FUNGAL BIOLOGY

Fungal Biology

Edited by:

Roksana Khalid

Delve Publishing

www.delvepublishing.com

Fungal Biology

Roksana Khalid

Delve Publishing
2010 Winston Park Drive,
2nd Floor
Oakville, ON L6H 5R7
Canada
www.delvepublishing.com
Tel: 001-289-291-7705
001-905-616-2116
Fax: 001-289-291-7601
Email: orders@arclereducation.com

ISBN: 978-1-77361-205-8 (Hardcover)

About the Ediror

ROKSANA KHALID

Roksana obtained her MSc from McMaster University in 2016. Her interests span from microbiology to neuroscience and she is currently working on the gut-brain axis as a visiting researcher at the Karolinska Institutet.

Contents

Preface

Fungi is a diverse kingdom comprising a vast range of eukaryotic organisms. It has some similar and different characters to both animals and plants. Therefore it was not easy to recognize where they should be classified in among living things. Because of a few important characteristics, fungi were separated from different eukaryotic systems of plants and animals. It was different from Protista, bacteria and plants due to presence of chitin in their cell wall. They are similar to animals because they are heterotrophs, they secrete enzymes in the environment that causes breakdown of food and converts them into dissolved food particles which the fungus absorbs. Fungi are major decomposers in their ecosystem. All these and some other differences urge to place fungi in a separate kingdom which is known as Eumycota (True Fungi). In biology the discipline which deals with fungus is known as mycology. It was in past considered as branch of botany, but recently due to progress in modern genetics it has proved that fungi is closer to animals than plants.

In term of presence, nearly 10,000 fungal species have been known, however its worldwide biodiversity is has not been understood completely. Fungal kingdom has been expected to have nearly 1.5 million species, on the base of interpretations of the number of plants species to the number of fungal species in particular atmosphere. Although according to recent research estimation there are 5 million species of fungi.Fungi are classified according to fruiting structure or shape and size of spores, and physiological and biochemical characteristics. Fungi can be found in vast range of habitats, comprising worst habitats like hot deserts, high salt concentrations, sediments of deep sea and ionizing radiations, e.g., cosmic radiations and intense UV. They also have huge difference in their sizes according to environmental conditions, types, etc. They can be found as microscopic organisms, for example, yeast and penicillium to macroscopic organisms, such as, mushrooms. Even the largest organism present on the earth is also a fungus known as "Honey fungus" from genus Armillaria.

Reproduction of fungus is also very complex due to wide variety of organisms It has been observed that about 1/3rd of fungi reproduce with the help of same method of reproduction called propagation, in two distinct stages, that are the anamorph and telomorph. They can reproduce through both types of reproduction, sexually and asexually. Asexual reproduction happens through mycelial fragmentation or via vegetative pores. Few fungi don't have observable sexual cycle, known as "Fungi imperfecti" or Deuteromycota. Sexual reproduction can be seen in all phyla of fungus except Glomeromycota, but genetic investigation put forward meiosis in Glomeromycota also.

Both sexual and asexual spores which can be called as sporangiospores are sometimes actively dispersed from their reproductive structures through forcible ejection. This discharge guarantees leaving of spores from the reproductive structures and also their travelling over long distances. Beside these processes some fungi might exchange their genetic material through para-sexual processes. However the process of sexual reproduction is greatly differ from animals and plants. Alterations also present among groups of fungus and can be used to differ species through morphological alterations in reproductive strategies and sexual reproduction. Some modern developments are also discussed in this book.

Current progress in genomics, molecular genetics and fungal genetics are discussed with deep examples, which comprises the processes of heritable information in fungi. Filamentous fungi and yeasts are greatly used as exemplary organisms for the purpose of genetic research of eukaryotes, comprising gene regulation, genetic recombination, chromatin structure and cell cycle regulation. One chapters includes growth and physiology of fungi. Growth occurs with the help of hyphae on or within the solid substrate or in the form of single cells in aquatic environments is adjusted for adequate withdrawal of nutrients. Some fungi even penetrate the substrate, Magnaporthe grisea form a structure known as appressorium that invade tissue of plants. A section of a book deals with pathogenic effects of fungus on animals and plants.

Some examples has been discussed in this book, Cryphonectria parasitica responsible for chestnut blight, Cryptoccocus cause immune-deficiencies in humans, and Paecilomyces lilacinus are predators of nematodes. Mostly fungi make biologically active compounds, few of which are toxic for plants and animals and that's why known as mycotoxins. Different methods of preventing from pathogenic fungi also been discussed in this book. As some organisms are parasites of fungus and known as mycoparasitic organisms. Oomycetes from genus Pythium are used as bio control agents in contradiction of various fungi.

One section deals with symbiotic associations of fungi with animals and plants. Mycorrhizal association is present among fungus and plant, and it is prehistoric dating to at least 400 million years ago. Lichen is an association among photosynthetic algae and fungi. Some insects cultivate fungi in order Agaricales as their main source of food. Last chapters have knowledge about role of fungi in humans and its significance in human life. Human use fungi for their benefit, for the purpose of preservation or food preparation and different other purposes has extended account. Fungi are commonly used for current chemotherapeutics, for example, penicillin's, for folk and traditional medicines, for pest control, and in food industry. This book has been written for students of graduate level who study mycology and are in search for knowledge related to fungi.

CHAPTER

1

INTRODUCTION-FUNGAL BIOLOGY

There are evaluated to remain no less than 1.5 million exclusive forms of organisms, though only 75,000 species have been represented up to the present time. For link, there are estimated to be 420,000 seed plant and around 4.9 million specie of arthropods.

FUNGI ON TAXONOMIC TREE

The Tree of Life Web Project aims to interface all the fundamental creatures on Earth as indicated by their common phylogenetic connections. The expectation is that this will lead us nearer to the very foundation of life on earth, which is right now assessed to be exactly 3.6–3.8 billion years prior (1 billion = 1000 million years, 109 years). In any case, growths of fungus arrived considerably later on the scene. The most seasoned known fungus fossils date to the time of Ordovician, in the vicinity of 460 and 455 million years prior which is a period when the biggest land plants are probably going to have been bryophytes (mosses and liverworts). [1] This treaties strikingly well with current investigations of phylogenetic, in light of examinations of groupings genes, talked about underneath.

Carl Woese from the University of Illinois at Urbana Champaign in USA, has stand up for the utilization of sub-atomic phylogenetics. The premise of this is to distinguish qualities that are available in every living being and that have a fundamental part, so they are probably going to be exceptionally rationed, amassing just little changes (transformations and back transformations) over huge ranges of developmental time. [2] Examinations of these groupings would then be able to demonstrate the connections between various life forms. There are restrictions and vulnerabilities in this approach, due to the potential for horizontal quality exchange amongst species and on the grounds that there are known to be variable rates of quality advancement between various gatherings of living beings. However a few profoundly moderated qualities and quality families can be utilized to give relative information. [3]

Most phylogenetic examinations are construct principally with respect to the qualities that code for the creation of ribosomal RNA. Ribosomes are basic parts of every single living life form since they are the locales of protein blend. They happen in vast numbers in all cells, and they are made out of a blend of RNA atoms (which have a basic part in the ribosome) and proteins. In prokaryotes (non-nucleate cells) the ribosomes contain three distinctive size groups of ribosomal RNA (rRNA), characterized by their sedimentation rates (S esteems, otherwise called Svedberg units) amid centrifugation in a sucrose arrangement. These three rRNAs are named 23S, 16S, and 5S. In eukaryotes (nucleate cells) there are likewise three rRNAs (28S, 18S, and 5.8S). The qualities encoding these rRNAs are found in numerous duplicates in the genome, and the distinctive rRNA qualities can be utilized to determine contrasts between living beings at various levels.

For most phylogenetic investigations the qualities that code for 16S rRNA (of prokaryotes) and the equal 18S rRNA (of eukaryotes) are utilized. These little subunit rDNAs contain enough data to recognize life forms over the phylogenetic range. Utilizing this approach, a few distinctive phylogenetic trees have been created, however large portions of them are basically comparable, both when all is said in done terms and particularly identifying with organisms.

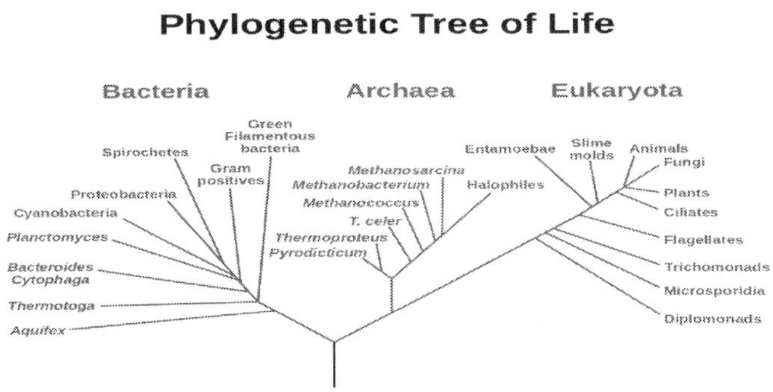

Figure- 1.1: Universal Phylogenetic Tree Showing Eukarya Position.

- Ribosomal DNA arrangement investigation shows that there are three developmentally particular gatherings of creatures, over the level of kingdom. These three gatherings, the Eucarya, Archaea, and Eucarya (eukaryotes) are named spaces and the contrasts between them are coordinated by numerous distinctions in cell structure and physiology.
- Beneath the level of spaces, there is still vulnerability about the ordered positions that ought to be doled out to living beings. Plants, creatures, and growths are all around viewed as independent kingdoms. Be that as it may, apparently, this status could likewise apply to the numerous "kingdoms" of microscopic organisms, particularly the tremendous Proteobacteria kingdom which incorporates most Gram-negative microbes. [4] Furthermore, it could be contended that the many separate gatherings of unicellular eukaryotes (flogs, ooze molds, amoebae, and so forth.) ought to likewise be viewed as kingdoms, in light of their clearly long haul division as judged by rDNA grouping disparity. Be that as it may, a considerable lot of these lower eukaryotes are still ineffectively examined, so they are frequently alluded to aggregately as "protists," pending further determination of their connections.
- The major multicellular living beings such as the creatures, plants, and organisms, shape a bunch at the extremely top of the Eucarya Domain, so they are regularly named the "crown eukaryotes". The fascinating component of these gatherings is that they appear to have separated from each other at generally

a similar time, and afterward experienced a noteworthy, fast extension and broadening [5]. The time when this happened, generally a large portion of a billion years prior, corresponds with the period when the land surfaces were colonized by primitive plants, for example, bryophytes (liverworts and greeneries) and when there were just three noteworthy mainland arrive masses: a land mass including present-day North America and Europe, situated close to the equator, some portion of current Siberia, towards the north, a land mass comprising of present-day South America, Africa, Antarctica, India, and Australia in the southern side of the equator. [6]

- Having put forth the defense for a long haul relationship amongst growths and land plants, we have to redress a generally held confusion: there is presently solid proof that organisms are more firmly identified with creatures than to plants. The organisms developed as an early branch from the creature genealogy, and both gatherings most likely have a typical inception in one of the basic unicellular eukaryotes. [7] At present it is trusted that the doubtlessly normal progenitor of both the parasitic and the creature kingdoms is a protozoan of the gathering named choanoflagellates, otherwise called the neckline lashes. These take after both the soonest branch of creatures (the wipes) and the most punctual branch of growths (the chytrids). It is a lowering suspected that people ought to have developed from something like this! [8]

COMPONENTS OF FUNGI

To start this area we should make an essential refinement between the genuine parasitic fungi and a scope of growth like living beings that have customarily been contemplated by mycologists, yet are in a general sense not the same as organisms. [9] Here we will concentrate on the genuine growths, regularly named the Mycota or Eumycota. All genuine growths have a scope of components that obviously isolate them from different life forms and that serve to characterize the parasitic kingdom (Mycota). [10] These elements are sketched out beneath:

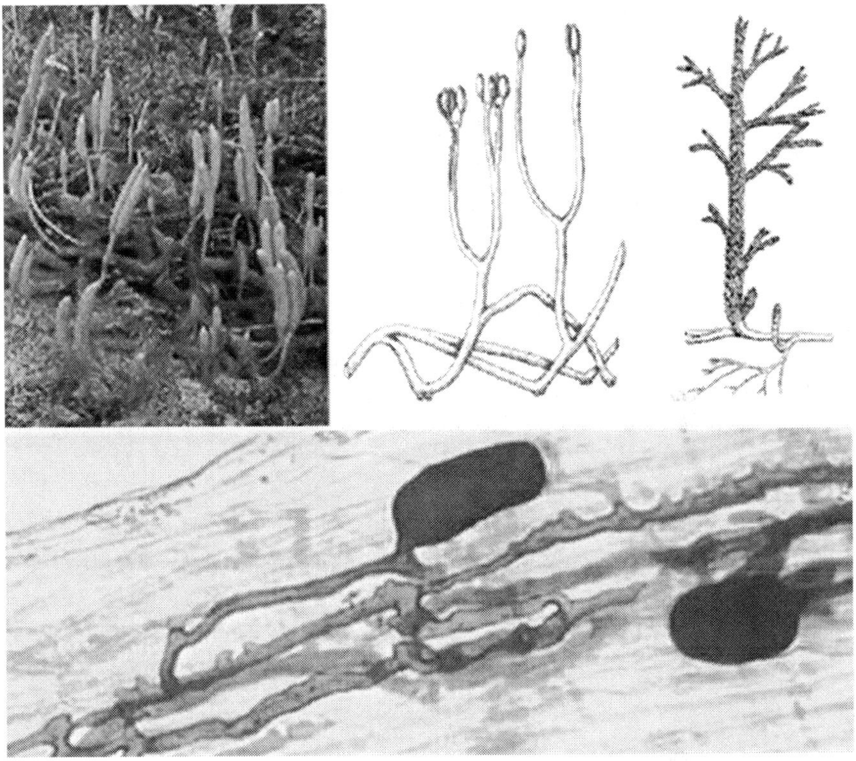

Figure- 1.2: (A) shows present-day Club moss (B, C) shows fossil pteridophytes (D) Swollen mycorrhizal fungus vesicles.

- All growths are eukaryotic. As such, they have layer bound cores containing a few chromosomes, and they have a scope of membranebound cytoplasmic organelles (vacuoles, mitochondria, and so forth.). [11] Different characterisitics, shared by all eukaryotes, include: cytoplasmic gushing, DNA that contains noncoding locales named introns, films that normally contain sterols, and ribosomes of the 80S sort as opposed to the 70S ribosomes of microorganisms ("S" alludes to Svedberg units, as specified prior). [12]
- Fungi commonly develop as fibers, named hyphae (particular: hypha), which broaden just at their outrageous tips. [13] Thus, parasitic fungi show apical development as opposed to numerous different filamentous life forms (e.g. filamentous green growth) which develop by rehashed cell divisions inside

a chain of cells (intercalary development). Parasitic hyphae branch over and over behind their tips, offering ascend to a system named a mycelium. Be that as it may, a few parasitic fungi develop as single-celled yeasts (e.g. Saccharomyces cerevisiae) which repeat by sprouting, and some can switch between a yeast stage and a hyphal stage because of ecological conditions. These dimorphic organisms (with two shapes) incorporate a few species that are not kidding pathogens of people. They frequently develop as yeast-like cells for multiplication in the body liquids however change over to hyphae for attack of the tissues. [14]

- Fungi are heterotrophs (chemo-organotrophs). At the end of the day, they require preformed natural mixes as vitality sources and furthermore as carbon skeletons for cell blend. The cell divider keeps parasitic fungi from overwhelming nourishment by phagocytosis, so growths ingest basic, dissolvable supplements through the divider and cell film. [15] By and large this is accomplished by emitting catalysts at the hyphal tips to debase complex polymers and after that engrossing the straightforward, solvent supplements discharged by depolymeraze (polymer-corrupting) compounds. [16]

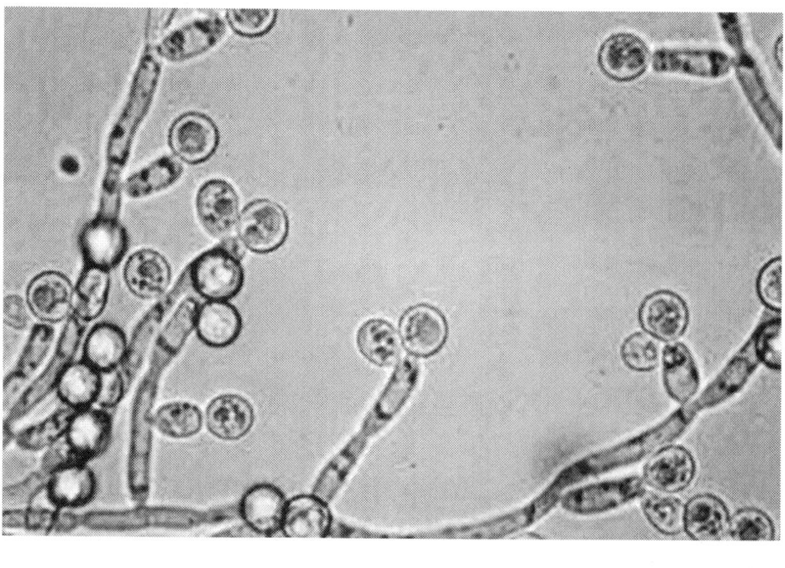

Figure- 1.3: Candida albicans (common dimorphic fungus).

- Fungi have a particular scope of divider segments, which ordinarily including chitin and glucans (polymers of glucose with dominatingly β-1, 3 and β-1, 6 linkages). [17] Short lengths of cellulose (a β-1, 4-connected polymer of glucose) have been identified in some parasitic dividers, particularly in a portion of the primitive growths. However parasitic fungi contrast from plants since they don't have cellulose-rich cell dividers.[18]
- Fungi have a trademark scope of solvent starches and capacity mixes, including mannitol and other sugar alcohols, trehalose (a disaccharide of glucose), and glycogen. These mixes are like those of a few creatures – strikingly the arthropods – however are not quite the same as those of plants.[19]
- Fungi normally have haploid cores are critical distinction from every single other eukaryote. Be that as it may, contagious hyphae frequently have a few cores inside each hyphal compartment, and many maturing yeasts are diploid. These distinctions in atomic status and atomic game plans have critical ramifications for parasitic hereditary qualities. [20]
- Fungi duplicate through asexual and sexual means, and commonly deliver spores. Contagious spores shift immensely fit as a fiddle, measure and different properties, identified with their different parts in dispersal or torpid survival.

Significant Activities of Fungi

As we have as of now observed, all growths require natural supplements for their vitality source and as carbon supplements for cell blend. [21] Be that as it may, an expansive qualification can be made by how these supplements are gotten:

- By developing as a symbiont in relationship with another life form.
- By developing as a saprotroph (saprophyte) on nonliving materials.
- By developing as a parasite (or a pathogen, a malady causing specialist) of another living life form.

Table- 1.1: Comparison of few fungal features with plants and animals

Character	Fungi	Animals	Plants
Growth habit	Hyphal tip growth or yeast budding	Not hyphal	Multicellular tissues
Nutrition	Heterotrophic, absorb soluble nutrients	Heterotrophic, ingest food.	Photosynthetic
Cell wall	Typically contains chitin	Chitin absent	Mainly cellulose
Nuclei	Usually haploid	Typically diploid	Diploid
Histones	Histone 2B	Histones 2B	Plant histones
Microtubules	Sensitive to benzimidazoles and griseofulvin	Sensitive to colchicine	Sensitive to colchicine
Membrane sterols	Ergosterol	Cholestrol	Sitosterol and different plant sterols

PARASITIC EFFECT OF FUNGUS ON PLANTS

A substantial number of organisms are adjusted to develop as parasitic fungi of plants, getting a few or the greater part of their supplements from the living tissues of their host. A considerable lot of these affiliations are very particular on the grounds that the growth taints just a single kind of host, and some of the time it is specific to the point that the organism can't develop at all in research center culture, it is a commit parasite that can become just in the host tissues. [22] Numerous cases of this are found among the rust organisms and fine mold growths, while different cases are found in the parasite like fleece molds, and the plasmodiophorids. [71] These host-particular organisms are named biotrophic parasitic fungi (profiles = life, trophy = bolstering) on the grounds that they encourage from living host cells without murdering them, frequently by creating exceptional nutrientabsorbing structures to tap the host's stores. At the flip side of the range are numerous basic parasitic fungi thataggressively assault plant tissues. They are named necrotrophic parasitic fungi (necros = demise) since they murder the host tissues as a major aspect of the sustaining procedure, for instance by delivering poisons or degradative chemicals. [23]

A typical illustration is the organism Botryotinia fuckeliana (all the more usually known by its previous name, Botrytis cinerea) which quickly obliterates delicate natural products, for example, strawberries, raspberries, and grapes, covering the natural product surface with its dim sporing structures. The contagious (or organism like) parasitic fungi of plants are colossally critical, representing over 70% of all the significant yield infections, and for some staggering pestilences. [24] To refer to only a couple of cases:

Potato scourge caused by the parasitic fungi like life form Phytophthora infestans obliterated the potato harvests of Ireland in the 1840s, prompting the starvation of up to one million individuals, and huge scale resettlement to whatever remains of Europe and the USA. [25] Indeed, even today the control of P. infestans and its nearby relatives, the fleece mold organisms, represents around 15% of world fungicide deals. The Advance of the Fungi by E. C. Huge in 1940 gives an interesting and exceedingly comprehensible record of potato scourge and its inheritance.

Dutch elm malady, caused by *O. ulmi*, and Ophiostoma novo-ulmi, has crushed the greater part of the normal elm (Ulmus procera) trees in Britain and Western Europe over the most recent 30 years, as it did in North America before in the 1900s. Essentially, chestnut curse caused by the organism Cryphonectria parasitica has crushed the local American chestnut (*Castanea dentata*) populace in the USA which is a plague that can be followed to the primary recorded infected chestnut tree in the New York Zoological Garden in 1904. [26] Furthermore, at the season of composing, another types of Phytophthora (*P. ramorum*) is causing sudden oak demise in southwestern USA and has officially spread to a few sections of Europe as pioneer colonizers of natural surroundings where no different life forms can develop, including rock surfaces and shaky, parched mineral soils. [27]

Mycorrhizas are personal relationship amongst parasitic fungi and the roots or other underground organs of plants. There are many sorts of mycorrhizal parasitic fungi, which have advanced autonomously of each other and which serve distinctive parts. In all cases these organisms rely upon the plant for a supply of carbon supplements, while the plants rely upon the growths for a supply of mineral supplements (phosphorus, nitrogen) from the soil. [28] Phosphorus is frequently the basic constraining element for plant development, since soil phosphates quickly frame insoluble edifices with natural issue or with divalent cations ($Ca2+$, $Mg2+$) and can't without much of a stretch diffuse to the plant roots. Mycorrhizal growths help to ease this issue by giving a broad hyphal system to catching mineral supplements and transporting them back to the roots. Notwithstanding, some other mycorrhizal parasitic fungi serve a very extraordinary part. Orchids and some nonphotosynthetic plants are completely reliant on organisms for all or part of the vegetation's, on the grounds that the plant nourishes on sugars provided by fungal growth. [29]

Lichens and mycorrhizas are by all account not the only cases of advantageous interaction. Lately many plants have been found to harbor symptomless endophytic organisms inside the plant dividers or intercellular spaces. [72] These organisms evidently do no mischief to the plants. Rather they can be valuable since they help to actuate plant resistance qualities and create bug against feedant mixes, for example, the ergot alkaloids. In any case, this is a twofold edged sword, on the grounds that the poisons can make genuine harm nibbling creatures, for example, stallions, cows, and sheep.

Mechanism of Symbiosis for Plants

Numerous parasitic fungi frame harmonious relationship with plants, in which both of the accomplices are probably going to profit. [69] The two most essential cases are lichens and mycorrhizas. Lichens are private relationship between two creatures which are photosynthetic accomplice (a green alga or a cyanobacterium) and a growth, which together deliver a thallus that can withstand the absolute most unfriendly situations on Earth. Commonly, the parasite encases and ensures the photosynthetic cells, and furthermore assimilates mineral supplements from follow levels in the earth, while the photosynthetic accomplice furnishes the organism with carbon supplements. [30] There are around 13,500 lichen species over the globe, and they assume fundamental parts as pioneer colonizers of natural surroundings where no different living beings can develop, including rock surfaces and flimsy, bone-dry mineral soils.

Mycorrhizas are cozy relationship amongst growths and the roots or other underground organs of plants. There are many sorts of mycorrhizal parasitic fungi, which have developed freely of each other and which serve diverse parts. In all cases these organisms rely upon the plant for a supply of carbon supplements, while the plants rely upon the parasitic fungi for a supply of mineral supplements (nitrogen, phosphorus) from the soil. [31] Phosphorus is regularly the basic restricting variable for plant development, since soil phosphates quickly shape insoluble buildings with natural issue or with divalent cations ($Ca2+$, $Mg2+$) and can't without much of a stretch diffuse to the plant roots. Mycorrhizal growths help to reduce this issue by giving a broad hyphal system to catching mineral supplements and transporting them back to the roots. Be that as it may, some other mycorrhizal parasitic fungi serve a very extraordinary part. [32] Some non-photosynthetic plants and orchids are totally reliant on parasitic fungi for all or part of the vegetation's, on account of the plant nourishes on sugars provided by fungi in soil. [70]

Mycorrhizas and lichens are not by any means the only cases of beneficial interaction. Lately many plants have been found to harbor symptomless endophytic parasitic fungi inside the plant dividers or intercellular spaces. These parasitic fungi evidently do no mischief to the plants. Rather they can be gainful in light of the fact that they help to enact plant barrier qualities and create bug hostile to feedant mixes, for example, the ergot alkaloids. Be that as it may, this is a twofold edged sword, in light of the fact that the poisons can make genuine harm nibbling creatures, for example, cows, sheeps and steeds.

Pathogenic Effect of Fungi on Humans

Rather than the numerous contagious parasitic fungi of plants, there are just somewhere in the range of 200 growths that taint people or other warm-blooded creatures. Truth be told, people have a high level of intrinsic invulnerability to parasitic fungi, except for the dermatophytic growths which regularly cause contaminations of the skin, nails, and hair. [33] Nonetheless, the circumstance changes radically when the resistant framework is bargained, and this is getting to be noticeably normal in patients with AIDS, transplant patients whose safe framework is deliberately stifled, patients experiencing tumor or propelled diabetes, and patients experiencing delayed corticosteroid treatment. In any of these conditions there is a critical possibility of contamination from organisms that represent no genuine danger to sound individuals. For instance, the broad and to a great degree regular airborne growth Aspergillus fumigatus ordinarily develops on fertilizers and in soil, however it has turned out to be a standout amongst the most huge obtrusive organisms in profound surgical techniques, and the survival rate can be as low as 30%. Numerous other growths that can develop at 37°C have spores that are sufficiently little to enter the lungs and achieve the alveoli. These parasitic fungi were practically obscure up to this point however basic reasons for disease in immunodeficient patients are currently to a great degree. [34] For instance, the parasite like life form *Pneumocystis jiroveci* (beforehand named P. carinii) normally causes pneumonia in patients contaminated with the human immunodeficiency infection (HIV). The onset of this ailment in patients with HIV is viewed as one of the "Guides characterizing" side effects. [35]

Just a modest bunch of antifungal medications are accessible to treat the real human mycoses, and most should be controlled at low measurements over a drawn out time to keep away from exorbitant harmfulness. Large portions of these medications are costly so they offer little would like to the poorest individuals in the creating scene.

FUNGAL AS SPECIALIST OF BIOLOGICAL CONTROLLING

Organisms parasitize many sorts of host, including other growths (mycoparasites), bugs (entomopathogens), and nematodes (nematophagous growths). Previously, such parasitic fungi may have been viewed as interests, however now they are perceived as being huge populace controllers of their hosts and as potential organic control (biocontrol) operators of real bugs or plant pathogens. [36]

Saprotrophic Fungus

Saprotrophs (saprophytes) are living beings that feast upon dead natural issue (sapros = demise, trophy = sustaining). Growths assume a noteworthy part in this regard in light of the fact that they produce an extensive variety of catalysts that debase complex polymers, for example, starch, cellulose, proteins, chitin, avionics lamp oil, keratin, and even the most complex lignified materials, for instance, wood. [37] Truth be told, there are few actually happening natural aggravates that can't be debased by some growth. One of only a handful couple of special cases is sporopollenin, the very safe polymer found in the dividers of dust grains.

Organisms are especially essential in the disintegration of cellulose, which speaks to around 40% of plant cell divider material and is the most bounteous regular polymer on Earth. Touching creatures (ruminants) likewise devour critical measures of cellulose, yet this is separated in the rumen (as a result, a huge anaerobic aging vessel) and the rumen parasitic fungi are thought to assume a noteworthy part in the decay procedure. [38] The breakdown of polymers by parasitic fungi is personally connected to hyphal development which gives both entering power and the planned arrival of extracellular catalysts and resulting reabsorption of the enzymic breakdown items. Be that as it may, diverse parasitic fungi are capable at debasing distinctive sorts of polymer, so contagious saprotrophs frequently develop in unpredictable, blended groups mirroring their diverse enzymic abilities. [68]

Despite the fact that the decomposer organisms assume indispensable parts in the reusing of real supplements, they can likewise be noteworthy deterioration operators. A notable illustration is the dry-spoil growth, Serpula lacrymans, which is a noteworthy reason for timber rot in structures. Essentially the "dirty molds" that ordinarily develop on kitchen and restroom dividers are to a great degree hard to annihilate. They use the solvent cellulose gels that are utilized as stabilizers in emulsion paints or as backdrop glues. These normal parasitic fungi incorporate types of Sydowia, polyspora, Alternaria, and Cladosporium (beforehand called Aureobasidium pullulans) which stain the dividers in view of their dimly pigmented hyphae and spores. Be that as it may, their normal living space is the surface of leaves or the rotting stalk tissues of plants, and they happen in structures simply because they find comparable conditions (and substrates) to those in their regular habitat. [67] General source specialists are currently giving careful consideration to security in the work environment, and especially to the potential parts of organisms in "wiped out building disorder," which has been connected (questionably) to newborn child bunk demise. The conditions in underventilated structures can absolutely advance the development of molds, including Stachybotrys chartarum, another normal dingy shape. Be that as it may, there is no authoritative confirmation to connect these growths to wipe out building disorder. [39]

Some saprotrophic growths represent a genuine danger to human and creature welfare by developing on put away nourishment items and delivering mycotoxins. These are a differing scope of parasitic auxiliary metabolites, regularly found in shamefully put away materials. For instance, aflatoxins are ordinarily created in groundnuts and cottonseed dinner. They are among the most powerful known cancer-causing agents and are firmly embroiled in hepatomas. Thus, the poisons created by a few Fusarium animal varieties on grain crops are ensnared in esophageal malignancy in Africa, and in kidney carcinomas. The pathways prompting the generation of these mixes are examined in one section of book, the upkeep of safe stockpiling conditions is shrouded in one section of book.

Role of Fungi in Biotechnology

Growths have numerous customary parts in biotechnology, additionally some novel parts, and there is significant degree for their future business advancement. Some of these parts are illustrated beneath.

Nourishment and Sustenance Flavorings

In 1994 the aggregate world generation of eatable mushrooms was assessed to be more than 5 million tons, with an estimation of US $14 billion. [40] A great part of the mushroom-developing industry depends on strains of the normal developed mushroom Agaricus bisporus (or A. brunnescens). Be that as it may, Lentinula edodes (the Shiitake mushroom, which is developed on logs), Volvariella volvacea (the padi straw mushroom, which is developed on rice straw), and Pleurotus ostreatus (the shellfish mushroom) are customarily developed in Japan and southeast Asia, and are presently broadly accessible in western grocery stores. [66]

Organisms are utilized to create a few customary nourishments and refreshments, including mixed beverages (ethanol from the yeast Saccharomyces cerevisiae) and bread, where the yeast produces CO2 for raising the mixture. Penicillium roqueforti is utilized as a part of the later phases of generation of the blue-veined cheeses, for example, Stilton and Roquefort, to which it gives a trademark enhance. *P. camemberti* is utilized to deliver the delicate cheeses, for example, bries and Camembert, it develops on the cheddar surface, framing an "outside layer," and creates proteases which logically debase the cheddar to give the delicate consistency. Less outstanding yet similarly critical is the part of growths in the aging of customary nourishments around the globe. [41] For instance, Rhizopus oligosporus is utilized to change over cooked soybean "corn meal" to a nutritious staple sustenance, called tempeh. This includes just a short (24–36 hour) hatching time, amid which the growth debases a portion of the fat and furthermore corrupts a trypsin inhibitor in soybeans, so that the normally high protein substance of this yield is all the more promptly accessible in the eating regimen, and a "tooting variable" is separated amid this procedure. [65] The nourishment named gari is a piece of the staple eating regimen in southern Nigeria, it is created from the high-yielding root edit, cassava, maybe better known in its handled frame, custard. Crude cassava contains a poisonous cyanogenic glycoside named linamarin, which is evacuated amid a drawn out and to a great extent uncontrolled aging in town groups. Quite a bit of this procedure includes microscopic organisms, yet the growth Galactomyces geotrichum (agamic stage: Geotrichum candidum) gives the item its coveted flavor.

A noteworthy advancement as of late has been the presentation of a totally new kind of nourishment, named Quorn™ mycoprotein. This is delivered industrially by growing a growth (*Fusarium venanatum*) in extensive maturation vessels, at that point reaping the contagious hyphae and handling them into meat-like lumps and different stove prepared suppers. [42] Quorn (as it is presently called) is broadly accessible in British and European general stores. It has a practically perfect healthful profile, with a high protein content, low fat substance, and nonattendance of cholesterol.

Fungal Metabolites

Metabolites can be assembled into two general classifications:

- ***Primary metabolites***: the intermediates or final results of the basic metabolic pathways of all living beings (glycerol, natural acids, amino acids, and sugars and so on.) and which are fundamental for the typical cell elements of growths. [64]
- ***Secondary metabolites***: a different scope of mixes shaped by particular pathways of specific living beings, they are not fundamental for development, in spite of the fact that they can give preference to the life forms that create them (e.g. parasitic poisons, anti-infection agents, and so forth.). [43]

Table- 1.2: Valueable secondary metabolites made commercially from fungi

Metabolite	Fungal source	Application
Penicillins	Penicillium chrysogenum	Antibacterial
Cephalosporins	Acremonium chrysogenum	Antibacterial
Griseofulvin	Penicillium griseofulvum	Antifungal
Fusidin	Fusidium coccineum	Antibacterial
Ciclosporins	Tolypocladium spp.	Immunosuppressants
Zearalenone	Gibberella zeae	Cattle growth promoter
Gibberellins	Gibberella fujikuroi	Plant hormone
Ergot and related compounds	Claviceps purpurea and related fungi	Effects comprising, antimigraine, vasoconstriction, vasodilation, anti-Parkinson, etc.

A few metabolites of both gatherings are created industrially from contagious societies. One of the best cases of a contagious essential metabolite is citrus extract, with an expected worldwide generation of 900,000 tons in the year 2000. [44] Citrus extract created on this immense scale is the backbone of the soda pops industry (lemonade, and so on.) on the grounds that it has a tart taste and furthermore improves enhance, decreases sweetness, and has cell reinforcement and additive qualities. [62] Uncommonly chose, overproducing strains of Aspergillus niger are utilized for the business generation of citrus extract, yet a few different conditions are vital such as the way of life must contain elevated amounts of promptly metabolizable sugars (up to at least 20%) and the grouping of either phosphate or nitrogen must be kept low, to restrict the measure of parasitic development. In these conditions at least 80% of the sugar provided to the way of life is changed over into citrus extract, which is then sent out from the cells and aggregates in the way of life medium. The impact of this is to bring down the pH of the way of life medium to 3.0 or less, which the parasite endures well. This emission of the corrosive is a critical element, in light of the fact that parasitic cells firmly manage their inside pH. Late confirmation shows that cells of *A. niger* keep up their intracellular pH at 7.7 when the cells are presented to outer pH levels running from 1.5 to 6.

Figure- 1.4: Commercial culture of the shiitake mushroom, Lentinula edodes, on inoculated logs.

Other natural acids are created economically by parasitic maturations. Gluconic corrosive (evaluated yearly worldwide generation of 50,000–

100,000 tons) is utilized essentially as a nourishment added substance, and is created by particular strains of *A. niger*, developed at typical pH. [45] This corrosive is delivered by the immediate oxidation of glucose, catalyzed by the chemical glucose oxidase. Itaconic corrosive (worldwide creation 70,000–80,000 tons) is delivered by Aspergillus terreus and is utilized as a co-polymer in fabricate of cements, paints, and so forth. [63]

In a few regards the creation of citrus extract and itaconic corrosive is like the generation of ethanol by Saccharomyces spp., the premise of the mixed beverages industry. Both sorts of item aggregate in the way of life medium when development is limited by some element yet when the biochemical apparatus keeps on working, similar to the motor of an auto removed from outfit. [61] For instance, ethanol collects as a metabolic endproduct when yeast is developed in a sugar-rich medium favoring digestion, yet in anaerobic conditions that breaking point cell development. [46]

Rather than the mass metabolites specified over, a huge scope of auxiliary metabolites are created by growths, and they incorporate a few high-esteem items with pharmaceutical applications. The best-known cases are the penicillins which is a gathering of fundamentally related β-lactam anti-microbials that are combined actually from little peptides. [47] The normally happening penicillins, for example, penicillin G (delivered by Penicillium chrysogenum) have a moderately limit range of action. Yet, an extensive variety of different penicillins can be delivered by compound change of the regular penicillins. Every cutting edge penicillin are semisynthetic mixes, they are acquired at first from maturation societies however are then basically changed for particular alluring properties. Schmidt in 2002 explored the produce and remedial parts of β-lactam anti-infection agents, including the cephalosporins which are basically identified with the penicillins. [48] Amazingly, in spite of their age (the penicillins were first delivered industrially in the late 1940s), the β-lactam anti-microbials still offer half of the world market for systemic anti-toxins, with deals in 1998 worth about US $4 billion for penicillins and about US $7 billion for the all the more as of late created cephalosporins.

A few non-β-lactam anti-infection agents are likewise delivered by organisms. They incorporate griseofulvin (from the parasite P. griseofulvum) which has been utilized for quite a while to treat dermatophyte contaminations of the skin, nails and hair of people, albeit as of late it has been supplanted by less dangerous medications. Fusidic corrosive (from different parasitic fungi) has been utilized to control staphylococci that have turned out to be impervious to penicillin, and there is restored enthusiasm for a scope of other common contagious items for treating the systemic contagious contaminations of people. Ciclosporins from different parasitic fungi (however chiefly from types of Tolypocladium) are utilized as immune-suppressants to forestall organ dismissal in transplant surgery. [49] Truth be told, 17 diverse parasitic taxa are accounted for to deliver ciclosporins. Another intense immunosuppressant is the anti-microbial gliotoxin (from Trichoderma virens), which is better known for its part in natural control of plant pathogenic organisms. As a last case, the ergot alkaloids and related poisons of the Claviceps purpurea, and ergot parasite, have numerous essential pharmacological applications. The fourmembered ring structure of the d-lysergic corrosive subsidiaries of ergot alkaloids emulate the ring structures of neurotransmitters (dopamine, serotonin, and epinephrine (adrenaline)). Be that as it may, at display a number of the ergot subordinates are excessively nonspecific in their methods of activity to meet their actual potential in treating human issue. [50]

Indeed, even these couple of illustrations bring up intriguing issues about the parts of contagious auxiliary metabolites. What capacities do they serve in organisms and what upper hand do they give? [60] As of late a considerable lot of the qualities encoding the optional biosynthetic pathways have been recognized and sequenced. This should lead both to a comprehension of their parts and to the potential development of transgenic strains that overproduce significant metabolites.

A portion of the polysaccharides of growths have potential business esteem. Pullulan is a α-1, 4-glucan (polymer of glucose) created as an extracellular sheath by Sydowia polyspora (once Aureobasidium pullulans), one of the dingy molds. [51] This polymer is utilized as a part of Japan to make a film-wrap for sustenances. A potential new market could create from the revelation that parasitic divider polymers or their incomplete breakdown items can be capable elicitors of plant guard reactions so they may be utilized to "inoculate" plants. For instance, the β-glucan parts from dividers of the yeast *S. cerevisiae* have this impact. So too does chitosan, the de-acetylated type of chitin in contagious cell dividers. At exhibit, chitosan is utilized on an extensive scale in Japan for clearing up sewage, however the wellspring of this chitosan is scavanger shells. Growths are an option, effortlessly inexhaustible wellspring of this and different polymers.

ENZYMIC TRANSFORMATIONS AND PROTEINS

Saprotrophic parasitic fungi and some plant-pathogenic growths deliver a scope of extracellular proteins with critical trade parts. The pectic compounds of growths are utilized to elucidate natural product squeezes, a parasitic amylase is utilized to change over starch to maltose amid breadmaking, and a contagious rennet is utilized to coagulate drain for cheddar making. [52] A solitary organism, *Aspergillus niger*, represents very nearly 95% of the business generation of these and other mass compounds from growths, albeit particular strains of the parasite have been chosen for specific purposes. The methanol-using yeasts (Pichia pastoris, Hansenula polymorpha, Candida lipolytica) have potential business esteem since they deliver a lot of liquor oxidase, which could be utilized as a fading operator in cleansers. [59] The wood-decaying organism Phanerochaete chrysosporium is to a great degree dynamic in debasing lignin, it can possibly be created for delignification of horticultural squanders and results of the wood-pulping industry, so that the cellulose in these materials could be utilized as a shoddy substrate for generation of fuel liquor by yeasts. [53]

Table- 1.3: Commercially produced fungal enzymes

Enzyme	**Fungal source**	**Application**
α – Amylase	*Aspergillus niger, A. oryzae*	Starch conversions
Amyloglucosidase	A. *Niger*	Starch syrups, dextrose foods
Pullulanase	*Aureobasidium pullulans*	Debranching of starch
Glucose oxidase	A. *Niger*	Food processing
Hemicellulase	A. *Niger*	Baking, gums

Lipases	*Mucor, Aspergillus, Penicillium*	Dairy industry, detergents
Glucose isomerase	*Mucor, Aspergillus*	High fructose syrups
Rennet	*Mucor spp.*	Milk coagulation
Pectinases	*Aspergillus, Rhizopus*	Clarifying fruit juices
Invertase	*Yeasts*	Sucrose conversions
Proteases (alkaline, neutral, acid)	*Aspergillus spp.*	Baking, Brewing
Glucose aerohydrogenase	A. *Niger*	Production of gluconic acid

Notwithstanding these cases of "mass" proteins, parasitic fungi have numerous inside catalysts and enzymic pathways that can be abused for the bioconversion of mixes, for example, pharmaceuticals. [58] For instance, organisms are utilized for the bioconversion of steroids, in light of the fact that contagious catalysts perform exceptionally particular dehydrogenations, hydroxylations and different adjustments of the complex sweet-smelling ring frameworks of steroids. [54] Forerunner steroids are bolstered to a growth, held at low supplement level either in culture or joined to a latent bed, so that the steroid is retained, changed and afterward discharged into the way of life medium from which it can be recovered.

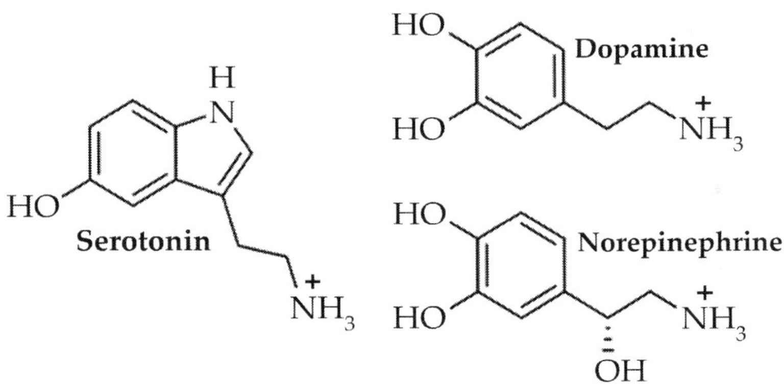

Figure- 1.5: Similarities among structures of serotonin, noradrenaline, and dopamine.

Heterologous Products of Gene

Hereditary building of fungus, especially *Saccharomyces cerevisiae*, has created to the phase where the cells can be utilized as plants to deliver pharmaceutical items, by the presentation of outside (heterologous) qualities, as we effectively noted for the hepatitis B antibody. [57] There are a few focal points in utilizing yeast to integrate such items. *S. cerevisiae* is as of now developed on a huge modern scale, so organizations know about its way of life. It is a GRAS life form, i.e. "by and large viewed as sheltered." Its genome was the first to be sequenced, and its hereditary qualities and sub-atomic hereditary qualities are very much investigated. Besides, yeast has a very much portrayed secretory framework for sending out quality items into a culture medium. [55] Cases of heterologous quality items that have been delivered tentatively from yeast incorporate epidermal development figure (required in wound recuperating), interferons (with antiviral and antitumor movement), atrial natriuretic component (for administration of hypertension), and α-1-antitrypsin (for potential help from emphysema). There are, nonetheless, inconveniences in utilizing *S. cerevisiae*. Specifically, this growth is hereditarily very not quite the same as other parasitic fungi and different eukaryotes, including its utilization of various codons for some amino acids, so it doesn't generally accurately read the presented qualities. Therefore consideration has changed to some other organisms, for example, the parting yeast *Schizosaccharomyces pombe* and the filamentous fungus *Emericella nidulans*, for both of which the genomes have now been sequenced. [56]

SUMMARY

- Fungi are the heterotrophic eukaryatic organisms with absorptive mode of nutrition. It includes organisms like yeast, molds, morels, truffles, smuts, puffballs, smut, mushroom and rusts, also many other organisms which are not well known. About 70,000 fungi species have been defined, but few accounts suggest that nearly 1.5 million fungi specie exists. Fungi are considered as separate kingdom from animals and plants. They have animals like absorptive mode of nutrition. They secrete enzymes that convert food into absorbable particles that is present in environment not in stomach.

- Fungi that grow on agar plates were observed by most biologists, which are filamentous and dense, however in nature filament becomes larger rather that the density of fungal colonies. When a filament comes in contact with food then whole colony become mobilize and relocates the food source in order to exploit new food. While the spores and filaments of fungi are microscopic, the colony can be very much large in size even larger than biggest plants or animal.
- Pheromones are the chemical messengers through which two fungi contact with each other before the mechanism of sexual reproduction. Atleast one pheromone is present in each phyla, and they might sort from carotenoids pathway derivatives in zygomycetes and chytridiomycetes to the oligopeptides in basidiomycetes and ascomycetes.
- Fungi are considered as major decomposers of plants and animals. Most of the fungus species are free living saprobes that are present in woody substrates, animal exudates, dead animals, leaf litter, and soil. Wood-decaying fungi creates large cavities by eating living trees and provide place to live for different animals. Some fungi are dikaryokic (they have two different and separate nuclei in same hyphal portion.

REFERENCES

1. Jennings, D. H., & Lysek, G. (1999). *Fungal biology: Understanding the fungal lifestyle*. Oxford: BIOS Scientific Publishers.
2. British Mycological Society. (1998). *Mycological research: An international journal of fungal biology*. Cambridge: Cambridge University Press.
3. Oliver, R., & Schweizer, M. (1999). *Molecular fungal biology*. Cambridge: Cambridge University Press.
4. ScienceDirect (Online service). (2010). *Fungal biology^*. New York: Elsevier.

5. Hudson, H. J. (1992). *Fungal biology*. Cambridge: Cambridge Univ P V.
6. Deacon, J. W. (2013). *Fungal Biology*. New York, NY: John Wiley & Sons.
7. Zhongguo jun wu xue hui. (2010). *Mycology: An international journal on fungal biology*. Abingdon: Taylor & Francis, on behalf of the Mycological Society of China.
8. Moore, D. (2013). *Fungal biology in the origin and emergence of life*. Cambridge: Cambridge University Press.
9. International Congress of Acarology, & Rodriguez, J. G. (1979). *Recent advances in acarology*. New York: Academic Press.
10. In Shefferson, R. P., In Jones, O. R., & In Salguero-Gómez, R. (2017). *The evolution of senescence in the tree of life*.
11. Worapong, J., & Montana State University. (2001). *Taxonomy, molecular phylogeny and taxol production in ed genera of endophytic fungi*. (Dissertation Abstracts International, 62-6.)
12. Symposium on Collection Building and Natural History Studies in Asia, Matsuura, K., Kokuritsu Kagaku Hakubutsukan (Japan), & Symposium on Collection Building and Natural History Studies in Asia. (2000). *Proceedings of the First and Second Symposia on Collection Building and Natural History Studies in Asia*. Tokyo: National Science Museum.
13. United States., & United States. (2016). *Comparison of pectin-degrading fungal communities in temperate forests using glycosyl hydrolase family 28 pectinase primers targeting Ascomycete fungi*. Washington, D.C: United States. Dept. of Energy. Office of Science.
14. Roeper, R. A. (1972). *Biology of symbiotic fungi associated with Ambrosia beetles of western United States*.
15. Kress, W. J., & Erickson, D. L. (2012). *DNA barcodes: Methods and protocols*. New York: Humana Press/Springer.
16. Dewdney, A. K. (1998). *Hungry Hollow: The story of a natural place*. New York, NY: Copernicus.
17. Cranbrook, G. G.-H., & International Union for Conservation of Nature and Natural Resources. (1988). *Malaysia*. Burlington: Elsevier Science.

18. Jørgensen, S. E., & Fath, B. D. (2008). *Encyclopedia of ecology*. Amsterdam, Netherlands: Elsevier.
19. Hanson, J. R. (2008). *The chemistry of fungi*. Cambridge, UK: Royal Society of Chemistry.
20. Maheshwari, R. (2005). *Fungi: Experimental methods in biology*. Boca Raton, FL: CRC/Taylor & Francis.
21. Carlile, M. J., Watkinson, S. C., & Gooday, G. W. (2001). *The fungi*. San Diego: Academic Press.
22. Hall, I. R., & Invermay Agricultural Research Centre. (1991). *A summary of the features of endogonaceous taxa*. Mosgiel, N.Z: Invermay Agricultural Centre.
23. Money, N. P. (2016). *Fungi: A very short introduction*.
24. Carlile, M. J., Watkinson, S. C., & Gooday, G. W. (2001). *The fungi*. San Diego: Academic Press.
25. In Horwitz, B. A. (2013). *Genomics of soil- and plant-associated fungi*.
26. Moore, D., Robson, G. D., & Trinci, A. P. J. (2011). *21st century guidebook to fungi*. Cambridge: Cambridge University Press.
27. Blanchette, R. A., & Biggs, A. R. (1992). *Defense mechanisms of woody plants against fungi*. Berlin: Springer-Verlag.
28. University of California, Berkeley., United States., & United States. (2015). *Comparative analysis of secretomes from Ectomycorrhizal fungi with an emphasis on small-secreted proteins*. Washington, D.C: United States. Dept. of Energy. Office of Science.
29. Hopwood, D. A. (2012). *Natural product biosynthesis by microorganisms and plants: Part C*. Amsterdam: Elsevier/Academic Press.
30. In Hurst, C. J. (2016). *The Rasputin Effect: When Commensals and Symbionts Become Parasitic*.
31. Flórez, L. V., & Friedrich-Schiller-Universität Jena. (2016). *Burkholderia as bacterial symbionts of Lagriinae beetles: Symbiont transmission, prevalence and ecological significance in Lagria villosa and Lagria hirta (Coleoptera: Tenebrionidae)*. Jena.
32. Hynson, N. (2010). *The Ecophysiology and Evolution of*

Mycoheterotrophic Plants in the Tribe Pyroleae (Ericaceae). Berkeley, CA.

33. Esser, K., & Deising, H. B. (2009). *The Mycota: A Comprehensive Treatise on Fungi as Experimental Systems for Basic and Applied Research.* Berlin, Heidelberg: Springer Berlin Heidelberg.
34. Khan, N. A. (2008). *Microbial pathogens and human diseases.* Enfield (NH: Science Publishers.
35. Esser, K. (2016). *The Mycota: A comprehensive treatise on fungi as experimental systems for basic and applied research.*
36. Domer, J. E., & Kobayashi, G. S. (2004). *The Mycota: A comprehensive treatise on fungi as experimental systems for basic and applied research.* Berlin: Springer-Verlag.
37. Seifert, H. S., & DiRita, V. J. (2006). *Evolution of microbial pathogens.* Washington, D.C: ASM Press.
38. Shaw, K. J. (2002). *Pathogen genomics: Impact on human health.* Totowa, N.J: Humana Press.
39. In Casadevall, A. (2015). *Human fungal pathogens: A subject collection from Cold Spring Harbor perspectives in medicine.*
40. Nester, E. W. (2012). *Microbiology: A human perspective.* New York: McGraw-Hill.
41. Austin, B. (1991). *Pathogens in the environment.* Oxford: Blackwell Scientific Publications.
42. Maheshwari, R. (2012). *Fungi: Experimental methods in biology.* Boca Raton: CRC Press.
43. Bellows, T. S., & Fisher, T. W. (1999). *Handbook of biological control: Principles and applications of biological control.* San Diego: Academic Press.
44. Boland, G. J., & Kuykendall, L. D. (1998). *Plant-microbe interactions and biological control.* New York: M. Dekker.
45. Merillon, J. M., & Ramawat, K. G. (2012). *Plant defence: Biological control.* Dordrecht: Springer.
46. Ehlers, R.-U. (2011). *Regulation of biological control agents.* Dordrecht: Springer.
47. Davies, K. G., & Spiegel, Y. (2011). *Biological control of plant-parasitic nematodes: Building coherence between microbial ecology*

and molecular mechanisms. Dordrecht: Springer.

48. Mérillon, J. M., & Ramawat, K. G. (2012). *Plant Defence: Biological Control*. Dordrecht: Springer Netherlands.
49. Bernier, J. (1995). *A study of fungal leaf decomposition in relation to biological control of the apple scab pathogen, Venturia inaequalis*.
50. Stirling, G. R. (2014). *Biological control of plant-parasitic nematodes: Soil ecosystem management in sustainable agriculture*. Wallingford, Oxfordshire: CABI.
51. Deacon, J. W. (2006). *Fungal biology*. Malden, Mass. [u.a.: Blackwell Publishing.
52. Dix, N. J., & Webster, J. (1995). *Fungal Ecology*. Dordrecht: Springer Netherlands.
53. Newsham, K. K., Frankland, J. C. J., Boddy, L., & Ineson, P. P. (1992). *Effects of dry-deposited sulphur dioxide on fungal decomposition of angiosperm tree leaf litter. I. Changes in communities of fungal saprotrophs*.
54. Ainsworth, A. M., & English Nature (Agency). (2004). *Developing tools for assessing fungal interest in habitats: 1*. Peterborough: English Nature.
55. Lawrence Berkeley National Laboratory., & United States. (2015). *Convergent losses of decay mechanisms and rapid turnover of symbiosis genes in mycorrhizal mutualists*. Berkeley, Calif: Lawrence Berkeley National Laboratory.
56. Carlile, M. J., Watkinson, S. C., & Gooday, G. W. (2005). *The fungi*. San Diego: Academic Press.
57. Schaechter, M. (2012). *Eukaryotic microbes*. Amsterdam: Elsevier/ Academic Press.
58. Moore, D., Robson, G. D., & Trinci, A. P. J. (2011). *21st century guidebook to fungi*. Cambridge: Cambridge University Press.
59. Raghukumar, S. (2017). *Fungi in coastal and oceanic marine ecosystems: Marine fungi*.
60. Lawrence Berkeley National Laboratory., United States., & United States. (2013). *Comparative genome analysis of Basidiomycete fungi*. Washington, D.C: United States. Dept. of Energy. Office of Science.

61. Gadd, G. M. (2001). *Fungi in bioremediation*. Cambridge: Published for the British Mycological Society [by] Cambridge University Press.
62. Ray, R. C., & Ward, O. P. (2006). *Microbial biotechnology in horticulture*. Enfield, NH: Science Publishers.
63. Ray, R. C. (2005). *Microbial biotechnology in agriculture and aquaculture*. Enfield, (NH: Science Publishers.
64. Salem, F., & Abdel-Azeem, A. (2014). *Screening of Anticancer Metabolites Produced by Endophytic Fungi: Of Some Medicinal Plants in Saint Katherine Protectorate*. Saarbrücken: LAP LAMBERT Academic Publishing.
65. Drauz, K., Waldmann, H., & Wiley InterScience (Online service). (2002). *Enzyme catalysis in organic synthesis: A comprehensive handbook*. Weinheim: Wiley-VCH.
66. Nord, F. F. (2009). *Advances in Enzymology and Related Areas of Molecular Biology: 10*. Hoboken: John Wiley & Sons.
67. Kuo, T. M., & Gardner, H. W. (2002). *Lipid biotechnology*. New York: Marcel Dekker.
68. Engels, W.J.M. (n.d.). *Volatile and non-volatile compounds in ripened cheese: their formation and their contribution to flavour*. Engels.
69. Czygan, F.-C. (1980). *Pigments in plants*. Stuttgart: G.F. Verlarg.
70. Gilmour, Sarah J., Zeevaart, Jan A. D., Schwenen, Ludger, & Graebe, Jan E. (1986). *Gibberellin Metabolism in Cell-Free Extracts from Spinach Leaves in Relation to Photoperiod 1*.
71. McDonald, T., Kim, K., & ALBERT EINSTEIN COLL OF MEDICINE OF (YESHIVA UNIV) BRONX NY. (2003). *Novel Lishmania and Malaria Potassium Channels: Candidate Therapeutic Targets*. Ft. Belvoir: Defense Technical Information Center.
72. In Rose, A. H., & In Harrison, J. S. (1993). *The yeasts: Volume 5*. London: Academic Press.

CHAPTER

2

DIVERSITY OF FUNGI AND FUNGUS LIKE ORGANISMS

In this section we concentrate on the significant gatherings of fungus like life forms, and parasitic fungus, comprising the entire traverse of fungus differences in its comprehensive logic. We will utilize chose cases to outline key components of the fungal gatherings, and their organic criticalness. There will be a few amazements in store. For instance, we will observe that the absolute most destroying pathogens of plants are not fungi actually by any means, but rather have a place with an altogether isolate kingdom. We also will observe a portion of the life forms once thought to be among the most "primitive" like as, diplomonads, trichomonads, and the microsporidia, that are gotten from fungus by the loss of components which they had once, for example, mitochondria. We will likewise perceive how the improvement of atomic strategies for deciding the connections among life forms has upgraded our fungal familiarity in various regards, yet there is still no agreement on the most ideal approach to develop phylogenetic trees.

FUNGI LIKE LIVING-BEINGS AND FUNGI

Every one of the fungal growths and parasitic fungus like creatures that are right now thought to be organisms in the broadest sense. By far most are genuine growths, some of the time named Eumycota (eu = "genuine"). [1] Up to this point, all the genuine organisms were appointed to four phyla, Basidiomycota, Chytridiomycota, Ascomycota, and Zygomycota. In any case, in 2001, a fifth phylum was raised which is the Glomeromycota (arbuscular mycorrhizal growths and their relatives). These had beforehand been incorporated into the Zygomycota. Glomeromycota were related with the most punctual land plants are still connected with by far most of plants today.

Inside the Kingdom Fungi, the Basidiomycota and the Ascomycota have many components in like manner, directing plainly toward a typical lineage. The phylum Chytridiomycota has customarily been portrayed on the premise of motile cells with a solitary back flagellum. This phylum was re-imagined as of late, in light of succession examination of the atomic qualities encoding little subunit (SSU) ribosomal RNA (18S rDNA). This has uncovered that some nonmotile parasitic fungi, beforehand alloted to the Zygomycota, are firmly identified with the Chytridiomycota and must be reassigned. An illustration is the fungal organism *Basidiobolus ranarum*, now exchanged to the Chytridiomycota. The status of Zygomycota (as of now characterized in the wake of barring the Glomeromycota) is as yet vague. Some of its individuals may should be isolated into new gatherings. In any case, the present view is that all living beings inside the Kingdom Fungi constitute a monophyletic gathering (all gotten from a typical predecessor), imparting a few elements to creatures. [2] Quality grouping investigations give the premise to a characteristic phylogeny, particularly when information for the SSU rDNA are upheld by succession examination of other quality families, for example, the genes of actin and tubulin.

Kingdom	Mycota
Phylum	Chytridiomycota
Phylum	Zygomycota
Phylum	Glomeromycota
Phylum	Ascomycota

Phylum	Basidiomycota
Kingdom	Straminipila
Phylum	Oomycota
Phylum	Hyphochytridiomycota
Phylum	Labyrinthulomycota
Fungus like organisms	
Phylum	Myxomycota
Phylum	Plasmodiophoromy-cota
Phylum	Dictyosteliomycota
Phylum	Acrasiomycota

The Kingdom Straminipila (straminipiles, or stramenopiles) is currently all around perceived as being particular from the genuine growths. It comprises of one vast and critical phylum, the Oomycota, and two little phyla, Labyrinthulomycota (with around 40 species) and the Hyphochytridiomycota (with around 25 species). [3] The Phylum Oomycota is astounding from multiple points of view. It incorporates the absolute most wrecking plant pathogens, including Phytophthora ramorum (sudden oak demise in California), Phytophthora cinnamomi (the scourge of extensive tracts of Eucalyptus woodland in Australia), Phytophthora infestans (potato curse), and numerous other vital plant pathogens, including Pythium and Aphanomyces spp. In any case, maybe most noteworthy of all is the way that Oomycota have developed a way of life that looks like that of the genuine growths in practically every regard. [4]

The growth like life forms of unverifiable proclivity incorporate four sorts of living being: the acrasid cell ooze shape, the dictyostelid cell ooze shape, the plasmodial sludge molds (Myxomycota), and the plasmodiophorids. [5] For a large portion of their life these creatures need cell dividers, and they develop as either a stripped protoplasmic mass or as amoeboid cells, changing over to a walled shape at the onset of sporulation. There is no proof that they are identified with growths, yet they have customarily been contemplated by mycologists, and they have a few intriguing components, which are talked about towards the finish of this section. [6]

TRUE FUNGI

Chitridiomycota

The Chytridiomycota, normally named chytrids, number around 1000 species (Barr 1990) and are thought to be the most punctual branch of the genuine growths, going back to around 1 billion years prior. They have cell dividers made primarily out of chitin and glucans (polymers of glucose) and numerous different components average of growths. [7] In any case, they are remarkable in one regard, since they are the main genuine organisms that create motile, lash zoospores. Commonly, the zoospore has a solitary, back whiplash flagellum, however a portion of the chytrids that develop in the rumen of creatures have a few flagella, and some different chytrids (e.g. Basidiobolus ranarum, as of late exchanged to the Chytridiomycota in view of SSU rDNA investigation), have no flagella. This gives a decent case of the estimation of DNA sequencing in deciding the genuine phylogenetic connections of life forms. [8]

Biology and Essentialness

Most chytrids are little, subtle creatures that develop as single cells or primitively fanned chains of cells on natural materials in wet soils or amphibian situations. They are considered to assume huge parts as essential colonizers and degraders of natural issue in these conditions. [9] Two normal illustrations are Allomyces arbuscula and Rhizophlyctis rosea (an unequivocally cellulolytic organism that is basic in characteristic soils).

Frequently, the Chytridiomycota are moored to their substrates by tight, decreasing rhizoids, which work like hyphae in discharging compounds and engrossing supplements. Now and again the rhizoidal framework is broad, or the thallus (the "body" of the organism) looks like a series of dots, with expanded cells emerging at interims along a rhizoidal system. A case of this is the parasitis fungi *Catenaria anguillulae*. [10] A couple of chytrids develop as commit intracellular parasitic fungi of plants (e.g. Olpidium brassicae), of little creatures (e.g. Coelomomyces), of green growth, or of parasitic spores. In any case, not very many chytrids can be thought to be financially vital, the most eminent case is *Synchytrium endobioticum*, which causes potato wart sickness, where the potato tubers grow unattractive nerves that render them unmarketable. [11]

Having said this, the hugeness of chytrids lies fundamentally in their captivating science. Anyone who has viewed a chytrid zoospore creeping like an one-celled critter along the body of a nematode, looking for the best site to encyst, and afterward twisting in its flagellum, encysting and entering the host will always remember the experience. [12] Every one of these occasions can be followed in straightforward glass chambers on magnifying instrument slides.

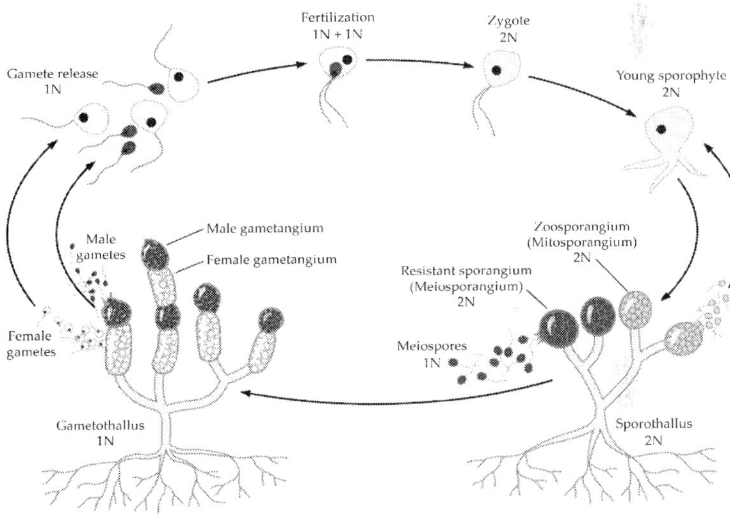

Figure- 2.1: Chytridiomycota (Life cycle of Allomyces).

The Chytridiomycota are hard to seclude by standard techniques, for example, weakening plating of soil onto agar plates. In any case, they can be discovered effectively by suspending little "trap" particles on the surface of characteristic waters or in dishes of overwhelmed soil. In these conditions, chytrid zoospores amass on lures, for example, cellulose, chitin, keratin, bug exoskeleton, or dust grains. [13] At that point they encyst and deliver rhizoids for port to the substrate. There is solid test prove that chytrid zoospores collect specifically on various sorts of goad. [14] For instance, when bits of cellulose or cleansed crab-shell chitin were added to zoospore suspensions, the zoospores supposedly encountered the snares aimlessly, yet then changed their swimming example, making incessant irregular turns, and frequently encysted inside 3–5 minutes. Zoospores of A. javanicus and Allomyces arbuscula collected and encysted on both cellulose and chitin, while zoospores of Chytridium confervae encysted specially on chitin, and zoospores of Rhizophlyctis rosea amassed and encysted just on cellulose. The in all probability clarification is that the zoospores have surface-found receptors that perceive diverse basic polymers, are marvel surely understood in zoospores of the parasitic fungus like Oomycota. [15]

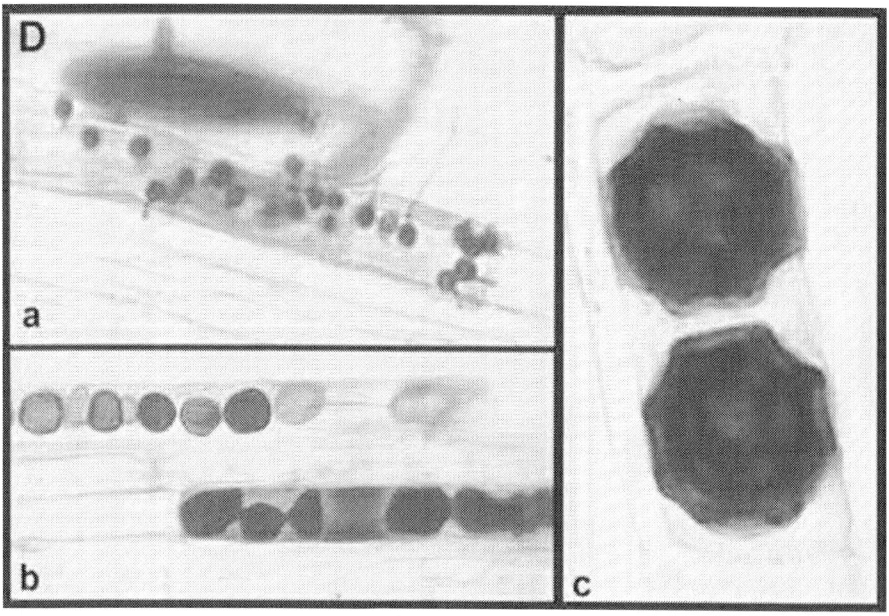

Figure- 2.2: Olpidium brassicae (a) Zoospore cysts (b, c) Thick walled spores (d) Binucleate Zoospores.

Taxonomy and Relationships

As of now, the Phylum Chytridiomycota is subdivided into five requests (Neocallimastigales, Spizellomycetales, Monoblepharidales, Blastocladiales, and Chytridiales) construct to a great extent in light of ultrastructural elements of the zoospores, which appear to show preserved examples of advancement in the diverse genealogies. [16] For instance, the zoospores of all individuals from the Order Blastocladiales (counting *Catenaria anguillulae*) have a prominent atomic top that encompasses the core and is loaded with ribosomes, yet unique game plans of the organelles are found in the other chytrid orders. A number of these distinctions identify with the course of action of microtubular components related with mooring of the flagellum, and the game plans of mitochondria and lipid stockpiling holds that are basic for zoospore motility. [17] We ought to likewise take note of that the Order Neocallimastigales is special in being obligately anaerobic. These living beings have as of late been appeared to have a hydrogenosome, equal to a mitochondrion, for creating vitality. Be that as it may, there is still uncertainty about whether the Neocallimastigales is a characteristic phylogenetic gathering. SSU rDNA examination of a more extensive scope of chytrids should clear up their connections. [18]

Reproduction and other features of Chytridiomycota

Most genuine organisms have a haploid genome however a few types of Allomyces can interchange amongst diploid and haploid eras. This is likewise valid for a few types of Blastocladiella. There are distinctive examples of the life cycle inside Allomyces spp. be that as it may, one of these examples, exemplified through A. macrogynus and *A. arbusculus* and, includes a dominatingly diploid stage. Sporangia are created on the diploid provinces, and cytoplasmic cleavage inside the sporangium prompts the arrival of diploid zoospores. [19]

These encyst and after that sprout to create promote diploid provinces. This procedure can proceed as long as the natural conditions are appropriate, however at the onset of troublesome conditions the growth creates thick-walled resting sporangia. Meiosis at that point happens inside the resting sporangia, prompting the arrival of haploid zoospores, which encyst and afterward grow to create haploid provinces. [20] These settlements deliver male and female gametangia, which discharge male and female haploid gametes. The gametes of inverse mating sort circuit to shape a diploid zygote, which encysts and after that grows to rehash the diploid period of the life cycle. [21] In different Allomyces spp. there is no different gametophyte era, rather this is most likely spoken to by a sore, which sprouts to deliver a further abiogenetic diploid province. [22]

A standout amongst the most charming elements of chytrid zoospores is their capacity to experience delayed amoeboid creeping, by pseudopodium-like augmentations of the cell. This happens both on glass surfaces and on the surfaces of potential has, for example, nematodes which are looking conduct for finding appropriate locales for encystment. [23] At that point the zoospores round-up and twist in their flagellum by turning most or the majority of the cell substance.

GLOMEROMYCOTA

Parasitic fungal fossils taking after the normal and financially essential arbuscular mycorrhizal (AM) growths were related with plants in the Rhynie chert stores of the Devonian time around 400 million years back (mya). [24] They presumably assumed an imperative part in the foundation of the land vegetation. Late itemized examination of these organisms, in view of SSU rDNA groupings, has demonstrated convincingly that they are particular from all other major parasitic gatherings, so they have been doled out to another phylum, the Glomeromycota. [25] The connections between this gathering and other organisms are as yet hazy, however, extremely solid bootstrap bolster for connecting these AM parasitic fungi.

The arbuscular mycorrhizal growths are considered, so we will examine them just quickly here. They share a few trademark and very striking elements:

- They are discovered developing inside the underlying foundations of most by far of plants, but they can't be developed autonomously, without a plant.
- When they enter the roots they develop dominatingly between the root cortical cells and regularly (yet not generally) create vast, swollen vesicles which are accepted to work as sustenance stockpiling saves. [26]
- The AM parasitic hyphae infiltrate singular root cortical cells to frame unpredictably fanned arbuscules (tree-like expanding frameworks). Be that as it may, they don't execute the root cells, and rather they set up a cozy sustaining relationship, which appears to profit both accomplices. The degree of this kind of relationship, in roots cleared of plant protoplasm, at that point recolored with trypan blue. [27]

As of late found fossil hyphae and spores from Wisconsin, USA, go back to the Ordovician, 460–455 mya and in this way pre-date the vascular plants (i.e. plants with water-directing tissues). They are amazingly like the AM contagious spores, despite the fact that it is stressed that they were not related with plants. In any case, they can give adjustment focuses in phylogenetic investigations. [28]

One last point that is maybe the most astounding, was the disclosure of a novel sort of beneficial interaction, first revealed in 1996 and still known from just a couple of characteristic destinations in Germany. In this advantageous interaction, AM organisms are joined to plant attaches and grow up to the dirt surface, where the AM growth produces straightforward bladders. [29] These bladders inundate and disguise the cells of a cyanobacterium, Nostoc punctiforme, so that the parasite acquires sugars from the cyanobacterial accomplice. The subsequent living being is called Geosiphon pyriforme, an individual from the Geosiphonaceae (Glomeromycota). [30]

ZYGOMYCOTA

Five noteworthy elements serve to describe the phylum Zygomycota:

- Cell dividers made out of a blend of chitin, chitosan (an inadequately or non-acetylated type of chitin) and polyglucuronic corrosive.
- Hyphae that normally need cross dividers, so every one of the cores are contained inside a typical cytoplasm (a coenocytic mycelium).
- The generation of a thick-walled resting spore, the zygospore, which if framed by a sexual procedure including the combination of two gametangia.
- The generation of abiogenetic spores by cytoplasmic cleavage inside a sporangium.
- A haploid genome. [31]

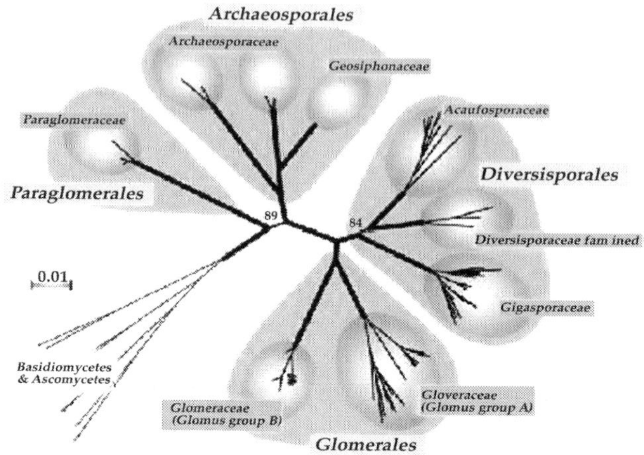

Figure- 2.3: General taxonomic structure of AM fungi.

Ecology and Significance

The Zygomycota contains a various scope of animal varieties, with various sorts of conduct, yet for comfort we begin with the most well-known cases in the Order Mucorales. This incorporates a few normal types of the genera Mucor, Rhizopus, Phycomyces and Thermomucor (a thermophilic organism that can grow up to around 60°C). [32] Every one of these parasitic fungi develop for the most part as saprotrophs in soil, on creature fertilizer, on manures on account of thermophilic species, or on different substrates, for example, finished ready natural products. They develop quickly, frequently covering an agar plate in 24–36 hours, and they are among the commonest parasitic fungi found on soil weakening plates. They have been named "sugar organisms", to indicate the way that they frequently rely upon basic, dissolvable carbon supplements. Be that as it may, some natural product spoiling species, for example, Rhizopus stolonifer create pectic catalysts that can quickly melt apples and delicate organic products. [33] What's more, a few types of Mortierella corrupt chitin by delivering chitinases (as likewise do a few chytrids). Chitin is a typical and essential polymer, since it is a noteworthy segment of parasitic dividers and of the exoskeletons of bugs and different spineless creatures. [34]

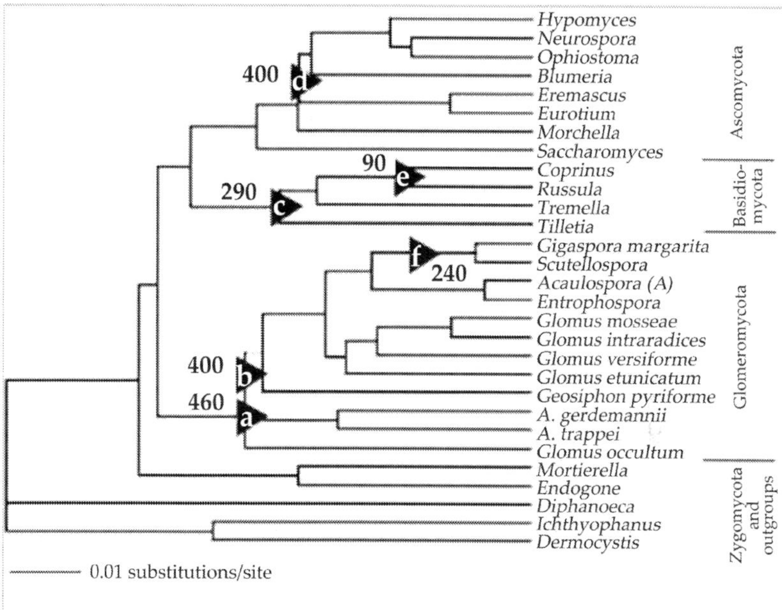

Figure- 2.4: A phylogenetic tree based on small subunit ribosomal RNA gene sequences.

After a period of fast substantial (vegetative) development, they deliver erect aeronautical branches (sporangiophores) behind the province edge, at that point a thinwalled sporangium is framed at the tip and the sporangial substance are divided around the individual cores to create numerous uninucleate spores (sporangiospores). [35] These spores are dimly pigmented, with melanin in their dividers. They are discharged and as a rule twist scattered by breakdown of the thin sporangial divider, however in a few animal types they are water-scattered. The tip of the sporangiophore ventures into the sporangium as a columella, which stays after the spores have been shed. Sporangiospores can develop quickly in reasonable conditions, rehashing the agamic cycle. [36]

Inside the Zygomycota there are a few varieties in the ways that sporangia and sporangiospores are delivered. A few animal groups create sporangia at the tips of since quite a while ago, unbranched sporangiophores yet the thermophilic species regularly create extended sporangiophores. [37] Thamnidium elegans is a psychrophilic (frosty adoring) animal types which can develop on meat in cool stockpiling, and at one phase was protected for softening steak. It delivers a huge sporangium at the tip of a long sporangiophore, however the sporangiophore additionally creates complex fanning structures that end in numerous few-spored sporangioles. Different varieties are found in various Orders and Families of the Zygomycota. In this regard perceive that the spore-bearing structures of growths have practical hugeness, their part is to amplify the odds of scattering spores to locales where the parasite can set up new states. Numerous individuals from the Mucorales additionally have a sexual stage. In suitable culture conditions elevated branches, named zygophores, develop towards each other and swell at their tips to deliver progametangia.

At that point the progametangia build up an entire septum, which removes every gametangium from the subtending hypha. The gametangia at that point breaker to frame a solitary cell that forms into a zygospore which is an ornamented resting spore and thick-walled. Combined with this, the terminal parts of the zygophores can swell, to deliver bulbous suspensors. [38] Zygospores display a time of constitutive lethargy, yet in the long run grow to create a sporangium, discharging haploid spores. Numerous species in the Mucorales are heterothallic (outcrossing) and create zygospores just whenstrains of inverse mating sort (assigned in addition to and less) are restricted on agar plates. Atomic combination and meiosis for the most part happen inside the zygospore. [39]

Notwithstanding the Order Mucorales, the Zygomycota contains an extensive variety of growths that parasitize other little creatures, including nematodes, other little spineless creatures, and other organisms. For instance, the family Piptocephalis, in the Order Zoopagales, contains a few species that parasitize other Zygomycota and infrequently other fungal growths.

Piptocephalis is one of a few genera that create few-spored sporangia, named merosporangia, with straightly masterminded spores. In nature the Piptocephalis spp. presumably develop as commit parasitic fungi on other organisms, particularly other Zygomycota, in spite of the fact that they can be developed in unadulterated culture if provided with the fundamental vitamins. [40] They enter the host hyphae without murdering them, by creating particular, supplement engrossing haustoria, and along these lines they tap the host's supplements to help their own particular improvement. This kind of collaboration is by all accounts administered by particular acknowledgment elements (lectins).

ZYGOMYCOTA

This fungal growth produces common place sporangia on long sporangiophores, yet complex stretching structures emerge from the sporangiophore and end in little, few-spored sporangioles. The Order Entomophthorales is eminent on the grounds that it incorporates numerous growths that parasitize creepy crawlies, and that can cause dynamite populace accidents of their bug has. Many are have particular, commonly contaminating just a thin scope of creepy crawlies, however others have moderately expansive host ranges. [41]

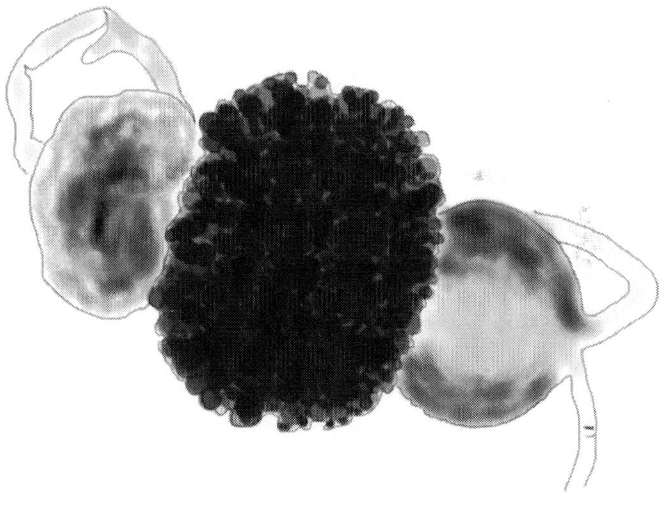

Figure- 2.5: Thick walled Zygospore.

Strategies have been created to mass deliver some of these organisms in strong or fluid culture frameworks, raising the possibility of utilizing them as pragmatic bio-control operators of creepy crawly bugs. The Entomophthorales are striking for the route in which they are scattered. [42] The hyphae rising up out of a dead creepy crawly deliver sporangiophores, each bearing a vast sporangium at its tip. The tip of the sporangiophore has a columella, which swells into the spore and is twofold walled (one divider being the tip of the columella, and the other being the mass of the sporangium). [43]

The dynamic develop of turgor weight makes the sporangium be shot abruptly from the columella, anticipating the sporangium to a separation of around 4 cm. The sporangium (conidium) is sticky and must hold fast to a creepy crawly keeping in mind the end goal to contaminate. In the event that it doesn't arrive on a bug it will sprout to deliver an auxiliary conidium, which is likewise scattered, and this procedure will be rehashed until the point that the conidium arrives on an appropriate host or until the point that the supplement holds are depleted. [44]

Some other parasitic fungi of the Entomophthorales show a dispersal methodology like that depicted previously. One illustration is Conidiobolus coronatus, a typical saprotrophic organism in soil. Another case is Basidiobolus ranarum, which is found in the gut, the excrement or the digestion tracts of creepy crawly eating reptiles or creatures of land and water (e.g. frogs, from which the species name ranarum is inferred). [45] This growth has as of late been exchanged to the Chytridiomycota, in view of SSU rDNA sequencing. A third parasite, Pilobolus in the Order Mucorales, demonstrates a parallel method of dispersal. This organism is regular in creature excrement and produces an extensive sporangium which is without shot from the manure, however the sporangium thusly separates to discharge numerous individual spores.

Zygomycota as Pathogens of Humans

A few basic individuals from the Mucorales, particularly Rhizopus arrhizus, Absidia corymbifera and Rhizomucor pusillus (a thermophilic species) can cause genuine, life-undermining diseases of people. Comparative contaminations can incidentally be caused by Conidiobolus (Entomophthorales) and Basidiobolus ranarum. [46] All in all, these illnesses are named zygomycosis (contaminations caused by individuals from the Zygomycota). In most by far of cases the disease is related with inclining elements. For instance, diabetics with ketoacidosis are among the most powerless against contamination, which regularly creates in the rhino-cerebral range (the nasal entries, advancing into the mind). Be that as it may, numerous other inclining variables can be included, incorporating extreme malnourishment in kids, serious consumes, disease, lymphoma, and immunodeficiency or insusceptible concealment.

These parasitic contaminations can grow quickly by spread through the corridors and intrusion of the encompassing tissues. [47] They regularly require surgical extraction of the influenced zones. All things considered, the guess is poor as the general death rate is around half and can ascend to 85% in patients with the rhino-cerebral type of the malady. The zygomycoses fall plainly into the class of pioneering diseases, since they are related with debilitated host safeguards. They are among the most troublesome contaminations to control on account of their quick, intrusive development.

ASCOMYCOTA

The phylum Ascomycota contains around 75% of the considerable number of parasitic fungi that have been depicted to date. It is the most essential phylum, as well as the most various, and a significant number of the connections inside this gathering still can't seem to be settled by present day atomic techniques. The one component that portrays all individuals from this phylum is the ascus which is a cell in which two good haploid cores of various mating sorts meet up and breaker to frame a diploid core, trailed by meiosis to create haploid sexual spores, named ascospores. [48] In numerous species the meiotic division is trailed by a solitary round of mitosis, prompting the creation of eight ascospores inside every ascus. In the more propelled individuals from the gathering, numerous asci are delivered inside a fruiting body, named an ascocarp. This can take different structures as like a jar formed perithecium with a pore at its tip, a container molded apothecium, a shut structure that separates at development, named a cleistothecium, or a pseudothecium which is typically inserted in a stack of tissue, named a stroma. In different cases, the asci are created separately and are not encased in a fruiting body, for instance, this is valid for the sprouting yeast Saccharomyces cerevisiae (which delivers just 4 ascospores in an exposed ascus) and the splitting yeast Schizosaccharomyces octosporus (eight-spored asci). [49]

Figure- 2.6: A developing perithecium of Sordaria macrospora.

By far most of Ascomycota likewise create agamic spores by mitosis. These mitospores (instead of meiospores, determined by meiosis) are to a great degree normal and capacity in dispersal. Actually, a generous number of Ascomycota are known just as their mitosporic (abiogenetic) stages. [50] Their sexual stages stay to be found and now and again may have been surrendered, however SSU rDNA arrangement investigation can be utilized to interface them to the Ascomycota. To give only one illustration, the to a great degree basic growth, Aspergillus fumigatus, has never been appeared to create a sexual stage. This parasite is winding up noticeably progressively normal as a noteworthy reason for death in healing facility patients experiencing profound surgical strategies, and it additionally causes the genuine respiratory sickness named aspergillosis. [51] In cases, for example, this it is critical to know the wellsprings of hereditary variety that is there a sexual stage that we presently can't seem to find?

In the same way as other individuals from the Ascomycota, *N. crassa*, has a sexual stage, prompting the creation of asci containing ascospores. N. crassa is heterothallic, requiring strains of various mating sorts (named An and a) for sexual improvement. The female sex organ, named an ascogonium, is frequently a wound, multinucleate hypha with a responsive trichogyne. [52] It is treated either by contact with a spermatium (a little uninucleate cell that can't sprout) or by a conidium.

At that point the ascogonium produces an ascogenous hypha that will in the long run offer ascent to the asci.

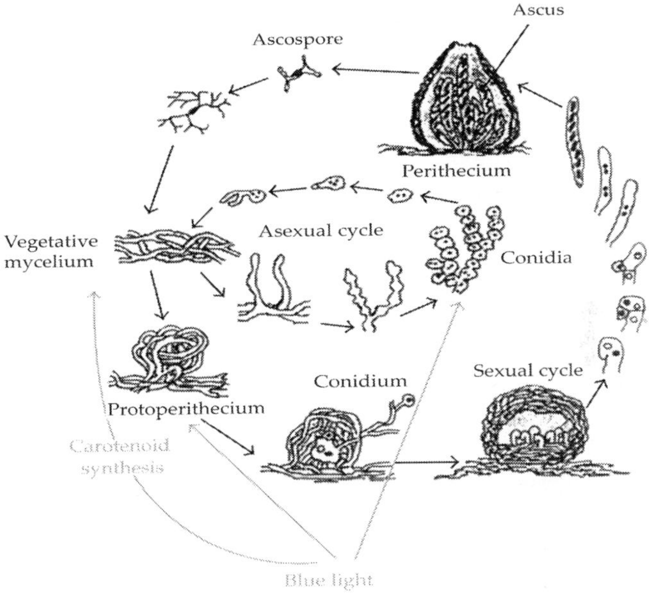

Figure- 2.7: Life cycle of Neurospora crassa.

Scientific Classification and Connections

The Ascomycota is thought to be a monophyletic gathering (all individuals sharing a typical family) that goes back to the coal age, no less than 300 million years prior. [53] This gathering is firmly identified with the Basidiomycota (the phylum containing mushrooms, and so on.) as a sister assemble that offers comparable components. As of now, the Ascomycota is partitioned into three subgroups, in spite of the fact that the status of a portion of the early separating individuals is as yet indistinct:

- Archaeascomycetes, including the splitting yeast, Schizosaccharomyces pombe, and some primitive plant pathogens, for example, Taphrina spp.
- Hemiascomycetes – the genuine yeasts, for example, Saccharomyces cerevisiae, used to deliver liquor and for bread-production. These parasitic fungi don't deliver ascocarps.

- Euascomycetes, involving the lion's share of species that create hyphae and ascocarps.

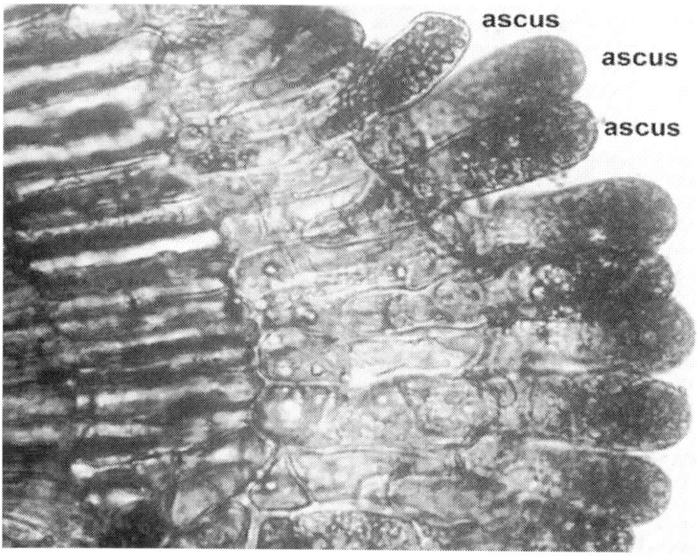

Figure- 2.8: A parasitic member of Archaeascomycota, Taphrina populina.

Notwithstanding the fundamental recognizing highlight, the ascus, the Ascomycota are portrayed by cell dividers made basically out of chitin and glucans, by hyphae that have cross dividers (septa, solitary septum) at standard interims yet with a focal pore that permits the entry of cores and other cell organelles, and by the creation of non-motile abiogenetic spores (mitospores, named conidia) that are delivered in different ways, yet never by cytoplasmic cleavage in a sporangium. [54]

The Life Cycle of Ascomycota

We will utilize Neurospora crassa for instance to outline noteworthy components of the life cycle of Ascomycota. Neurospora develops as an extended system of hyphae with punctured septa, and produces airborne hyphae that end in expanded chains of conidia (abiogenetic spores). [55] These are shaped by a bud-like process including swelling and rehashed septation. Whenever develop, the conidia build up a pink tinge and are effortlessly expelled from the hyphae via air streams. This cycle of sporulation, dispersal and spore germination happens over and again when substrates are promptly accessible.

There are a few varieties in sexual conceptive frameworks of the Ascomycota, which we require specify just quickly. For instance, *N. crassa* and *N. sitophila* are heterothallic (out-intersection) yet a few species (*Neurospora tetrasperma*, *Podospora anserina*) create just 4 binucleate ascospores in every ascus, and are homothallic. A solitary ascospore can give rise a state that will create the sexual stage. [56]

Ecology and Significance

Since the Ascomycota is a huge and critical phylum, numerous parts of its science are shrouded in the later sections of this book. The gathering all in all incorporates numerous financially imperative plant pathogens, for example, the fine buildup growths of many yield plants, the vascular wither fungal organisms, *Ophiostoma ulmi* and *O. novo-ulmi*, which cause the staggering Dutch elm ailment that cleared over and again crosswise over Europe and the USA in the most recent century, the similarly wrecking chestnut curse illness in the USA, caused by Cryphonectria parasitica and a few toxigenic growths, for example, Claviceps purpurea (ergot of grains).

The Ascomycota likewise incorporates a few pathogens of people, tamed creatures and animals. For instance, the omnipresent dermatophytic (ringworm) growths are assessed to taint about portion of the aggregate human populace, particularly in the tropics and subtropics, additionally in created nations, where contaminations, for example, "competitor's foot" and "nail organism" are normal. [57] The ascomycetous yeast, *Candida albicans*, is a typical commensal life form in the gut and on other mucosal layers of people, making disturbance individuals who wear dentures and to ladies amid period or pregnancy. What's more, some ascomycetous parasitic fungi (*Blastomyces dermatitidis*, *Histoplasma capsulatum*) can cause life-debilitating maladies of people whose safe framework is insufficient or weakened.

In an alternate setting, a portion of the Ascomycota shape mycorrhizal relationship with backwoods trees (e.g. Tuber spp., the truffles) and an expected 96% of the 13,500 referred to types of lichens have Ascomycota as the parasitic accomplice. [58] Not very many of these lichenized parasitic fungi develop in a free-living state. As far as numbers, the ascomycetous lichen organisms most likely speak to the biggest single part of biodiversity in the whole contagious kingdom. Numerous ascomycetous yeasts (Saccharomyces, Candida oleophila, and so forth.) ordinarily develop as saprotrophs on leaf and natural product surfaces, and some are currently being showcased as organic control specialists to avoid organic product decays. Numerous different ascomycetous organisms debase cellulose and other auxiliary polymers, e.g. Xylaria, Chaetomium in soil and fertilizers, and, Lulworthia on wood in estuarine conditions, and Hypoxylon as operators of wood rot, Ascobolus and Sordaria in herbivore manure.

BASIDIOMYCOTA

The Basidiomycota contains around 30,000 portrayed species, which is 37% of the depicted types of genuine fungal growths. In spite of the fact that the most usual place cases of this gathering are the toadstools and mushrooms, there is a huge differences of species, including basidiomycetous yeasts, numerous imperative plant pathogens, and some genuine human pathogens. [59]

The one single element that portrays the gathering is the basidium in which meiosis happens, prompting the generation of sexual spores (basidiospores) that generally are created remotely on short stalks named sterigmata.

A few different components are found in the Basidiomycota. The hyphae frequently have a complex dolipore septum that keeps cores from moving between various hyphal compartments. Its cell wall ordinarily is made out of chitin and glucans (however chitin and mannans in the yeast frames).

The cores commonly are haploid, yet all through the greater part of the life of a state each hyphal compartment contains two cores, speaking to two diverse mating similarity gatherings. A mycelium of this sort is named a dikaryon. The criticalness of this will be clarified presently. [60]

Scientific Classification and Connections

The Basidiomycota is a monophyletic gathering (the majority of its individuals having a typical family line) and is a sister gathering to the Ascomycota. At the end of the day, both of these gatherings have a typical progenitor. Three noteworthy sub-bunches are perceived inside the Phylum Basidiomycota:

- Urediniomycetes, including the rust organisms (Uredinales) which are monetarily huge plant pathogens of many harvests and wild plants.
- Ustilaginomycetes, including the muck parasitic fungi (Ustilaginales) some of which again are noteworthy plant pathogens, and pick up their name from their dark, dingy spores.
- Hymenomycetes, including mushrooms, puffballs, and jam parasitic fungi.

Nonetheless, the phylogenetic connections between and inside these three gatherings are as yet vague. **Noteworthy components in the life-cycle of Basidiomycota:**

We will start by considering the summed up life-cycle of a commonplace mushroom. Basidiospores, each containing a solitary haploid core, grow and develop into hyphal provinces that have a solitary core in each hyphal compartment. [61] This stage is named a monokaryon (implying that it has one atomic sort in each hyphal compartment). The monokaryon can deliver little oidia that either go about as treating components or they can grow to create promote monokaryotic provinces.

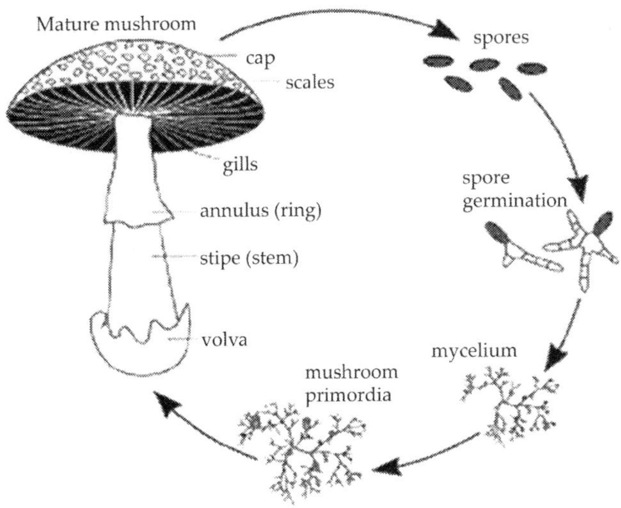

Figure- 2.9: Life cycle of mushrooms.

The following phase of improvement happens when the hyphae of two monokaryons of various mating similarity bunches intertwine with each other, or when an oidium of one mating similarity aggregate pulls in a hypha of an alternate mating similarity gathering. [62] This causes a solid "homing reaction" however the synthetic attractants causing it have not been distinguished. Once the two good strains have melded by the procedure named plasmogamy, all consequent development happens by dikaryotic hyphae (with two cores such as one of every similarity assemble, in each hyphal compartment). [63]

The parasite can develop for a long time, months or even years in this frame, creating a broad system of dikaryotic hyphae, however because of ecological signs it will deliver a mushroom or other kind of fruitbody. Every one of the tissues of the mushroom are made out of dikaryotic hyphae, however at a generally late stage being developed of the mushroom the gills wind up noticeably fixed with (dikaryotic) basidia. Inside every basidium the two cores breaker to frame a diploid core (karyogamy) which in this manner experiences meiosis. Sterigmata at that point create from the highest point of every basidium, and the four haploid cores move into the creating basidiospores. [64] Once in a while just two spores are delivered, rather than four such as a typical case being the developed mushroom, Agaricus bisporus. This is on account of, after meiosis in the basidium, 2 cores enter every spore, and the growing basidiospores create a dikaryotic mycelium that is homothallic (self-rich). Be that as it may, by far most of mushroom-delivering organisms are heterothallic, around 90%. [65]

The path in which the basidiospores are discharged is not completely seen, but rather just before their discharge a little bead, named the hylar bead, is seen at the base of the spore, at that point the spore is "flew" from the sterigma into the space between the gills, where it falls vertically, free from the fruitbody. After a time of lethargy the haploid basidiospores develop to create monokaryotic hyphae, which rehash the entire cycle.

A standout amongst the most fascinating parts of this life cycle is the way in which the dikaryon creates. The dikaryotic tip cell broadens and blends protoplasm until the point that it accomplishes a basic cytoplasmic volume. At that point a little backwardly-anticipating branch emerges, and the two cores partition synchronously and is one of the cores separating along the hub of the hypha, and the other isolating so that one of the cores enters the branch. The branch is then cut off by the improvement of a septum (cross-divider) and a septum likewise creates in the fundamental pivot of the hypha. The impact of this is to make a tip cell with two cores of various similarity sorts, a penultimate cell with a solitary core, and a branch with a solitary core, segregated by septa. [66] The branch at that point wires with the penultimate cell, by disintegration of the divider, and the core moves into the penultimate cell, reestablishing the dikaryotic condition. This procedure happens each time the cores separate. The subsequent "knocks" on the hyphae are effortlessly observed by light microscopy, and are named clasp associations. Nonetheless, not all individuals from the Basidiomycota create brace associations, but then despite everything they develop as dikaryons, recommending that other administrative systems are included.

- In the event that we allude back to the advancement of asci, we see a striking closeness in the conduct of cinch associations and the improvement of asci in the Ascomycota. In both cases the atomic conveyance is kept up by a hypha becoming back upon itself and combining with the cell behind it. What's more, in both cases there is a dikaryotic stage, in spite of the fact that it happens just incidentally in the ascus initials. These lines of proof fortify the view that the Basidiomycota and Ascomycota are firmly related. [67]

Nature and Importance

The Phylum Basidiomycota displays the most extraordinary variety of any gathering of growths. A portion of the yeasts on plant surfaces are basidiomycetous, in spite of the fact that they are far dwarfed by the ascomycetous yeasts. A typical case is the salmon-hued yeast, Sporobolomyces roseus, which develops on senescent leaf surfaces. It isolates by growing additionally has a sexual stage, in which single yeast cells deliver a sterigma and discharge a basidiospore, clearly by an indistinguishable instrument from in mushrooms. *S. roseus* is a typical tenant of leaf surfaces, and the spore populace noticeable all around can be sufficiently high to cause respiratory sensitivities. Be that as it may, a couple of basidiomycetous yeasts are a great deal huger and can be life-debilitating pathogens of people, particularly of individuals with impeded safe guards. [68]

The most striking cases are Blastomyces dermatitidis, Cryptococcus neoformans and Histoplasma capsulatum, all of which can create sexual stages when strains of various mating sorts are crossed in research center conditions, yet the sexual stages are rarely found in nature. [69] The wellsprings of inoculum that start disease in the lungs are regular ecological examples, for example, winged animals' droppings and even the bark of eucalypt trees. The imperative lesson from this is even the clearly immaterial saprotrophic growths of plant surfaces and different materials can significantly affect human wellbeing.

Probably the most monetarily essential plant pathogens are individuals from the Basidiomycota. They incorporate the numerous species and genera of rust parasitic fungi (Uredinales) that are specific to taint specific sorts of plant. One of the best-known cases is Puccinia graminis, which causes dark stem rust of wheat. These growths can have complex life cycles, with up to four distinct sorts of spore being created at various phases of the life cycle. Another imperative gathering of pathogens, the muck growths (Ustilaginales), are uncommon on the grounds that they become just as yeasts in culture however as mycelia in their hosts. [70]

The most recognizable parasitic fungi in the Basidiomycota have a place with the class Hymenomycetes, and incorporate all the bigger growths, for example, the mushrooms (generally known as toadstools), woody sections, puffballs, earth-stars, sprinkle mugs, and jam like structures. The vast majority of these "higher" Basidiomycota corrupt polymers, for example, lignin, hemicelluloses and cellulose. [71,72] They are found in manures (e.g. Coprinus), leaf litter (e.g. Mycena), the cover of dead leaf sheaths in old fields (e.g. Marasmius oreades, which produces "pixie rings" and in woody substrates, where they assume a noteworthy part in wood rot. Some of these parasitic fungi cause significant tree infections, e.g. *Armillaria spp.* in hardwood stands and *Heterobasidion annosum* in conifers.

Many other mushroom-producing fungi form mycorrhizal associations with forest trees, although this applies mainly to the cooler regions of the world, because most tropical and subtropical trees have arbuscular-mycorrhizal fungi. Common examples of mycorrhizal fungi found in the cooler regions include species of Lactarius, Amanita, Hebeloma, Boletus, Russula, Cortinarius, etc. They play major roles in facilitating the uptake of mineral nutrients from soil. But, whereas the arbuscular mycorrhizal fungi of most land plants seem to be involved primarily in the uptake of phosphates from soil, the ectomycorrhizal fungi of forest trees (excluding the tropics and subtropics) seem to be most significant in degrading soil proteins and providing the trees with nitrogen.

Numerous other mushroom-generating parasitic fungi frame mycorrhizal relationship with woodland trees, in spite of the fact that this applies basically to the cooler locales of the world, on the grounds that most tropical and subtropical trees have arbuscular-mycorrhizal organisms. Basic cases of mycorrhizal parasitic fungi found in the cooler areas incorporate types of Lactarius, Amanita, Hebeloma, Cortinarius, Russula, Boletus, and so forth. They assume real parts in encouraging the take-up of mineral supplements from soil. Be that as it may, though the arbuscular mycorrhizal parasitic fungi of most land plants appear to be included fundamentally in the take-up of phosphates from soil, the ectomycorrhizal organisms of woodland trees (barring the tropics and subtropics) appear to be most huge in corrupting soil proteins and furnishing the trees with nitrogen.

A few individuals from the Basidiomycota are developed industrially as mushroom crops (e.g. *Agaricus bisporus*, *Volvariella volvacea*, *Lentinula edodes*), while others are "developed" by creepy crawlies to give a nourishment source to the bug states. Two great cases of this are found in the "fungal growth gardens" of the leaf-cutting attine ants (*Atta* species) and in a sub-group of termites that need cellulolytic microbes in their guts. In both cases, the bugs get their principle nourishment source by bringing bits of leaf or other material into the homes and vaccinating this with spores of basidiomycetous organisms.

The creepy crawly province at that point encourages on the fungal parasitic hyphae, and removes any contaminant organisms. Cladistic investigation has followed the cause of these mutualistic relationship to around 50–60 million years back. Different individuals from the Basidiomycota, for example, Amanita phalloides, "the crushing heavenly attendant" which deliver savage poisons that dilemma to the cytoskeletal proteins in cells. These poisons have turned out to be imperative devices in cell science for examining cell progression.

Basidiocarps Diversity produced by Basidiomycota

As a methods for outlining the significant assorted qualities of shape and capacity of basidiocarps, a scope of cases are appeared, with explanations.

Figure- 2.10: Basidiocarps of birds nest fungi.

MITOSPORIC FUNGI

At long last, among the genuine parasitic fungi, we consider the vast number of ascomycetous growths that create conidia, yet whose sexual stages are truant, uncommon or obscure. Previously, these growths have differently been named "Deuteromycota" (Deuteromycotina) or "Parasitic fungi Imperfecti" yet the more fitting term is mitosporic organisms, showing that the spores are created just by mitotic atomic division. The share of these organisms are probably going to be allocated to genera of the Ascomycota by quality arrangement examinations. Be that as it may, at show, the genera and species with no known sexual stage are given temporary non specific names, named "frame genera." When a sexual stage is found the parasite must be renamed, and portrayed by the elements of its sexual regenerative stage, the alleged "impeccable state." (Other terms utilized incorporate anamorph for the abiogenetic stage, and teleomorph for the sexual stage) Some exceptionally normal and essential organisms are known just by their agamic stages.

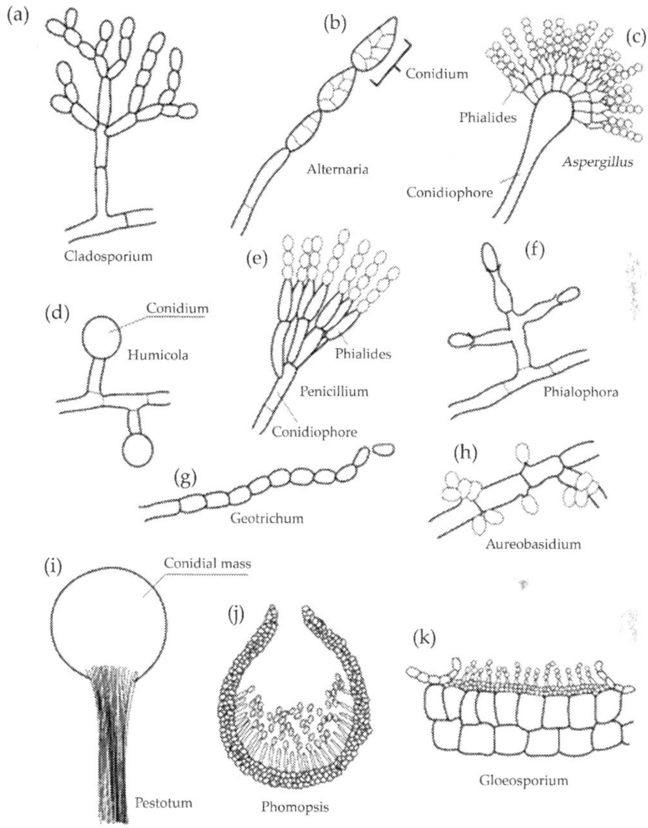

Figure- 2.11: Conidium production in mitosporic fungi.

Cases incorporate growths of the shape genera Alternaria (albeit one species has a sexual stage, named Lewia infectoria), Aspergillus (albeit two species have a sexual stage: Emericella nidulans for Aspergillus nidulans and Fenellia flavipes for Aspergillus flavipes), Phialophora, Penicillium, Cladosporium, Humicola, and so on. The fascinating inquiry is: the reason have such a variety of growths generally or totally surrendered sexual propagation? The appropriate response may be that they have created elective methods for hereditary recombination, for example, a "parasexual cycle" that has been depicted in (mitosporic) genera, for example, Penicillium and Aspergillus.

The Methods of Conidial Production

The mitosporic organisms, similar to the agamic phases of known Ascomycota, deliver conidia in different ways, however never by cytoplasmic cleavage in a sporangium. It has been recommended that the fundamental sorts of conidium advancement can be either thallic (basically by a fracture procedure including septation, as in the growth Geotrichum candidum) or blastic, including swelling of the conidium initials before they are isolated by septa (e.g. Alternaria and Cladosporium). [73]

Most sorts of conidium advancement would fall into the "blastic" sort. For instance, the conidia of Aureobasidium pullulans, now named by its sexual (teleomorph) arrange, Sydowia polyspora, deliver conidia by growing at or close to the septa. The conidia of Humicola spp. are delivered by a swelling of the tip of a conidiophore branch. The conidia of numerous organisms are created by expelling spores from jar formed cells named phialides (Penicillium, Aspergillus, and Phialophora).

Different varieties are found in the ways that conidiophores are masterminded. For instance, Penicillium spp. ordinarily have brush-like conidiophores, ending in phialides, and the progressive conidia collect in chains. Aspergillus spp. normally have a conidiophore with phialides orchestrated on a swollen vesicle, and again they create chains of conidia. The conidiophores can be amassed into a stalk named a synnema or coremium (e.g. Pesotum) or the conidia can emerge from a stack of tissue (an acervulus as in Gloeosporium) or they can be created in a flagon formed pycnidium (e.g. Phomopsis).

Environment and Importance

As will be clear from the remarks above, mitosporic parasitic fungi are to a great degree normal and create inexhaustible conidia which are scattered by wind, rain or creatures. A considerable lot of these fungal growths are basic saprotrophs, with vital parts in supplement reusing. A few have conidia that are sufficiently vast 20–40 μm to affect in the nostrils, causing roughage fever indications (e.g. Alternaria). Others have conidia that are sufficiently little (around 4 μm distance across) to escape impaction in the upper respiratory tract, and achieve the alveoli where they can cause intense hypersensitive reactions or now and again (e.g. Aspergillus, fumigatus) set up long haul foci of disease in the lungs. Mitosporic growths, for example, Aspergillus Penicillium, and Fusarium spp. can likewise be real deterioration specialists of put away nourishment items, and deliver a few unsafe mycotoxins (e.g. the aflatoxins of Aspergillus flavus). A few mitosporic parasitic fungi are real plant pathogens, including the vascular shrivel growths Fusarium oxysporum and Verticillium dahliae. Others parasitize creepy crawlies, for instance, the many strains of Metarhizium and Beauveria that have potential for natural control of bugs. Mitosporic parasitic fungi have numerous other useful parts and are abused financially for their tremendous scope of valuable metabolites (natural acids, anti-infection agents, immunosuppressants, proteins, and so forth.).

Fungus-like different life Forms

Notwithstanding the genuine growths, a few parasitic fungi like living beings have customarily been considered by mycologists. This heritage goes back quite a while and it is faultless, in as much as we perceive that the growth like living beings are not genuine parasitic fungi. The principle parasite like living beings that can be considered as organisms in a more extensive sense are:

- The Oomycota, which have a place with the Kingdom Straminipila (stramenipiles) and are firmly identified with the brilliant dark colored green growth and diatoms.
- The acrasid and dictyostelid sludge molds (Dictyosteliomycota and Acrasiomycota).
- The plasmodial sludge molds (Myxomycota).
- The plasmodiophorids (Plasmodiophoromycota).

OOMYCOTA

The Oomycota are the most financially vital growth like life forms and assume greatly critical parts in numerous ecological procedures. They are especially critical as plant pathogens, and cause probably the most pulverizing plant sicknesses, for example, potato scourge (caused by Phytophthora infestans), sudden oak demise (caused by Phytophthora ramorum), numerous seedling and "decrease" ailments caused by Pythium spp., and a few wool buildup ailments.

A standout amongst the most striking components of Oomycota is that they act precisely like the true fungal parasitic fungi. Their hyphae display apical development and enter plant cells by creating cell-divider debasing catalysts. They additionally utilize comparable disease procedures to those of growths. Along these lines, by a striking level of concurrent advancement the Oomycota have built up a way of life identical to that of growths. By the by, they are completely random to the genuine growths. Aside from their absence of photosynthetic shades, they have many plant-like components, including: Cell walls composed primarily of glucans, including cellulose-like polymers.

- Cell dividers made basically out of glucans, including cellulose-like polymers.
- Diploid cores, as opposed to the haploid cores of generally growths.
- Cell layers made out of plant sterols, as opposed to ergosterol, the trademark contagious sterol.
- Energy stockpiling mixes like those of plants, as opposed to the polyols (sugar alcohols) and trehalose found in generally parasitic fungi.
- A scope of ultrastructural highlights more firmly identified with those of plants than of growths, including stacked, plate-like golgi cisternae (the parasitic fungi have tubular golgi cisternae), and tubular mitochondrial cisternae (the organisms have plate-like or circle like mitochondrial cisternae.
- Different sensitivities to a scope of antifungal operators.

Components of Life-Cycle of Oomycota

The significant life-cycle elements of Oomycota, spoken to by the plant-pathogen, Phytophthora infestans, and water shape, for example, Saprolegnia spp. regularly, and the substantial (vegetative) stages are wide, quickly developing hyphae with diploid cores. The hyphae need cross-dividers (septa) aside from when they create finish, unperforated septa to disconnect the conceptive structures. Agamic generation includes the creation of a multinucleate sporangium, which is isolated by a septum.

The sporangial substance are then divided around the individual cores, bringing about the creation of diploid zoospores, each with an anteriorly coordinated tinsel-sort flagellum and a trailing whiplash flagellum. The zoospores are discharged when the tip of the sporangium separates, however in a few animal categories (e.g. Phytophthora infestans) the entire sporangium can be discharged and works as a dispersal spore. Contingent upon natural conditions (particularly temperature) this can grow by creating a hypha or it can experience cytoplasmic cleavage and discharge zoospores.

The critical fleece mold fungal pathogens of a few yield plants discharge sporangia that are wind-scattered and develop by a hypha to contaminate their hosts. Sexual proliferation regularly includes the generation of a male sex organ (antheridium) and a female sex organ (oogonium) that may contain one or a few eggs. Meiosis happens inside these sex organs, and treatment is accomplished by the exchange of a solitary haploid core through a preparation tube to every haploid egg. In some Phytophthora species (e.g. P. cactorum) the antheridia are appended along the side to the oogonium. In any case, in others (e.g. P. infestans) the oogonial hypha becomes through the antheridium and after that swells to shape the oogonium. Regardless, preparation prompts the advancement of at least one thick-walled, diploid oospores. These more often than not have a constitutive torpid period before they sprout to create either diploid hyphae or a sporangium that discharges diploid zoospores. There are numerous minor departure from these fundamental topics.

For instance, a portion of the Oomycota are homothallic (self-rich) and some heterothallic. A few animal types (e.g. Pythium oligandrum) distinctively create oospores parthenogenetically, without a sexual procedure. In a portion of the water molds (Saprolegniales) the zoospores discharged from the sporangia encyst promptly and afterward sprout to deliver an optional zoospore, which can swim for a long time. In Pythium spp. The sporangia don't discharge their zoospores "straightforwardly" by disintegration of the tip of the sporangium. Rather, the sporangium creates a short leave tube and the tip of this separates to frame a film bound vesicle. The substance of the sporangium are expelled into this vesicle, where the zoospores separate and after that are discharged by break of the vesicle.

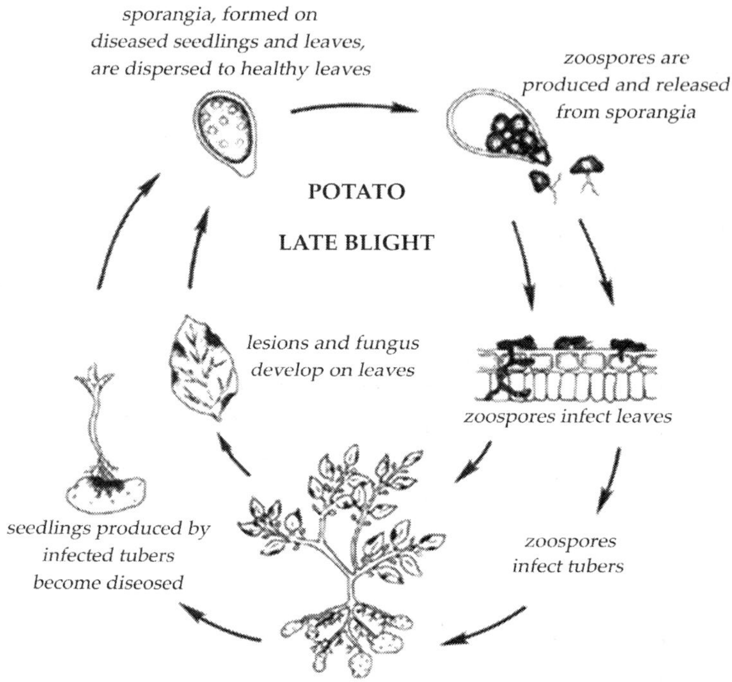

Figure- 2.12: Life cycle of Phytpphthora infestans.

Classification, Ecology, and Importance

There are assessed to be 500–800 species in the Oomycota, arranged in 5 or 6 Orders. Be that as it may, some of these Orders contain just a couple of creatures, which will be specified where suitable in later parts. Here we will concentrate on the two noteworthy Orders, Peronosporales and Saprolegniales.

The Saprolegniales, regularly known as water molds, are to a great degree normal in freshwater territories and in some harsh, estuarine situations. They can regularly be goaded by submerging hemp seeds in characteristic waters, yet it is hard to free the way of life from microorganisms. A few types of Saprolegnia and Achlya are normal saprotrophs in freshwater territories, and add to the reusing of natural issue, however *S. diclina* is an essential pathogen of salmonid angle. The related Aphanomyces spp. are maybe best-known as forceful root pathogens (e.g. *A. euteiches* on beans and pea). Be that as it may, *A. astaci* is a fungus pathogen of crawfish and has for all intents and purposes disposed of the local European crawfish (the illness being known as crawfish torment). The presented American crawfish, be that as it may, is impervious to the illness and has now supplanted the European crawfish in all locales.

The Order Peronosporales is by a long shot the biggest and most monetarily imperative gathering. It incorporates numerous true fungal plant pathogens. For instance, numerous Pythium spp. cause seedling ailments and assault the feeder underlying foundations of all harvest plants, however a couple of Pythium spp. parasitize other organisms and have potential for biocontrol of plant illnesses. The Peronosporales additionally incorporates all the Phytophthora spp., causing infections, for example, potato curse (*P. infestans*). A standout amongst the most forceful plant pathogens is *Phytophthora cinnamomi*, with a broad host run.

It causes root spoil of pines, eucalypts, organic product trees, and numerous different plants in hotter locales of the globe, and can attack common vegetation with wrecking outcomes. A likewise forceful animal groups, Phytophthora ramorum, causes sudden oak passing in California and is presently spreading crosswise over Europe, representing a danger to numerous local European trees. Additionally in this gathering are the essential wool mold pathogens (e.g. Plasmopara viticola on grape vines, and Bremia lactucae on lettuce,) that commit as biotrophic fungal parasitic fungi.

The rest of the Orders incorporate Leptomitales (a little gathering of amphibian organisms, for example, Leptomitis lacteus which is normal in sewage-contaminated waters). This gathering has a few intriguing and particular components, for example, fermentative as opposed to respiratory digestion, a necessity for sulfur-containing amino acids since they can't utilize inorganic sulfur sources, and hyphae that are tightened at interims, similar to chains of hotdogs. The dividers of a few animal varieties contain noteworthy measures of chitin (which is unordinary for Oomycota) and the hyphal choking influences are regularly stopped by cellulin granules, made generally out of chitin. [74]

The Order Lagenidales is a little gathering of basic symptomless parasitic fungi of plant roots (e.g. Lagena radicicola) or parasitic fungi of green growth, organisms, or spineless creatures (e.g. mosquito hatchlings,

Lagenidium giganteum on nematodes, and so on.). Frequently these can't be developed in culture without a host.

Cellular Slime Moulds

The cell sludge molds contain two gatherings of creatures, the dictyostelid cell ooze molds and the acrasid cell ooze molds. These two gatherings are naturally like each other, yet the dictyostelids appear to be a monophyletic gathering, though the acrasids appear to be more unique. Both sorts of living being develop and partition as haploid, unicellular amoebae, which inundate microbes and other nourishment particles by phagocytosis. They are normally found in sodden natural creature manure, leaf litter, rich soil, and comparable sorts of substrate. At the point when supplements end up plainly drained, the amoebae experience a complex formative grouping in which a large number of amoebae total and in the long run create a stalked fruiting body.

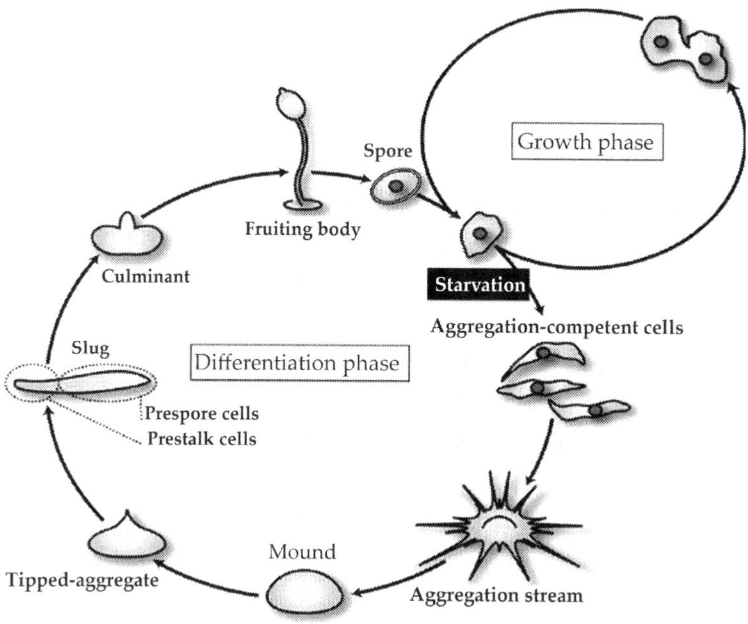

Figure- 2.13: Life cycle of Dictyostelium discoideum.

This procedure has been contemplated seriously in the dictyostelid sludge form, Dictyostelium discoideum, which has turned into a model arrangement of cell correspondence and separation. At the onset of starvation, a couple of amoebae go about as a total focus and deliver occasional beats of cyclic AMP (cAMP). This causes other amoebae to join the total by chemotaxis, and the amoebae at that point experience spilling along characterized tracks with the goal that they total as a hill. At that point the cells in the hill separate into two sorts which are the pre-stalk cells and the pre-spore cells, each in its characterized position where it will later turn into a stalk cell or a spore. At this stage the hill topples over, and the subsequent slug (grex) moves to the surface of the substrate where it separates into a fruiting body (sorocarp) with a stalk and a sporing head. The walled spores are consequently discharged and wind-scattered. [75]

A sexual cycle has additionally been portrayed, prompting the generation of thick-walled, diploid macrocystis, which experience meiosis taken after by rehashed mitotic divisions. At that point the cytoplasm is severed and numerous haploid amoebae are discharged to rehash the cycle.

MYXOMYCOTA

The plasmodial ooze molds are divider less life forms that comprise of a multinucleate system of protoplasm (the plasmodium) showing fast, musical surges of protoplasmic spilling. These living beings are normally found in the pre-winter on sodden decaying wood, on senescing parasitic fruit bodies and on comparative natural substrata where microscopic organisms are copious. They overwhelm microbes and other nourishment particles by phagocytosis. The plasmodium ordinarily creates inside a substrate however then moves to the surface in light of supplement consumption and the entire plasmodium changes over into fruiting structures (sporangia) containing numerous haploid spores that are wind-scattered. These spores sprout to deliver either amoeboid cells named myxamoebae or lash swarmers, which combine in sets, and the subsequent diploid cell develops into a plasmodium. Physarum polycephalum is maybe the best-known illustration. It has been examined seriously by morphogeneticists since it can be kept up on characterized (sans bacterium) media.

PLASMODIOPHORIDS

Plasmodiophorids are commit intracellular parasitic fungi of plants, green fungal growth, or organisms. They become just in a host living being, and can't be developed in research center culture. As a gathering, the plasmodiophorids demonstrate no undeniable relationship to parasitic fungi or to different growth like living beings, so their phylogenetic status stays vague. The most essential life form of this gathering is Plamodiophora brassicae, which causes the harming clubroot ailment of cruciferous products (the cabbage family). In this sickness the roots are extremely twisted, with multiplication of cell division caused by the generation of phytohormones, and this can prompt genuine yield misfortunes. A related life form, Spongospora subterranea, causes fine scab of potato tubers. Notwithstanding these, few basic plasmodiophorids, for example, Polymyxa spp., develop as symptomless parasitic fungi in the foundations of many plants, causing practically zero harm, yet they can go about as vectors of some monetarily vital plant infections. Noteworthy stages in the life cycle of *P. brassicae*, despite the fact that there is still uncertainty about the subtle elements of a few phases. The thick-walled resting spores can hold on in soil for a long time, and just a low extent of them will grow in any one year.

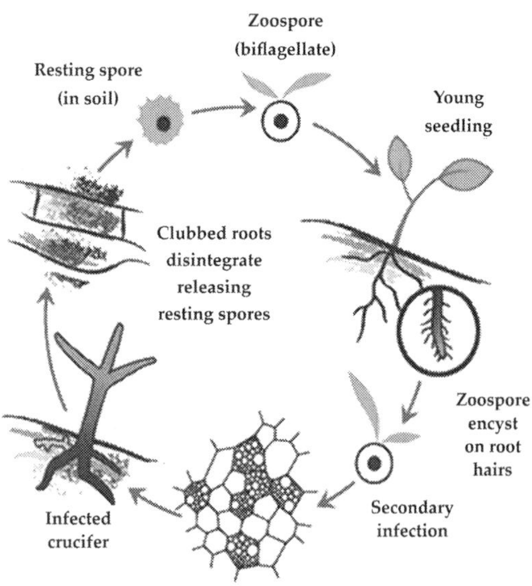

Figure- 2.14: life cycle of Plasmodiophora brassicae.

Thus, once this parasite is set up in a site it is practically difficult to annihilate. The resting spores in the end develop to discharge a motile, biflagellate zoospore, which finds a host root by chemotaxis and afterward encysts in a characterized introduction. The growth protoplast is then infused into a root hair by methods for a slug like structure and a root epidermal cell. [76] The protoplast at that point develops into a little essential plasmodium inside the contaminated cell, and believers into a sporangium, which discharges zoospores when the root cell bites the dust or when a leave tube is shaped. At this stage, the zoospores are accepted to meld in sets, and the subsequent optional zoospores contaminate root cortical cells, where they form into expansive, auxiliary plasmodia that will inevitably create thick-walled resting spores. There is confirm that meiosis happens just before the improvement of resting spores, proposing that the essential zoospores are haploid, and in this way that the auxiliary zoospores and zoosporangia are diploid.

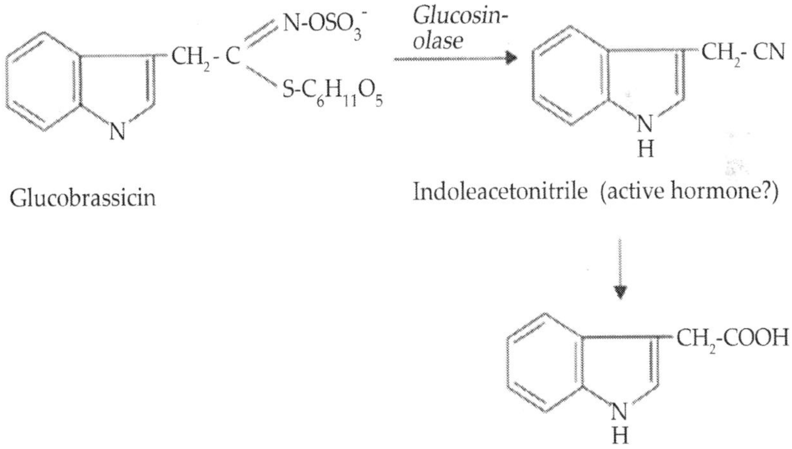

Glucobrassicin

Indoleacetonitrile (active hormone?)

Indoleacetic acid (1AA) (active hormone)

Figure- 2.15: Plasmodiophora brassicae convert glucobrassicin to the hormone of plants indoleacetic acid.

The fundamental harm caused by Plasmodiophora lies in the way that the root reacts to contamination of the cortex by experiencing quick cell extension (hypertrophy) and cell division (hyperplasia), prompting the advancement of vast nerves. The plant supplements occupied to these nerves seriously lessen the shoot development and yield of harvests, for example, cauliflower, while the nerves themselves make root crops unmarketable. The proposed clarification for bother advancement is that glucobrassicin, a trademark compound in cruciferous plants, is changed over to the plant hormone indolylacetic corrosive (IAA) by the activity of a chemical, glucosinolase.

As a gathering, the plasmodiophorids are hard to contemplate on the grounds that they can't be developed in research center culture, thus numerous essential parts of their science stay hazy. They demonstrate no conspicuous relationship to different parasite like creatures, so they appear to speak to a basal ancestry of the early eukaryotic living beings. [77] They are extremely normal, since when the underlying foundations of any grass plant are cleared of protoplasm and recolored with trypan blue the roots are as often as possible seen to contain resting spores of Polymyxa graminis or comparative life forms, for example, Ligniera spp. They don't cause distortion of the roots, and even Plasmodiophora brassicae appears to be ready to develop in the underlying foundations of numerous cruciferous plants, including the normal weed Capsella bursa-pastoris (shepherd's tote), without causing nerves, maybe in light of the fact that these plants don't contain large amounts of glucobrassicin or don't react to disease by delivering dynamic hormones.

SUMMARY

- Fungi are primary multicellular organisms that arose from protists ancestors and are grouped into supergroup Opisthokonta. Fungi are ecologically very important organisms and are pervasive. Body of fungus is commonly a mycelium consists of filaments that spread in order to absorb nutrients. Development of hyphae occurs through process that use cytoplasmic streaming and osmosis.
- Usually fungi can reproduce with the help of both processes sexually and asexually, also with the help of spore formation. Nearly all fungi have ability to make asexual spores, and in some species spores are the only source of reproduction. Mostly fungi can tooform spores through exclusive sexual mechanism. Fruiting bodies forms in sexual reproduction for the purpose of spore dispersal. Some single celled fungi, such as yeast, reproduce by budding asexually.

- Fungi is divided further into 5 main groups which are as follows, Chytrids, Zygomycetes, Ascomycetes, Arbuscular mycorrizal, and Basidiomycetes. Chytrids are the simplest and earliest fungi. It contributes to amphibian population decline universaly. Zygomycetes form zygospores by sexual reproduction. Arbuscular mycorrhizal live just with plants, and they form large spores through asexual reproduction. Basidiomycetes make club like arrangement amid sexual reproduction. Ascomycetes make sacs consisting of 8 sexual pores.

REFERENCES

1. Jones, E. B. G., & Pang, K. (2012). *Marine fungi and fungal-like organisms*. Boston: De Gruyter.
2. Webster, J. (2006). *Lower fungi and fungus-like organisms*. Göttingen: IWF Wissen-und Medien-gGmbH.
3. Taylor, T. N., Krings, M., & Taylor, E. L. (2015). *Fossil fungi*.
4. Stephenson, S. L. (2012). *The Kingdom Fungi: The Biology of Mushrooms, Molds, and Lichens*.
5. Piepenbring, M., & American Phytopathological Society,. (2015). *Introduction to mycology in the tropics*.
6. Webster, J., & Weber, R. (2006). *Mycology: Vol. 1*. Ludwigsburg: STUMM-FILM in cooperation with Wissen und Medien.
7. Webster, J., & Weber, R. W. S. (2006). *Mycology*. Ludwigsburg: STUMM-FILM in cooperation with Wissen und Medien.
8. Ownley, B. H., & Trigiano, R. N. (2016). *Plant Pathology Concepts and Laboratory Exercises, Third Edition*. Boca Raton: CRC Press.
9. Deacon, J. W. (2006). *Fungal biology*. Malden, Mass. [u.a.: Blackwell Publishing.
10. Quinn, P. J. (2011). *Veterinary microbiology and microbial disease*. Chichester: Wiley-Blackwell.
11. Alexopoulos, C. J., & Bold, H. C. (1967). *Algae and fungi*. New York: Macmillan.
12. Jones, E. B. G., & Pang, K. (2012). *Marine fungi and fungal-like organisms*. Boston: De Gruyter.

13. Sharma, P. D. (2005). *Fungi and allied organisms*. Oxford, U.K: Alpha Science International.
14. Farr, M. L. (1981). *How to know the true slime molds*. Dubuque, Iowa: W.C. Brown Co.
15. Ainsworth, G. C., Bisby, G. R., Kirk, P. M., & CABI Bioscience. (2008). *Ainsworth & Bisby's dictionary of the fungi*. Wallingford, Oxon, UK: CABI.
16. Rippon, J. W. (1988). *Medical mycology: The pathogenic fungi and the pathogenic actinomycetes*. Philadelphia: W.B. Saunders Co.
17. DiConsiglio, J. (2008). *There's a fungus among us: True stories of killer molds*. New York: Franklin Watts.
18. Jordan, M. (1995). *The encyclopedia of fungi of Britain and Europe*. London: David & Charles.
19. Soil Resource Management National Program (U.S.), & United States. (2006). *Development of efficient and practical methods for producing arbuscular mycorrhizal fungi ... annual report*. Washington, D.C: U.S. Dept. of Agriculture, Agricultural Research Service.
20. Schwarzott, Daniel. (2004). *Phylogenie und Divergenz der arbuskulären Mykorrhizapilze und Geosiphon pyriformis*.
21. Hruschka, D. J. (2010). *Friendship: Development, ecology, and evolution of a relationship*. Berkeley: University of California Press.
22. Gairn, L. (2008). *Ecology and modern scottish literature*. Edinburgh: Edinburgh University Press.
23. Blow, W. H. (1979). *The Cainozoic Globigerinida: A study of morphology, taxonomy, evolutionary relationships and the stratigraphical distribution of some globigerinida (mainly globigerinacea)*. Leiden: Brill.
24. Vellucci, S. L. (1995). *Bibliographic relationships among musical bibliographic entities: A conceptual analysis of music represented in a library catalog with a taxonomy of the relationships discovered*. Ann Arbor, Mich: UMI Dissertation Services.
25. Bloom, B. S., Krathwohl, D. R., & Masia, B. B. (1956). *Taxonomy of educational objectives: The classification of educational goals*. New York: D. McKay.

26. Mason, K. A., Losos, J. B., Singer, S. R., Raven, P. H., & Johnson, G. B. (2013). *Biology*. New York, N.Y: McGraw Hill.
27. Błaszkowski, J. A. (2012). *Glomeromycota*. Kraków: W. Szafer Institute of Botany, Polish Academy of Sciences.
28. Souza, F. A. (2005). *Biology, ecology and evolution of the family Gigasporaceae, Arbuscular Mycorrhizal Fungi (Glomeromycota).*
29. Mennes, C. B. (2015). *Antiquities of the rainforest: Evolution of mycoheterotrophic angiosperms growing on Glomeromycota.* Netherlands: publisher not identified.
30. Krüger, M. (2012). *Molecular phylogeny taxonomy & evolution of glomeromycota: DNA-based characterization and identification of arbuscular mycorrhizal fungi.* Saarbrücken, Deutschland: Sudwestdeutscher Verlag F.
31. Souza, T. (2015). *Handbook of Arbuscular Mycorrhizal Fungi.*
32. Chapman, M. J., & Margulis, Lynn. (2009). *Kingdoms and Domains.* Academic Press.
33. Gordon, D. P. (2012). *New Zealand inventory of biodiversity: Volume three.* Christchurch, N.Z: Canterbury University Press.
34. Fullerlove, G. (2002). *Two-photon fluorescence light microscopy to zygomycota.* London [u.a.: Nature Publ. Group.
35. Alencar, Y. B., Rios-Velasquez, C. M., Lichtwardt, R. W., & Hamada, N. (2003). *Trichomycetes (Zygomycota) in the digestive tract of arthropods in Amazonas, Brazil.* (Memórias do instituto Oswaldo Cruz, 98, 6, 799-810.)
36. Vojvodic, S. (2006). *The ecology and development of several trichomycete (zygomycota) fungi found in aquatic dipteran hosts.*
37. Nelder, M. P. (2003). *The ecology of trichomycete (zygomycota) fungi inhabiting black fly larvae (dipters: simuliidae) of coastal Alabama.*
38. Dickinson, M. (2003). *Molecular Plant Pathology.* UK: Routledge.
39. Jones, E. B. G., & Pang, K. (2012). *Marine fungi and fungal-like organisms.* Boston: De Gruyter.
40. Stern, K. R., Bidlack, J. E., Jansky, S., & Uno, G. (2006). *Introductory plant biology.* Boston: McGraw-Hill Higher Education.

41. Petrini, L. E., & Petrini, O. (2013). *Identifying moulds: A practical guide.*
42. Kües, U., & Fischer, R. (2006). *Growth, differentiation, and sexuality*. Berlin: Springer.
43. Watanabe, T. (2010). *Pictorial atlas of soil and seed fungi: Morphologies of cultured fungi and key to species*. Boca Raton: CRC Press/Taylor & Francis.
44. Rogers, K. (2011). *Fungi, Algae, and Protists*. Britannica Educational Publishing.
45. Sinclair, W. A., & Lyon, H. H. (2005). *Diseases of trees and shrubs*. Ithaca: Comstock Pub. Associates.
46. In McLaughlin, D. J., & In Spatafora, J. W. (2014). *Systematics and evolution: Part A*.
47. McLaughlin, David J. ; Spatafora, Joseph W. (2014). *Systematics and Evolution : Part A*. Springer Berlin Heidelberg.
48. Rogers, K. (2011). *Fungi, Algae, and Protists*. Britannica Educational Publishing.
49. Cappuccinelli, P., Ashworth, J. M., & European Molecular Biology Organization. (1977). *Development and differentiation in the cellular slime moulds: Proceedings of the international workshop held at Porto Conte, Sardinia on 12-16 April, 1977*. Amsterdam: Elsevier/North-Holland Biomedical Press.
50. Bonner, J. T. (1967). *The cellular slime moulds*. Princeton, N.J: Princeton University Press.
51. Cappuccinelli, P. (2014). *Development and Differentiation in the Cellular Slime Moulds*. Elsevier Science.
52. Stevens, R. (2002). *A computer model of the cellular slime mould Dictyostelium discoideum*.
53. Lackie, J. M. (1986). *Cell Movement and Cell Behaviour*. Dordrecht: Springer Netherlands.
54. In Baldauf, S., In Escalante, R., & In Romeralo, M. (2013). *Dictyostelids: Evolution, Genomics and Cell Biology*. Berlin, Heidelberg: Springer Berlin Heidelberg.
55. Peberdy, J. F. (1980). *Developmental Microbiology*. Boston, MA: Springer US.

56. Pickett-Heaps, J. (2006). *The kingdom protista: The dazzling world of living cells*. S.l.: Cytographics [prod..
57. Hardie, D. G., & Hanks, S. (1995). *The protein kinase factsbook*. London: Academic Press.
58. Bonner, J. T., & Indian Academy of Sciences. (1991). *Researches on cellular slime moulds: ed papers of J.T. Bonner*. Bangalore: Indian Academy of Sciences.
59. May, T. W. (2003). *Basidiomycota p.p. & Myxomycota p.p.* Canberra: CSIRO.
60. Krishnan, U. (1999). *Evolution of rna editing sites in mitochondria of the myxomycota*.
61. Orchard, A. E., May, T. W., & Australian Biological Resources Study. (2003). *Catalogue and bibliography of Australian fungi 2. Basidiomycota p.p. & Myxomycota p.p.* (Fungi of Australia.) Canberra: Australian Biological Resources Study.
62. Burns, G. W. (1974). *The plant kingdom*. New York: Macmillan.
63. In McLaughlin, D. J., & In Spatafora, J. W. (2014). *Systematics and evolution: Part A*.
64. Binion, D. (2008). *Macrofungi associated with oaks of Eastern North America*. Morgantown: West Virginia University Press.
65. Stern, K. R., Bidlack, J. E., Jansky, S., & Uno, G. (2006). *Introductory plant biology*. Boston: McGraw-Hill Higher Education.
66. Barron, G. L. (1999). *Mushrooms of northeast North America: Midwest to New England*. Edmonton, AB: Lone Pine Pub.
67. Sharma, O. P. (2011). *Fungi and allied microbes*. New Delhi: Tata McGraw Hill.
68. Browne, S. A. (2012). *Aquatic ecosystems*. New York: Nova Science Publisher's.
69. Gherbawy, Y., & Voigt, K. (2010). *Molecular identification of fungi*. Berlin: Springer.
70. Sime-Ngando, T., & Niquil, N. (2011). *Disregarded microbial diversity and ecological protentials in aquatic systems*. Dordrecht: Springer.
71. Dick, M. W. (2001). *Straminipilous Fungi: Systematics of the Peronosporomycetes including accounts of the Marine Straminipilous*

Protists, the Plasmodiophorids and similar organisms. Dordrecht [etc: Kluwer Academic Publishers.

72. Institute of Arable Crops Research (Great Britain). (1997). *Molecular studies of the Plasmodiophorids: Rothamsted's site collating molecular information on the Plasmodiophorales*. Rothamsted, UK: IACR.
73. Faggian, R., & RMIT University. (2002). *The application of molecular biology techniques for the detection and genetic differentiation of the plasmodiophorids*.
74. Hernandez, M. M. L. (2012). *Manipulating plant defences for improved control of diseases caused by plasmodiophorids: A thesis submitted in partial fulfilment of the requirements for the Degree of Doctor of Philosophy at Lincoln University*.
75. Neuhauser, Sigrid, Kirchmair, Martin, Bulman, Simon, & Bass, David. (2014). *Cross-kingdom host shifts of phytomyxid parasites*. (BioMed Central Ltd.) BioMed Central Ltd.
76. Burki, Fabien, Kudryavtsev, Alexander, Matz, Mikhail V, Aglyamova, Galina V, Bulman, Simon, Fiers, Mark, Keeling, Patrick J, ... Pawlowski, Jan. (2010). *Evolution of Rhizaria: new insights from phylogenomic analysis of uncultivated protists*. (BioMed Central Ltd.) BioMed Central Ltd.
77. Fuller, M., & Longcore, J. (n.d.). *Zoosporic Fungi Online*. Joyce Longcore.

CHAPTER

3

FUNGAL STRUCTURE

Parasitic fungus have numerous extraordinary elements as far as their cell segments, cell associations, and structure. [69] These elements are personally connected to the processes of contagious development and in this way to the numerous actions of parasitic fungus as decomposer life forms, human pathogen, and plant pathogen. In this section we study the primary ultrastructural and basic elements of parasitic fungus, and a portion of the effective systems that empower us to see the progression of subcellular segments in living hyphae of fungi, giving an understanding into the way that it develop.

FUNGAL HYPHAE STRUCTURE

The hypha is basically a tube with an inflexible divider, containing a moving slug of protoplasm. It is of vague length yet frequently has a genuinely consistent width, extending from 2 μm to 30 μm or all the more (typically 5–10 μm), contingent upon the species and development conditions. [1] Hyphae become just at their tips, where there is a decreased area named the expansion zone, this can be up to 30 μm long in the quickest developing hyphae, for example, Neurospora crassa which can stretch out at up to 40 μm every moment. Behind the developing tip, the hypha ages logically and in the most seasoned locales it might separate via autolysis or be separated by the compounds of different life forms (heterolysis). While the tip is developing, the protoplasm moves ceaselessly from the more seasoned areas of the hypha towards the tip. Along these lines, a parasitic hypha consistently reaches out toward one side and ceaselessly ages at the flip side, drawing the protoplasm forward as it develops. [2]

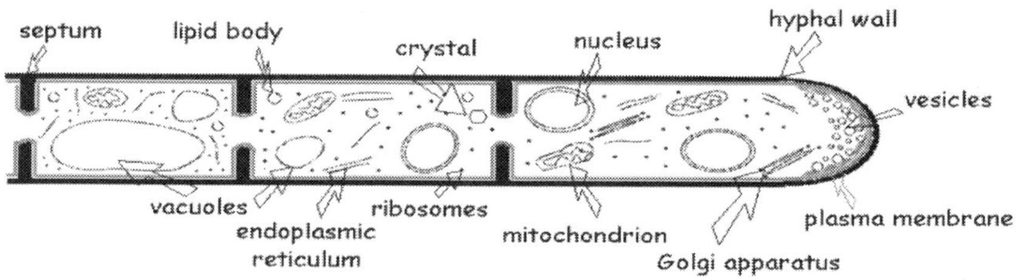

Figure- 3.1: Diagrammatic representation of fungal hyphae.

The hyphae of most growths have cross dividers (septa, solitary septum) at genuinely consistent interims, however septa are missing from hyphae of most Zygomycota and Oomycota, with the exception of where they happen as total dividers to disengage old or conceptive districts. [3] By and by, the utilitarian refinement amongst septate and aseptate organisms is not as incredible as may be thought, in light of the fact that septa have pores through which the cytoplasm and even the cores can relocate. Entirely, along these lines, septate hyphae don't comprise of cells however of interconnected compartments, similar to the compartments of manufacturing. As we will see later, this empowers cell parts to move either advances or in reverse along committed pathways, so that a hypha can work as a coordinated unit. [4]

All parasitic hyphae are encompassed by a mass of complex association, portrayed later. It is thin at the zenith (around 50 nm in

Neurospora crassa) however thickens to around 125 nm at 250 μm behind the tip. The plasma layer lies near the divider and is by all accounts immovably connected to it in light of the fact that hyphae are hard to plasmolyse. [70]

FUNGAL ULTRASTRUCTURE

Transmission electron microscopy of thin areas of contagious hyphae has been a standout amongst the most essential devices for understanding the conduct of parasitic hyphae, and for explaining the systems of apical development. [5] The underlying examinations on parasitic ultrastructure depended on synthetic obsession strategies: the hyphae were "settled" by drenching in aldehydes, for example, glutaraldehyde, at that point post-treated with electrondense substances (osmium tetroxide and uranyl acetic acid derivation) to give greatest differentiation.

Yet, a fresher method named solidify substitution was produced in the late 1970s, and this gives significantly more noteworthy determination so it has now turned into the standard. In this method, the living hyphae are dove into fluid propane (−190°C) causing practically momentary safeguarding of the hyphae, at that point exchanged to icy (−80°C) $CH_3)_2CO$ with osmium tetroxide and uranyl acetic acid derivation, and gradually taken back to room temperature. [6] Examination of the electron micrographs acquired by these two techniques ought to be made with alert, for no less than two reasons:

- The artificially settled hyphae have a place with a types of Pythium (Oomycota), which is not a genuine growth, and the example was recolored to give most extreme determination of the inside organelles, so the divider is ineffectively recolored. [71]
- The hyphae arranged by solidify substitution speak to a genuine parasite, Athelia (Sclerotium) rolfsii, and a portion of the organelles of growths are unique in relation to those of the Oomycota, as talked about later in this part.

Notwithstanding, it is remarkable that the level of conservation of the hyphae is greatly improved in the stop substituted material. [7] Specifically, the plasma film has a smooth profile contrasted and the numerous spaces seen with the concoction obsession technique, and the subcellular organelles are all the more obviously characterized.

Zonation of Organelles in Apical Compartment

All effectively becoming contagious hyphae that are including hyphae of the parasite like Oomycota, demonstrate an unmistakably characterized extremity in the game plan of organelles from the hyphal tip back towards the base of the apical compartment. [8] The outrageous hyphal tip contains a huge collection of film bound vesicles, yet no other real organelles. These vesicles indicate contrasts in electron thickness, proposing that they have distinctive substance. A large portion of the vesicles are believed to be gotten from golgi bodies (or what might as well be called golgi bodies in organisms) additionally back in the apical compartment, and to be transported to the pinnacle where they add to apical development. In help of this view, cytochemical recoloring of a portion of the vesicles has demonstrated that they contain chitin synthase, the catalyst in charge of creating the chitin part of parasitic dividers. [9]

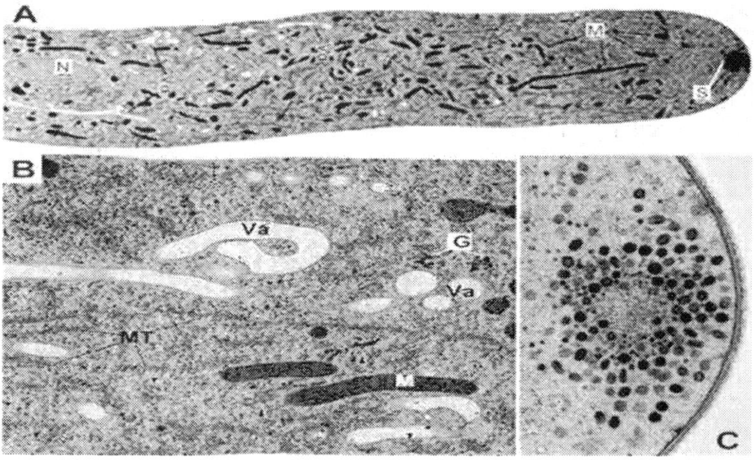

Figure- 3.2: Ultrastructure of the hyphae of Sclerotium rolfsii (a) M represents mitochondria, N abbreviate nucleus, S represent Spitzen-krper, (b) Va (vacuoles), MT mitochondria, G (golgi bodies), and M represents mitochondria.

The accumulation of vesicles at the hyphal tip is named the apical vesicle bunch (AVC). At high amplification in the genuine growths the focal point of the AVC is by all accounts a sans vesicle center comprising of a system of actin microfilaments, with microtubules going through this and now and then reaching out up to the extraordinary tip of the hypha. [10] Early light microscopists had perceived the presence of the AVC as a stage dull body in the hyphal tips of septate growths, when seen by phasecontrast microscopy. It was named the Spitzenkörper ("apical body"), and it was noted to vanish when development was halted by applying a gentle stun to the hyphal tip, and to return when development restarted.

Also, the Spitzenkörper supposedly shifted towards the side of a hyphal tip when hyphae altered their development course, and all the more as of late the Spitzenkörper has been appeared to part preceding the arrangement of a hyphal branch, with the goal that two hints are created from the first one. This confirmation focuses to a focal part of the Spitzenkörper in contagious tip development, a body that capacities as what might as well be called a "tip-development organelle." [11]

Behind the outrageous tip is a short zone with few or no real organelles. At that point commonly there is a zone rich in mitochondria. Through the activities of ATPase, this zone more likely than not produces the proton-thought process constrain (H+ inclination over the cell layer) that drives supplement take-up at the hyphal tip. The mitochondria of parasitic fungi have plate-like cisternae like those of creatures and plants, while the mitochondria of Oomycota have tubular cisternae. [72] Additionally back along the apical compartment there is expanding advancement of an arrangement of fanned tubular vacuoles. These reach out through the septa into the hyphal compartments additionally back, giving a vehicle framework to the development of metabolites, including phosphates which frequently can be hard to come by.

The appropriation of cores shifts between and inside parasitic gatherings. The aseptate organisms (e.g. Zygomycota) and parasite like creatures (e.g. Oomycota) contain numerous cores inside a typical cytoplasm, so these organisms are coenocytic. [12] Numerous septate organisms (e.g. mitosporic fungal parasitic fungi and Ascomycota) additionally have a few cores in the apical compartment, however at times just a single or two cores in compartments behind the summit. Nonetheless, cores can press through the septal pores, so the idea of a solitary core deciding the elements of a settled volume of cytoplasm does not by any stretch of the imagination apply to generally parasitic fungi. Rather, we should view numerous growths as having a few cores sharing a typical pool of cytoplasm. This has vital ramifications for the hereditary ease of parasitic fungi. [13]

Numerous individuals from the Basidiomycota (toadstools, mushrooms, and so forth.) have a standard plan of one core in every compartment of the monokaryon, however two cores (of various mating similarity gatherings) in every compartment of the dikaryon. [14] This course of action is guaranteed by a unique kind of septum, named the dolipore septum, which has just a restricted pore, too little to enable cores to go through.

Hyphae as Colony Part

Parasitic settlements ordinarily create from a solitary sprouting spore, which delivers a germ tube (a youthful hypha) that develops and branches behind the tip. As the first hypha and the primary request branches develop, they deliver additionally branches behind their tips. [15] These branches veer from each other until, in the end, the province builds up a trademark round framework. At that point, in the more seasoned parts of a province where supplements have been drained, numerous fungal growths deliver limit hyphal branches that develop towards each other as opposed to veering, and circuit by tip-to-tip contact, including confined breakdown of their dividers. This procedure of hyphal anastomosis makes a system for the pooling and remobilization of protoplasm to deliver chlamydospores or other, bigger separated structures. [16]

The marvels are striking for a few reasons:

- In supplement rich conditions, the hyphae and branches at the province edge dependably wander from each other, situating themselves with the goal that they develop in the spaces between existing hyphae. These situating instruments are extremely exact, however we have no firm proof of their causes, perhaps nearby slopes of carbon dioxide, or other development metabolites?
- Nutrient-poor conditions close to the focal point of a state make precisely the inverse impact, in which hyphae develop towards each other and wire at the purposes of contact.

[17] As found in, this again includes exact introduction of hyphal tip development, in light of the fact that the "homing reaction" is amazingly precise and dependably prompts tip-to-tip combination. We have no information of the elements included.

- Hyphal anastomosis just happens between individuals (states) of similar species, and from time to time, if at any point, between inconsequential species. Indeed, even inside a species, strains that are hereditarily diverse can intertwine, yet this ordinarily prompts fast passing of the combined hyphal compartments, a response administered by heterokaryon contradiction (het) loci. [73]
- The hyphae of Oomycota from time to time if at any time display "vegetative" anastomosis, however they do experience hyphal combinations when they deliver sexual hyphae of inverse mating sorts, the advancement of these is controlled by sex hormones.

Figure- 3.3: Phases of progress of individual colony of fungus.

STRUCTURE OF YEAST

A few organisms develop as growing, uninucleate yeasts, as opposed to as hyphae. Normal illustrations incorporate the bread yeast, Saccharomyces cerevisiae (Ascomycota), Cryptococcus spp. (Ascomycota), and Sporobolomyces roseus (Basidiomycota). [18] Some other parasitic fungi are distinctively dimorphic (with two shapes): they switch between a hyphal and a yeast frame in light of changes in natural conditions. A decent case is the class Candida (Ascomycota), including the basic species Candida albicans, which can be a critical pathogen of people.

Along these lines, yeasts are not in a general sense unique in relation to hyphal organisms, they only speak to an alternate development shape, adjusted to spread in supplement rich water movies and comparable sorts of condition. [74]

Every yeast cell has a solitary core and an average scope of cytoplasmic organelles, including a vast, prominent vacuole and generally a few stage splendid lipid bodies. The vacuole and the lipid bodies are regularly the main structures unmistakably obvious by light microscopy. Most yeasts partition by maturing from at least one areas on the phone surface. [19] Amid this procedure a little outgrowth shows up at the bud site, at that point the bud continuously extends before venturing into an adjusted frame, by union of new divider parts over the entire of the phone surface. At the point when the bud has almost achieved its last size, the core of the mother cell relocates towards the bud site and partitions, so that one core stays in the mother cell and a moment core enters the cell. [75]

The last partition of the two cells is accomplished by the improvement of a septum. In Saccharomyces this happens when a ring of chitin is delivered at the "neck" site, and this ring of chitin grows inwards until the point that it turns into a total chitin plate between the mother and female cell. At that point other divider materials are saved on each side of this chitin plate, and the cells isolate by enzymic cleavage of the divider between the chitin plate and the little girl cell. [20] This procedure leaves a bud scar on the mother cell and a birth scar on the little girl cell. These scars are unnoticeable by typical light microscopy, yet the chitin plate on the mother cell can be seen plainly if yeasts are treated with a fluorescent brightener, for example, Cellufluor (calcofluor), which ties to chitin and fluoresces blue when the cell is seen under a fluorescence magnifying lens. By utilizing fluorescent colors that predicament to chitin, we can check the quantity of bud scars on the cell surface. [21]

This uncovers Saccharomyces cerevisiae is a multipolar growing yeast, it generally buds from an alternate point on the cell surface, never from a past bud site, though some different yeasts display bipolar sprouting, the buds dependably emerge at similar positions, frequently at the posts of the cell. In principle, the cells of bipolar species are eternal, with no restriction to the quantity of times they can bud, though the cells of multipolar species would inevitably come up short on potential bud locales, computed to be up to 100, albeit just around 40 bud scars have been seen on a solitary cell. [22] Be that as it may, by and by this refinement is immaterial on the grounds that in exponentially developing societies (where one cell produces two of every a unit of time, two create four in the following unit of time, et cetera) one portion of the considerable number of cells will be first-time moms, one quarter will be second-time moms, and an infinitessimally modest number will be exceptionally old moms. Actually, the average chitin substance of a yeast populace from time to time surpasses over 1–2% of the aggregate divider material, and the greater part of this chitin is found in the bud scars, not in whatever is left of the yeast cell divider. Some critical advances have been made in understanding development extremity, in view of the utilization of cell-cycle mutants in the research facility of the Nobel Laureate, Sir Paul Nurse. [76]

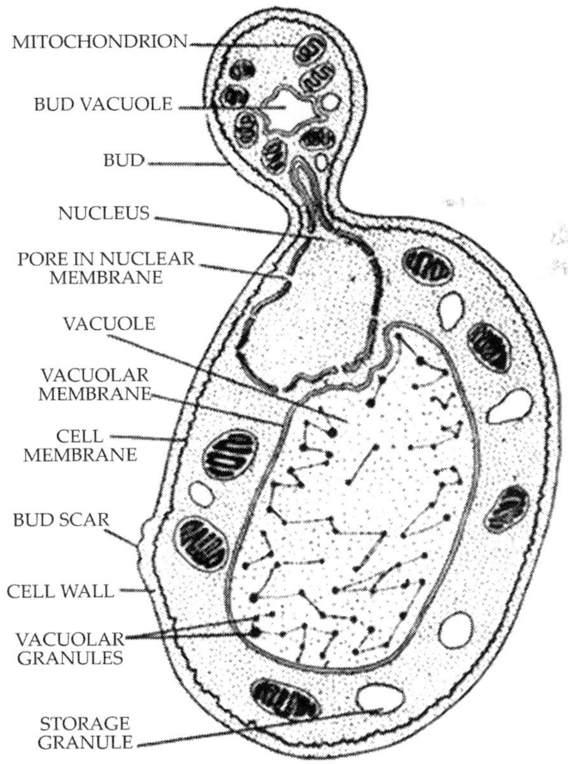

Figure- 3.4: Structure of Yeast.

The scientific categorization of yeasts is confused by the absence of clear morphological components, so as of late it has depended vigorously on sub-atomic methodologies. There are likely a huge number, if not thousands, of yeast species that presently can't seem to be unmistakably divided and portrayed. Late confirmation from the ascomycetous sprouting yeasts recommends that these are a monophyletic assemble, all with a typical predecessor, and they are not just diminished structures gotten from the mycelial Ascomycota. They are likewise particular from the "splitting yeasts, for example, Schizosaccharomyces species, which don't bud however rather shape fibers that section by septation into block like cells (arthroconidia or arthrospores). [23]

Fungal Walls Features

As of late it has turned out to be evident that parasitic dividers serve numerous essential parts, very separated from the conspicuous part of giving an auxiliary boundary. For instance, the route in which a parasitic fungi develops is whether as round and hollow hyphae or as yeasts, is dictated by the divider parts and the courses in which these are amassed and attached to each other. [24] The divider is likewise the interface between a parasite and its condition: it ensures against osmotic lysis, it goes about as a sub-atomic strainer directing the entry of substantial particles through the divider pore space, and if the divider contains colors, for example, melanin it can secure the cells against bright radiation or the lytic proteins of different creatures. Notwithstanding these focuses, the divider can have a few physiological parts. It can contain restricting locales for chemicals, in light of the fact that numerous disaccharides (e.g. sucrose and cellobiose) and little peptides should be corrupted to monomers before they can go through the phone film, and this is ordinarily accomplished by the activities of divider bound compounds. [25] The divider additionally can have surface parts that intercede the collaborations of parasitic fungi with different life forms, including plant and creature has. Every one of these elements require a point by point comprehension of divider structure and design.

Significant Segment Wall

The essential way to deal with examining the divider piece of parasitic fungi is to upset contagious cells and decontaminate the dividers by utilizing cleansers and other gentle synthetic medicines, at that point utilize acids, soluble bases and proteins to debase the dividers consecutively. Albeit moderately couple of organisms have been investigated in detail, these medications demonstrate that parasitic dividers are transcendently made out of polysaccharides, with lesser measures of proteins and different segments. The significant divider parts can be classified into two noteworthy sorts: (i) the basic (fibrillar) polymers that comprise prevalently of straight-chain atoms, giving auxiliary unbending nature, and (ii) the framework segments that cross-connect the fibrils and that coat and insert the basic polymers. [26]

Table- 3.1: Main components of fungal walls

Taxonomic group	**Fibrillar components**	**Matrix components**
Chytridiomycota	Chitin	Glucan
	Glucan	
Zygomycota	Chtitin	Polyglucuronic acid
	Chitosan	Mannoproteins
Ascomycota	Chitin	Mannoproteins
	b-(1→3), *b*-(1→6)-glucan	*a*(1→3)-glucan
Basidiomycota	Chitin	Mannoprotein
	b-(1→3), *b*-(1→6)-glucan	*a*(1→3)-glucan
Oomycota	*b*-(1→3), *b*-(1→6)-glucan	Glucan

The fundamental divider polysaccharides contrast between the major contagious gatherings. The Basidiomycota, Ascomycota, and Chytridiomycota, commonly have chitin and glucans (polymers of glucose) as their real divider polysaccharides. Chitin comprises of long, straight chains of β-1, 4 connected N-acetylglucosamine buildups, though the parasitic glucans are extended polymers, comprising for the most part of β-1,3-connected spines with short β-1,6-connected side chains. [27] The Zygomycota ordinarily have a blend of chitin and chitosan (an ineffectively acetylated or non-acetylated type of chitin, polymers of uronic acids, for example, mannoproteins, and glucuronic corrosive. The Oomycota (which are not true fungal growths) have little chitin, and rather they have a blend of a cellulose-like β-1, 4-connected glucan and different glucans. [28]

Having made these focuses, perceive that the divider creation of a parasite is not settled, but rather can change considerably at various phases of the life cycle. This is additionally valid for some dimorphic organisms, which can develop as either hyphae or sprouting yeast-like cells.

Wall Architecture

The significant divider parts of growths can be thought of as the blocks and mortar, yet it was just in 1970 that we started truly to comprehend the engineering of the contagious divider. For this, Hunsley and Burnett built up an elegently straightforward strategy which they named enzymatic dismemberment. [29] They mechanically upset contagious hyphae so that lone the dividers remained, and afterward treated the dividers with blends and groupings of various proteins, coupling this with electron microscopy to identify any progressions that the catalyst medicines had caused. On the off chance that, for instance, the surface appearance of the divider changed after treatment with a specific compound "X," at that point the substrate of "X" is probably going to be the peripheral divider segment. In this way, by utilizing different successions and blends of proteins it was conceivable to strip away the real divider segments and to see their connections to each other. Three organisms were utilized as a part of this work, however we will take Neurospora crassa for instance to show the fundamental elements. [30]

In develop areas of hyphae the mass of Neurospora was appeared to have no less than four concentric zones, these are appeared as discrete layers, yet in all actuality they review into each other. The peripheral zone comprises of shapeless glucans with transcendently β-1, 3 and β-1, 6 linkages, which are debased by the catalyst laminarinase. Underneath this is a system of glycoprotein inserted in a protein lattice. At that point there is a pretty much discrete layer of protein, and afterward a deepest district of chitin microfibrils implanted in protein. [31] The aggregate divider thickness for this situation is around 125 nm. Be that as it may, the divider at the developing tip is more slender (c. 50 nm) and less difficult, comprising of an internal zone of chitin inserted in protein and an external layer of mostly protein. So unmistakably the divider winds up noticeably more grounded and more intricate behind the broadening hyphal tip, as further materials are included or as further holding happens between the segments. [32]

Neurospora appears to have a surprisingly complex divider design, in light of the fact that a glycoprotein arrange has not been found in some other organisms. By the by, the general example of divider design of hyphae is genuinely steady: the principle, straight-chain microfibillar parts (chitin, or cellulose in the Oomycota) are discovered transcendently in the inward area of the divider, and they are overlaid by nonfibrillar or "lattice" segments (e.g. different proteins, glucans, and mannans) in the external district. Be that as it may, there is generous holding between the different parts, serving to reinforce the divider behind the zenith. Specifically, a portion of the glucans are covalently clung to chitin, and the glucans are fortified together close by chains. [33]

The Extrahyphal Matrix

Notwithstanding the primary auxiliary segments of the divider, a few yeasts can have a discrete polysaccharide case, and both hyphae and yeasts can be encompassed by a pretty much diffuse layer of polysaccharide or glycoprotein, effortlessly evacuated by washing or gentle synthetic treatment. These extracellular lattice materials can have vital parts in the connections of growths with different creatures. For instance, the yeast Cryptococcus neoformans is a critical pathogen of people, its polysaccharide container veils the antigenic parts of the cell divider so that the growth is not overwhelmed by phagocytes and can multiply in the tissues. [77] In an alternate setting, the growth Piptocephalis virginiana (Zygomycota) parasitizes other Zygomycota, for example, Mucor spp. on agar media, yet does not parasitize them in fluid culture where the host organisms create an extracellular polysaccharide. Truth be told, the generation of an extracellular grid is frequently affected by development conditions. Pullulan is delivered financially from Aureobasidium pullulans, for this situation its blend is supported by a bounteous sugar supply in nitrogen-restricting development conditions. A portion of the other gel-like materials around growths additionally have potential commerical parts. [34]

SEPTA

Septa (cross-dividers) are found at genuinely normal interims along the length of most hyphae in the Ascomycota, Basidiomycota, and mitosporic growths, in spite of the fact that the septa are punctured to permit coherence of the protoplasm. In the event that piece of the hypha is harmed, at that point a Woronin body or a fitting of coagulated protoplasm quickly seals the septal pore to limit the harm.

At that point the hypha can regrow from a recently shaped tip behind the harmed compartment, or at times another tip can develop into the harmed compartment. [78] Unmistakably, these "harm impediment" reactions help to preserve the uprightness of the hypha, and may regularly be important when hyphae are eaten by bugs, or are assaulted by parasitic organisms (mycoparasites), or when hyphae of various vegetative similarity bunches endeavor to anastomose. Nonetheless, there are a few parasitic fungi and growth like living beings that don't create septa in the ordinary vegetative hyphae, for instance the Zygomycota (Mucor, and so on.) and the organism like Oomycota (Phytophthora, Pythium, and so on.). These aseptate parasitic fungi are more powerless against harm. [35]

Septa may give basic help to hyphae, particularly in states of water push. Be that as it may, one of their principle parts is by all accounts to empower separation. By obstructing the septal pores, a parasitic hypha is changed from a ceaseless arrangement of compartments to various free cells or locales that can experience isolate improvement.

Septa can be seen by typical light microscopy, yet electron microscopy uncovers a few distinct sorts of septum. The Ascomycota and mitosporic parasitic fungi have a straight-forward septum with a generally extensive focal pore, extending from 0.05 to 0.5 μm width, which permits the entry of cytoplasmic organelles and even cores. [36] The advancement of these septa happens strikingly rapidly, normally being finished inside a couple of minutes. They create as an ingrowing ring from the horizontal dividers of the hypha, and this is related with confined changes of the parallel dividers themselves, including a limited multiplication of a glycoprotein reticulum in the dividers of Neurospora crassa.

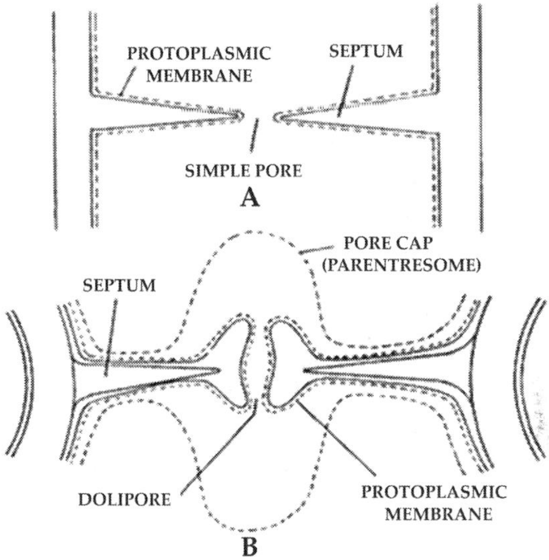

Figure- 3.5: Septal pores.

The Basidiomycota additionally have straightforward septa when they are developing as monokaryons (with one core in every cell). However, they frequently have a more intricate dolipore septum when strains of various mating similarity bunches circuit to shape a dikaryon, with two cores in every compartment. [37]The dolipore septum has a tight focal channel (around 100–150 nm breadth) limited by two ribs of undefined glucan. On either side of this septum are section formed membraneous structures named parenthosomes, which have pores to permit cytoplasmic progression however which keep the entry of significant organelles. In this manner, the Basidiomycota frequently have a more general course of action of cores contrasted and other organisms. Be that as it may, the septa are specifically corrupted when the Basidiomycota start to shape a complex fruiting body, for example, a mushroom. This empowers the mass redistribution of materials important to fuel the improvement of the fruiting structure.

From even this concise record, obviously septa assume a few imperative parts in parasitic science. [38]They can segregate compartments or they can permit the free entry of organelles through the septal por es, and they can be corrupted to permit the mass translocation of supplements and cytoplasmic segments to destinations of future improvement.

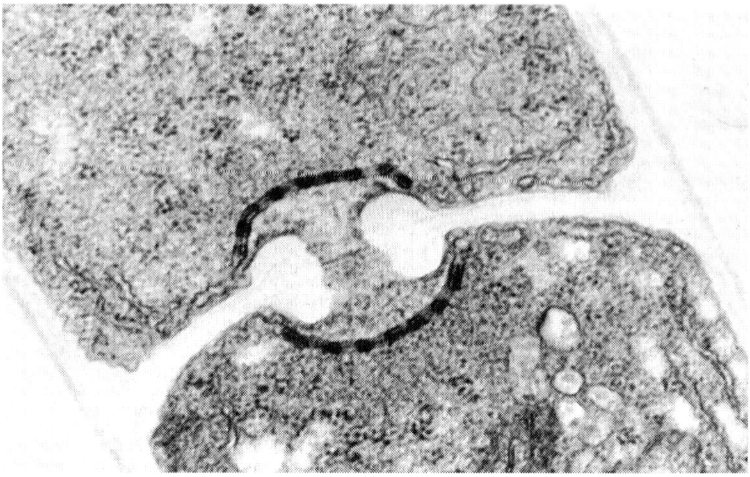

Figure- 3.6: The dolipore septum.

Nucleus of Fungi

Contagious cores are normally little (1–2 μm breadth) however astoundingly can be up to 20–25 μm distance across (e.g. Chytridiomycota, Basidiobolus ranarum,). They are encompassed by a two-fold atomic film with pores, as in all eukaryotes. Be that as it may, organisms are eminent for a few exceptional elements of their cores and atomic division. To start with, the atomic film and the nucleolus stay in place amid most phases of mitosis, though in most different living beings the atomic layer separates at an early stage amid atomic division. [39] A conceivable explanation behind this is the maintenance of an atomic film may forestall scattering of the atomic substance in hyphal compartments that contain a few cores and quickly gushing protoplasm. Second, in parasitic fungi there is no certain metaphase plate, rather the chromosomes appear to be haphazardly scattered, and at anaphase the little chromatids pull separated along two tracks, on shaft strands of various lengths. A third purpose of contrast is that growths have different sorts of shaft post bodies (additionally called microtubule-arranging focuses). [40]

They are in charge of microtubule get together amid atomic division. In Ascomycota and mitosporic organisms the shaft post bodies are circle formed, while in Basidiomycota they are frequently made out of two globular finishes associated by a scaffold. Be that as it may, the essentialness of these distinctions is misty. All parasitic fungi require a microtubule-arranging focus to guarantee that the chromosomes isolate accurately amid atomic division.

By far most of organisms are haploid with chromosome numbers extending from around 6 to 20. For instance there are six chromosomes in Schizophyllum cooperative (Basidiomycota), 16 in Saccharomyces cerevisiae (Ascomycota), eight in Emericella (Aspergillus) nidulans (Ascomycota), and 20 in Ustilago maydis (Basidiomycota) and seven in Neurospora crassa (Acomycota). A couple of organisms are actually diploid (e.g. the yeast Candida albicans, and individuals from the Oomycota). Also, a few growths can exchange amongst haploid and diploid eras, e.g. *S. cerevisiae*, and Allomyces spp. (Chytridiomycota) or exist as polyploid arrangement (Allomyces and a few Phytophthora spp.). [41]

Numerous parts of contagious hereditary qualities and parasitic fungal genomes, yet here we have to address a standout amongst the most wonderful components: growths are the main significant gathering of eukaryotic creatures that are haploid. All different eukaryotes are diploid, including plants, creatures, Oomycota, and the many single-celled life forms inexactly named "protists." The feasible explanation behind the haploid idea of parasitic fungi is that it gives a particular preferred standpoint. From one viewpoint, as we have seen, contagious hyphae for the most part contain a few cores in each hyphal compartment, and there is congruity of protoplasmic stream between the compartments. On the off chance that a passive transformation happens in one of the cores, at that point the growth will at present carry on as wild-sort, yet the latent change will be held, subsequently putting away potential changeability, the sign of diploid life forms. Then again, a transformed core can be presented to determination weight occasionally, when the parasite produces uninucleate spores or when a branch emerges and a solitary "author" core enters it. In this way, in any event a few regards, haploid parasitic fungi with multinucleate compartments have large portions of the focal points related with both a haploid and a diploid way of life. [42]

Cytoplasmic Organelles

A large portion of the cell segments of growths are practically like those of different eukaryotes, yet vary in some vital regards. The significant contrasts are talked about underneath.

Plasma Membrane

As in all eukaryotes, the parasitic plasma film comprises of a phospholipid bilayer with related transmembrane proteins, a hefty portion of which are included specifically or in a roundabout way in supplement take-up. [43] The layer likewise can grapple a few catalysts. Truth be told, the two fundamental divider engineered chemicals, chitin synthase and glucan synthase, are vital film proteins, they progress toward becoming moored in the layer such that they create polysaccharide chains from the external layer confront. The plasma film has a third essential part, in transferring signals from the outer condition to the phone inside, the procedure named flag transduction. [44]

The contagious plasma layer is novel in one vital regard, it normally contains ergosterol as the primary film sterol, rather than creatures, which have cholesterol, and plants which have phytosterols, for example, β-sitosterol. The Oomycota likewise have plant-like sterols. Be that as it may, some plant-pathogenic parasitic fungi, for example, Phytophthora spp., and Pythium can't combine sterols from non-sterol antecedents and rather should be provided with sterols from the host. Ergosterol is the essential focus of a few fungicides that are utilized to control plant-pathogenic organisms. Ergosterol is additionally an essential focus of a few antifungal medications that are utilized to treat human mycoses. [45]

Secretory System of Fungi

Parasitic fungi have a secretory framework, comprising of the endoplasmic reticulum (ER), the Golgi mechanical assembly (or Golgi equal), and layer bound vesicles. Proteins bound for send out from the cell are combined on ribosomes joined to the ER, at that point enter the ER lumen and are transported to the Golgi. Amid their dynamic transport through the Golgi cisternae, proteins experience different alterations, including fractional cleavage and reassembly, collapsing into a tertiary structure, and the expansion of sugar chains (glycosylation). [46]

At that point the proteins, or glycoproteins, are bundled into vesicles which bud from the developing face of the Golgi, and are transported to the plasma layer for discharge. This complicated postal framework sorts and conveys proteins to particular goals, including the compounds (pectinases, cellulases, proteases, and so forth.) that are bound for fare to debase polymers in the encompassing condition. The secretory framework is included in at any rate a few parts of parasitic tip development, since glycoproteins are just created in the Golgi and are transported in vesicles to the destinations of divider development. [47]

Furthermore, the Golgi is vital for the business creation, and discharge from the cell, of remote (heterologous) quality items. The capacity of a protein to enter the ER is dictated by a flag succession at the N-end, which is consequently evacuated. Without this succession the protein will stay in the cell. By and by, the parasitic secretory framework demonstrates some remarkable properties. The common Golgi mechanical assembly of creatures, plants, and Oomycota comprises of a heap of membraneous cisternae. Be that as it may, organisms have a Golgi proportionate which is ultrastructually very not the same as a commonplace Golgi as it comprises of frankfurter formed strings, dabs, and circles.[48]

CHITOSOMES

The vast majority of the vesicles seen in hyphal tips have not been artificially described, but rather a portion of the littler ones (microvesicles) take after particles that have been cleaned in vitro and are named chitosomes. These particles were at first found by homogenizing hyphae, expelling the divider material and major membranous organelles by centrifugation, and after that subjecting the supernatant to sucrose thickness centrifugation so that its segments isolated out as groups at various profundities in the tubes. When one of these groups was inspected by electron microscopy it was found to contain chitosomes, little spheroidal bodies, 40–70 nm breadth, each encompassed by a "shell" around 7 nm thick. At the point when the band containing these particles was hatched within the sight of an activator (a protease compound) and UDP-N-acetylglucosamine (the sugar-nucleotide from which chitin is made), every molecule supposedly produced a curl of chitin inside it, at that point the particles cracked to discharge strip like chitin microfibrils, each made out of a few chitin chains.[49]

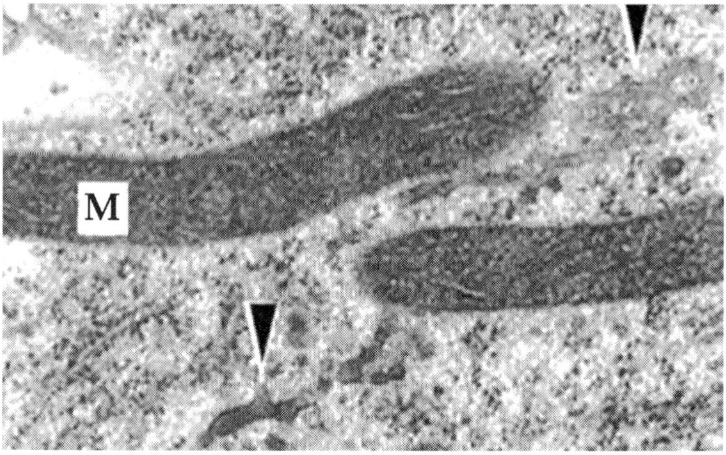

Figure- 3.7: Magnification of mithochondria in fungi.

On the off chance that the supernatant was first treated with digitonin (a saponin which solubilizes sterol-containing layers) and afterward centrifuged in a thickness angle, the chitosomes were never again found, rather, the expansion of activator and UDP-N-acetylglucosamine brought about the generation of chitin in an alternate band comparing to a lighter part of the homogenate. In any case, resulting expulsion of the digitonin caused the chitosomes to return, demonstrating that they are self-collecting totals of the catalyst chitin synthase, every molecule containing adequate compound atoms to deliver, after proteolytic initiation, a chitin microfibril made out of a few chitin chains. [50] Thus, this rich communitarian ponder by natural chemists and electron microscopists prompted the distinguishing proof of one of the sorts of vesicle regularly found in the hyphal pinnacle. One of the inquiries still to be settled is the manner by which these particles come to the hyphal tip after they have been amassed additionally back in the hyphae. One recommended component is that they are bundled into membrane bound multivesicular bodies, for example, those occasionally found in electron micrographs.

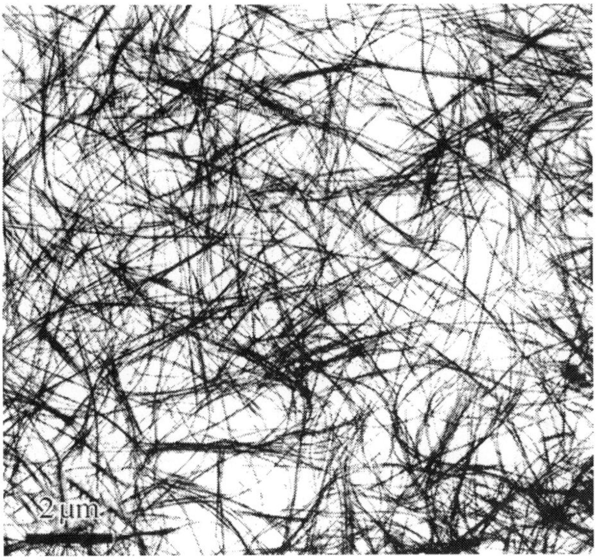

Figure- 3.8: Chitin microfibrils ribbon like masses.

VACUOLES

The vacuolar arrangement of growths has a few capacities, including the capacity and reusing of cell metabolites. For instance the vacuoles of a few organisms, including mycorrhizal species, collect phosphates as polyphosphate.[51] Vacuoles likewise appear to be real locales for capacity of calcium which can be discharged into the cytoplasm as a feature of the intracellular flagging framework. Vacuoles contain proteases for separating cell proteins and reusing of the amino acids, and vacuoles likewise have a part in the control of cell pH. All these vital physiological parts are notwithstanding the potential part of vacuoles in cell extension and (perhaps) in driving the protoplasm advances as hyphae stretch at the tips.

Vacuoles frequently are viewed as obvious, adjusted structures in the more established areas of hyphae, however late work has demonstrated that there is likewise a tubular vacuolar framework reaching out into the tip cells. [52] It is a greatly unique framework, comprising of restricted tubules which can enlarge and contract, as expanded components go along them in a peristaltic way. This was shown by concentrate the development of a fluorescent compound, carboxyfluorescein (CF), in living hyphae of a scope of parasitic fungi. The hyphae were treated with a nonfluorescent antecedent, carboxyfluorescein diacetate (CFDA), which is lipid-dissolvable so it enters the plasma film and is then quickly retained from the cytosol into the vacuoles. There the acetic acid derivation bunches are cut by esterases to yield CF, which is film impermeable so it stays in the vacuolar framework. Rees demonstrated that the tubular vacuoles which contain CF frame a pretty much consistent framework which goes through even the dolipore septa of Basidiomycota, transporting materials in reverse and advances in the hyphae. This has essential ramifications for the bidirectional translocation of materials inside single hyphae, and maybe particularly for mycorrhizal frameworks. [53]

Travelling of Vesicles and Endocytosis

Concentrates on plant cells, creature cells, and sprouting yeast propose that the take-up of substances through the cell layer may be accomplished by a dynamic procedure of endocytosis, the opposite of exocytosis, which is generally perceived to happen when vesicles intertwine with the plasma film and discharge their substance into the encompassing condition. Nonetheless, there is still some open deliberation about whether contagious hyphae have an endocytotic framework, and (on the off chance that they do as such) regardless of whether it is equal to the endosomal framework normally detailed in other cell sorts. [54]

The confirmation for endocytosis in contagious hyphae depends on the utilization of amphiphilic steryl colors, for example, FM4-64 and FM1-43, which are accounted for to be marker colors for endocytosis in a wide range of life forms. These key colors can be taken up by cells and are then envisioned by laser filtering confocal microscopy. Basically comparative marking designs with FM4-64 were discovered when hyphae of eight unique growths were tried, including agents of Basidiomycota, Ascomycota, Zygomycota, and mitosporic parasitic fungi. [55]The explanations behind the diverse marking examples of these colors are as yet misty, however the solid naming of the plasma film by FM4-64 is reliable with reports that this color progresses toward becoming intercalated between the external and inward layers of the cell layer. It at that point is by all accounts disguised by a dynamic procedure, since it can't latently cross the internal leaf of the plasma film. Comparative hyphae treated with a metabolic inhibitor (sodium azide) demonstrated no confirmation of color take-up. In a few occurrences a satellite Spitzenkörper supposedly originated at some separation behind the hyphal tip and it moved towards the tip, eventually combining with the principle Spitzenkörper. [56]

Albeit none of these lines of confirmation gives conclusive verification of the presence of an endosomal framework in filamentous growths, the proof is in any event perfect with an endosomal framework in which vesicle trafficking between various organelles gives an instrument to reusing of layers and their substance between various subcellular compartments. On this premise, a theoretical model of the association of a vesicle-trafficking system, including the conceivable inclusion of endosomes, has been proposed. This is, in any case, a region of debate in light of the fact that Torralba and Heath did not discover proof of endocytosis in Neurospora crassa hyphae. [57]

Concentrates with mutants might be expected to determine this issue. Numerous proteins related with endocytosis in Saccharomyces cerevisiae have homologs in the Neurospora genome arrangement and could be possibility for mutational examinations to test whether filamentous organisms have an equal endosomal framework.

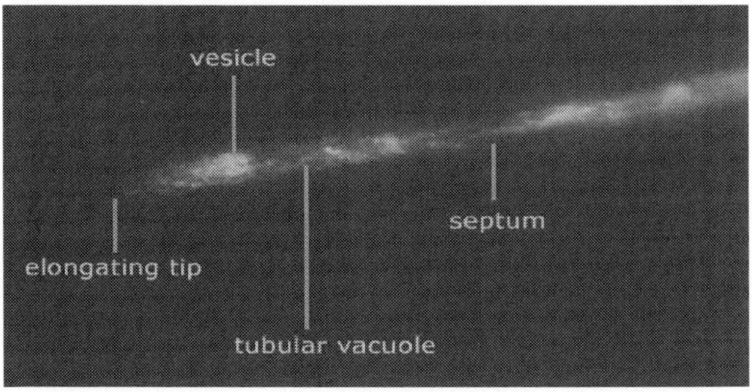

Figure- 3.9: Tubular vacuolar system of fungal hyphae.

Atomic Cytoskeleton and Motor

The cytoskeleton assumes a noteworthy part in the inner association of eukaryotic cells, giving a dynamic basic system to transporting organelles, for cytoplasmic gushing, and for chromosome partition amid cell division. [58] The three primary components of the cytoskeleton are: (i) microtubules, comprising of polymers of tubulin proteins, (ii) microfilaments, comprising of the contractile protein actin, and (iii) middle of the road fibers which give elasticity.

The appropriation of the cytoskeletal segments can be envisioned in living hyphae by utilizing fluorescent fundamental colors. These components are considered to assume a noteworthy part in organizing hyphal tip development. In electron micrographs of hyphae, the microtubules are viewed as long, straight tubules, around 25 nm breadth, happening either independently or as parallel exhibits. [59]

They are seen for the most part in the fringe areas of the hypha yet can likewise reach out up to the film at the outrageous tip. Microtubules are additionally observed to be intently connected with layer bound organelles, demonstrating that they could give a tramline framework to organellar development. Reliable with this, the benzimidazole fungicides that have been utilized generally to control plant pathogens fungus apply their antifungal activity by official to contagious microtubules, and this causes hyphae to quit developing. [60]The benzimidazole mixes comparatively square atomic division by authoritative to the axle microtubules. Microtubules are dynamic cell segments.

They depolymerize in light of medications, for example, cool stun, and on the other hand they can be balanced out by mixes, for example, taxol, the poison from yew trees (which is additionally poisonous to people). In regularly developing cells the microtubules are believed to be constantly corrupted and changed. Steady with this, microtubules can polymerize without anyone else's input get together in sans cell frameworks. [61] This is a two-arrange prepare: in the main stage, an atom of the protein α-tubulin joins with a particle of a sister protein β-tubulin to shape a dimer, at that point the dimers polymerize to frame tubulin chains. The microtubules shaped by this procedure collaborate with two mechanochemical compounds (engine proteins), kinesin and dynein, and likely help to transport organelles inside the contagious hypha.

Actin microfilaments are much smaller than microtubules, being around 5–8 nm breadth. In the ooze form Physarum polycephalum the actin microfilaments are known to work in cytoplasmic constriction, when actin partners with its engine protein, myosin. [62] This could likewise be valid for filamentous growths, where myosin-like proteins have as of late been identified. Be that as it may, the most convincing confirmation for a part of the actin cytoskeleton in sorting out parasitic development has originated from contemplates on the yeast, Saccharomyces cerevisiae. This growth has microtubules in any case, rather than mycelial parasitic fungi, the microtubules of yeast don't appear to be required in the vehicle of significant organelles, rather, they appear to work primarily in introduction of the mitotic shaft. However, actin links do assume a noteworthy part in guiding secretory vesicles to development locales in yeast, and by relationship actin may assume a comparative part in conveying the apical vesicles to destinations of divider development in hyphal tips. [63]

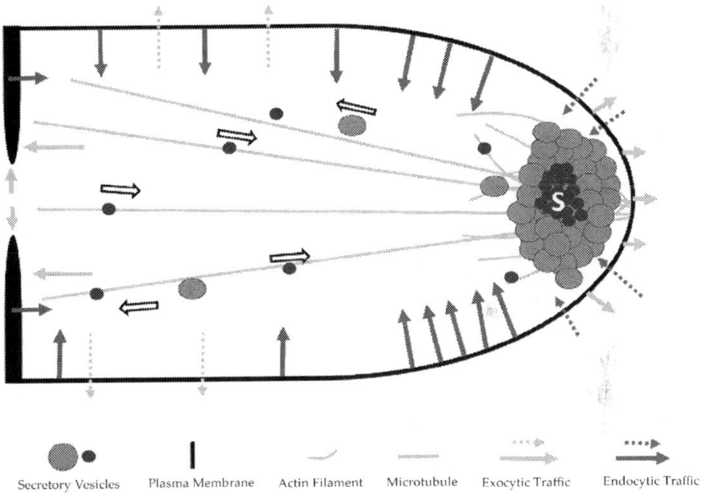

Figure- 3.10: Trafficking network of vesicle in growing hyphae.

In later sections we come back to the parts of the cytoskeleton in parasitic development, separation, and the conduct of contagious zoospores. Yet, we end this section by taking note of that parasitic tubulins contrast from those of plants and creatures. They are restrained by the anti-infection griseofulvin and by the benzimidazole fungicides, while plant and creature tubulins are unfeeling to these mixes. [64] This is the reason griseofulvin can be utilized to treat the dermatophytic (ringworm) parasitic contaminations of people, and why the benzimidazoles can be utilized to treat contagious diseases of plants. On the other hand, the contagious tubulins (except for Oomycota) are unaffected by colchicine (the poison from the harvest time crocus) which represses atomic division in plant and creature cells.

SUMMARY

- Fungi have exclusive structure in term of their features such as, cellular organization, cellular components, and structure. Fungi consists of long, slender, filaments known as hyphae. Hyphae has diameter alternating from 2 μm to 30 μm which dependent on growth and species condition. These hyphae may have cross walls and known as septate hyphae, while without cross walls is known as aseptate hyphae, for example in Zygomycota and Oomycota.[65]
- All the fungal hayphae thar are growing actively express clearly described polarity in the organization from hyphal tip to base of apical section. Hyphae collect vesicles from its tip, this phenomena is called as apical vesicle cluster. Amount of nuclei differ within and among groups of fungi. Fungi that have one nucleus are called monokaryon, and those having two nucleus are called dikaryon. Example of monokaryon is Basidiomycota. [66]
- Outermost zones of fungal cell wall consists of amorphous glucans having mainly β-1,3 and β-1,6 linkages, that might be tarnished with the help of enzyme laminarinase. In addition to this some fungi also have discrete polysaccharide capsule, which can be removed by mild chemical treatment and by washing. Nuclie of fungi is 1 to 2 μm in diameter, however in some conditions can be 20 to 25 μm in diameter. [67] It has unique cytoskeleton and molecular motors. Tubulin are also present in fungi but it is different from animals and plants.

REFERENCES

1. Hunter, B. B., Sylvester, M. A., & American Society for Microbiology. (1981). *Fungal morphology*. Washington, D.C: American Society for Microbiology.
2. Moore, D. (2002). *Fungal Morphogenesis*. Cambridge: Cambridge University Press.
3. Burnett, J. H. (2003). *Fungal population genetics and speciation*. Oxford: Oxford University Press.

4. In Martin, F. (2017). *Molecular mycorrhizal symbiosis.*
5. Samanta, I. (2015). *Veterinary Mycology*. Dordrecht: Springer.
6. Nicolau, D. V., University of Adelaide., Queensland University of Technology., & Society of Photo-optical Instrumentation Engineers. (2006). *BioMEMS and nanotechnology II*. Bellingham, Wash., USA: SPIE.
7. Xu, H., & University of Alberta. (2015). *Major intrinsic proteins of laccaria bicolor: Characterization, transcript profiling and functions in ectomycorrhizal associations with picea glauca.*
8. Peterson, R. L., Massicotte, H. B., & Melville, L. H. (2004). *Mycorrhizas: Anatomy and Cell Biology*. Boulder: NetLibrary, Inc.
9. Garcia, E. O., Deising, H. N., & Vögele, R. T. (2013). *Infection structure-specificity of [beta]-1,3-glucan synthase is essential for pathogenicity of Colletotrichum graminicola and evasion of glucan-triggered immunity*. Halle, Saale: Universitäts- und Landesbibliothek Sachsen-Anhalt.
10. Zabel, R. A., & Morrell, J. J. (1992). *Wood Microbiology: Decay and Its Prevention*. Oxford: Elsevier Science.
11. Jones, E. B. G., & Pang, K. (2012). *Marine fungi and fungal-like organisms*. Boston: De Gruyter.
12. Fuller, R., Lovelock, D. W., & Society for Applied Bacteriology. (1976). *Microbial ultrastructure: The use of the electron microscope*. London: Academic Press.
13. Kavanagh, K. (2011). *Fungi: Biology and Applications*. Chichester, UK: Wiley-Blackwell.
14. Sumbali, G., & Johri, B. M. (2005). *The fungi*. Harrow, Middlesex: Alpha Science International.
15. Aldrich, H. C., & Todd, W. J. (1986). *Ultrastructure techniques for microorganisms*. New York: Plenum Press.
16. Sharma, P. D. (2005). *Fungi and allied organisms*. Oxford, U.K: Alpha Science International.
17. Li, J. (1993). *Ultrastructure, taxonomy and phylogeny of the chytridiomycetous anaerobic gut fungi*. Ottawa: National Library of Canada = Bibliothèque nationale du Canada.

18. Kurylowicz, W. (1975). *Atlas of ultrastructure of Streptomyces in course of biosynthesis of antibiotics*. Warszawa: Polish Medical Publishers.
19. Gruhn, C. M. (1989). *Effect of a heavy metal on ecto- and vesicular-arbuscular mycorrhizal fungi: The physiology, ultrastructure, and ecology of copper stress and tolerance.*
20. Longcore, J. E. (1991). *Morphology, occurrence, and zoospore ultrastructure of three chytridialean fungi from Maine lakes*. Orono, Me.
21. Pierson, E S, Miller, D D, Callaham, D A, Shipley, A M, Rivers, B A, Cresti, M, & Hepler, P K. (n.d.). *Pollen tube growth is coupled to the extracellular calcium ion flux and the intracellular calcium gradient: effect of BAPTA-type buffers and hypertonic media.*
22. Miller, Deborah Denise. (1995). *The role of calcium and myosin in {\it Lilium longiflorum\/} pollen tube growth*. ScholarWorks@UMass Amherst.
23. Cinq-Mars, L. . (1949). *Interactions between Venturia inaequalis (Oke.) Wint. and Saprophytic fungi and bacteria inhabiting apple leaves.*
24. Koneman, E. W., Fann, S. E., & Astrin, J. (1971). *Clinical laboratory mycology, part I*. New York: Medcom.
25. Coulson, G. (Supervisor), & Cinq-Mars, Lionel, 1919-1973. (1949). *Interactions between Venturia inaequalis (Oke.) Wint. and Saprophytic fungi and bacteria inhabiting apple leaves*. McGill University.
26. Gerards, S., Klein Gunnewiek, P.J.A., Modderman, R., & De Boer, W. (1999). *Response of the chitinolytic microbial community to chitin amendments of dune soils*. 29, 170-177.)
27. Smith, J. M. (2005). *Powdery mildew (Podosphaera macularis Braun & Takamatsu) resistance in wild hop genetic r* Koehler, J. K., & Lawrence Radiation Laboratory. (1961). *The fine structure of normal and irradiated yeast cells and yeast ribosomes*. Berkeley, Ca: Lawrence Radiation Laboratory.
28. Fantes, P., & Beggs, J. (2000). *The yeast nucleus*. Oxford: Oxford University Press.

29. Prescott, H. P. (1870). *Strong drink and tobacco smoke: The structure, growth, and uses of malt, hops, yeast, and tobacco*. New York: W. Wood & Co.
30. Koschubs, T. (2010). *Structure and functional architecture of the mediator middle module from budding yeast*.
31. Valerius, O. (2001). *Chromatin structure and regulation of gene expresion at the histidine/adenin branch point in yeast and aspergillus*. Göttingen: Niedersächsische Staats- und Universitätsbibliothek.
32. Winton, A. L., & Winton, K. G. B. (1932). *The structure and composition of foods*. New York: J. Wiley & Sons.
33. Větvička, V., & Novák, M. (2013). *Biology and chemistry of beta glucan: Beta-glucan, structure, chemistry and specific application*. Oak Park, IL: Bentham Science Publishers.
34. Berrios, M. (1998). *Nuclear structure and function*. San Diego: Academic Press.
35. Whisstock, J. C., & Bird, P. I. (2011). *Serpin structure and evolution: Volume 501*. Amsterdam: Elsevier/Academic Press.
36. Elgin, S. C. R., & Workman, J. L. (2000). *Chromatin structure and gene expression*. Oxford: Oxford University Press.
37. In Buckeridge, M. S., & In De, S. A. P. (2017). *Advances of basic science for second generation bioethanol from sugarcane*.
38. Malhotra, S. K. (1983). *The plasma membrane: ed topics in structure and function*. New York: Wiley.
39. Grossmann, G. (2009). *Plasma membrane compartmentation in Saccharomyces cerevisiae*.
40. Fischer, R. (1989). *Molecular cloning of a human plasma membrane Ca^{2+}-ATPase*. Zürich.
41. Parham, P., & Janeway, C. (2015). *The immune system*.
42. Schaechter, M., Engleberg, N. C., DiRita, V. J., & Dermody, T. (2013). *Schaechter's mechanisms of microbial disease*. Philadelphia, PA: Wolters Kluwer Health/Lippincott Williams & Wilkins.
43. Gupta, P. K. (2009). *Cell and molecular biology*. Meerut, India: Rastogi Publications.
44. Jeon, K. W. (2014). *International review of cell and molecular biology*.

45. Cregg, J. M. (2007). *Pichia protocols*. Totowa, N.J: Humana Press.

46. Williams, A., Hussell, T., & Lloyd, C. (2012). *Immunology: Mucosal and body surface defences*. Chichester, West Sussex: Wiley-Blackwell.

47. In Polin, R. A., In Abman, S. H., In Rowitch, D. H., In Benitz, W. E., & In Fox, W. W. (2017). *Fetal and neonatal physiology*.

48. Vacante, V. (2015). *Mites of economic plants: Identification, bio-ecology and control*.

49. Gozalbo, D. (1993). *Characterization of chitosomes in Candida albicans protoplast*. S.l.: s.n..

50. Schwencke, L. J., Flores, M. A., & Vorisek, J. (1984). *Are chitosomes a type of general secretory vesicle?*. Francia: Elsevier Science Publishers B.V.

51. Ziman, M, Chuang, J S, & Schekman, R W. (1996). *Chs1p and Chs3p, two proteins involved in chitin synthesis, populate a compartment of the Saccharomyces cerevisiae endocytic pathway*.

52. De, D. N. (2000). *Plant cell vacuoles: An introduction*. Collingwood, Vic: CSIRO.

53. Kitching, J. A. (1956). *Food vacuoles*. Wien: Springer.

54. Hurley, R. R. (1977). *Studies on yeast vacuoles*.

55. Kitching, J. A. (1956). *Contractile vacuoles of protozoa*. Wien: Springer-Verlag.

56. Krantz, M. J. (1971). *Studies on gas vacuoles*.

57. Gu, M. (2010). *Synaptic vesicle recycling by clathrin-mediated endocytosis*.

58. Lai, M. M.-S. (2001). *Regulation of synaptic vesicle endocytosis by calcineurin and associated proteins*.

59. Wienisch, M. (2006). *Visualization of synaptic vesicle protein recycling during exo-endocytosis at individual hippocampal boutons*. S.l: s.n.

60. Rajappa, R. (2014). *On the role of synaptic vesicle proteins self-assembly during exo- and endocytosis*.

61. Kohli, B. M. (2006). *Mass spectrometry of APP-interacting protoeins in vivo identifies a protein network involved in vesicle*

turnover and a link to endocytosis-mediated nuclear signaling.

62. Schliwa, M. (2003). *Molecular motors*. Weinheim: Wiley-VCH.
63. Vallee, R. B. (1991). *Molecular motors and the cytoskeleton*. Academic Press Inc.
64. Lucy, J. A. (1978). *The plasma membrane*. Burlington, N.C: Scientific Publications Division, Carolina Biological Supply Co.
65. Russell, B. J., Phoenix Learning Group., Coronet (Firm), Bill Walker Productions., & BioMedia Associates. (2000). *The plasma membrane*. St. Louis, MO: Phoenix Learning Group.
66. Poskanzer, K., University of California, San Francisco., & University of California, San Francisco. (2006). *Regulation of synaptic vesicle endocytosis by synaptotagmin I*. (Dissertation Abstracts International, 67-2.)
67. Yarden, Y., & Tarcic, G. (2013). *Vesicle Trafficking in Cancer*. New York, NY: Springer New York.
68. Morgan, J. R. (2001). *Mechanisms of synaptic vesicle endocytosis: The functions of clathrin in the nerve terminal*.
69. Gu, M. (2010). *Synaptic vesicle recycling by clathrin-mediated endocytosis*.
70. Lai, M. M.-S. (2001). *Regulation of synaptic vesicle endocytosis by calcineurin and associated proteins*.
71. Wienisch, M. (2006). *Visualization of synaptic vesicle protein recycling during exo-endocytosis at individual hippocampal boutons*. S.l: s.n.
72. Rajappa, R. (2014). *On the role of synaptic vesicle proteins self-assembly during exo- and endocytosis*.
73. Kohli, B. M. (2006). *Mass spectrometry of APP-interacting protoeins in vivo identifies a protein network involved in vesicle turnover and a link to endocytosis-mediated nuclear signaling*.
74. Schliwa, M. (2003). *Molecular motors*. Weinheim: Wiley-VCH.
75. Vallee, R. B. (1991). *Molecular motors and the cytoskeleton*. Academic Press Inc.
76. Banting, G., Higgins, S. J., & Biochemical society (GB). (2000). *Molecular motors*. London: Portland Press.

77. Bloom, G. S. (2000). *Molecular motors*. Amsterdam: Elsevier.
78. Vallee, R. B. (1991). *Molecular motors and the cytoskeleton: [Pt. A]*. San Diego, Calif. [u.a.: Academic Press.

CHAPTER

4

FUNGAL GROWTH

In this part we concentrate on the processes of parasitic fungal development, with exceptional reference to the synthesis of wall at tip of hyphae and association of development, which is key to understanding the science of fungal growths. We also talk about the routes in which hyphal branches emerge and orientate themselves for greatest proficiency of supplement catch. Furthermore, we consider the energy of contagious development in clump culture and constant culture frameworks, relating this to mechanical scale forms, comprising the viable creation of Quorn™ mycoprotein which is a profoundly fruitful maturation framework for delivering "single-cell protein."

APICAL DEVELOPMENT OF FUNGAL HYPHAE

Apical development is the sign of organisms. Aside from the organism like Oomycota, which have embraced apical development by an amazing level of joined advancement, no different living beings develop consistently as a tube that reaches out at the outrageous tip by the restricted combination of divider parts, and, seemingly, no other gathering shows such extraordinary pliancy. [1] The hyphal summit can swell into an inflatable like structure, for example, a spore or yeast cell, or it can decrease to such an extent, to the point that it can infiltrate a layer of dormant gold film or the mass of a host plant by applying turgor weight alone. In different conditions, the parasitic hypha can offer ascent to complex tissues and contamination structures, when hyphae of Neurospora crassa are seen by setting a coverslip over the edge of a province on an agar plate. The succession of nine edges was assumed control over a 1-hour duration, beginning from the time when the coverslip was included. In the main casing (a) the hyphal tip was developing ordinarily, and two horizontal branches had emerged behind the developing tip. [2] Before long thereafter (b and c) the hyphal tips started to swell (a reaction to aggravation caused by the coverslip) and after that extended over and over from the tips before continuing a more typical example of apical development. By taking any helpful reference focuses, for example, the fanning focuses appeared as v1 and v2, it is seen that the length of hypha officially framed stays unaltered, and all new development happens from the first hyphal tip or from the branch tips. [3] Truth be told, the consolidate particle of new divider material is essentially limited to the outrageous tip. Developing hyphae were presented to a short beat of radio-categorized divider forerunners, for example, 3H-N-acetylglucosamine (from which chitin is orchestrated) or 3H-glucose (from which divider glucans are made) and after that autoradiographed. The radiolabel isincorporated maximally at the hyphal tip, and the rateof name fuse tumbles off pointedly finished the initial couple of micrometers, the apical arch of the hypha. Fast rates of hyphal tip augmentation, (for example, 40 μm min−1 in N. crassa) must be conceivable if the zenith is provided with vesicles and other cytoplasmic segments from behind. Thus, what we term apical development is really apical augmentation, in light of the fact that the genuine rate of development, characterized as increment in biomass per unit of time, is much slower. [4] The length of hypha expected to help an expanding zenith can be assessed by making a corner to corner cut over a state edge with a surgical tool, so that individual hyphae are disjoined at various separations from their tips. Hyphae that are sliced excessively near the tip bite the dust from physical harm. Hyphae slice additionally back keep on extending yet more gradually than expected, and in the end a point is come to at which the cut is so far back that it has no impact on the apical expansion rate. This separation is named the fringe development zone of a parasitic province, characterized as the length of hypha expected to keep up the most extreme augmentation rate of the main hyphae at the settlement edge, it differs between growths, from underneath 200 μm up to a few millimeters for the quickest expanding organisms. [5]

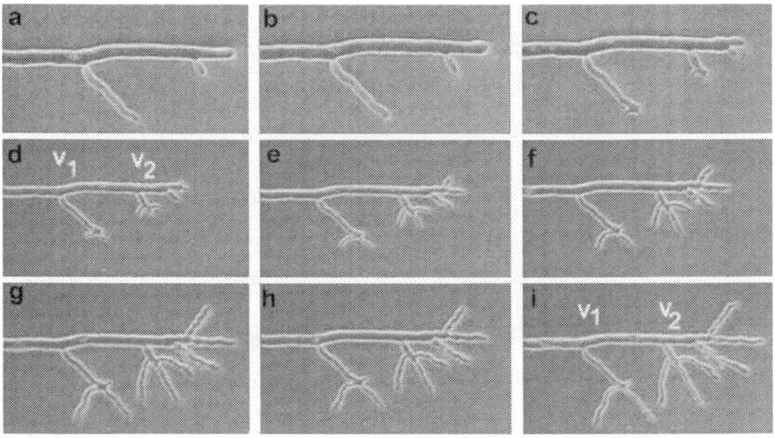

Figure- 4.1: Arrangement of fungal hyphae growth from videotape.

Experiments on Processes of Fungal Apical Growth

Robertson did a large portion of the key early tests on apical development of growths, utilizing to a great degree basic strategies combined with genuinely amazing knowledge. He developed provinces of Fusarium oxysporum or *N. crassa* on supplement rich agar, overwhelmed them with water and after that watched the conduct of the tips. [6] Generally 50% of the tips quit developing instantly and after that continued development inside a moment, yet from a smaller pinnacle than some time recently. Alternate tips quit developing for a few minutes, swelled into a precious stone shape amid this time, and in the long run regrew by creating at least one limited tips simply behind the first summit. He at that point rehashed the investigations, again flooding the provinces with water yet supplanting this inside 40 seconds by an answer of an indistinguishable osmotic potential from the first agar (an isotonic arrangement). This made every one of the tips stop for a few minutes, however they swelled amid this time and in the end regrew from limit subapical branches. [7]

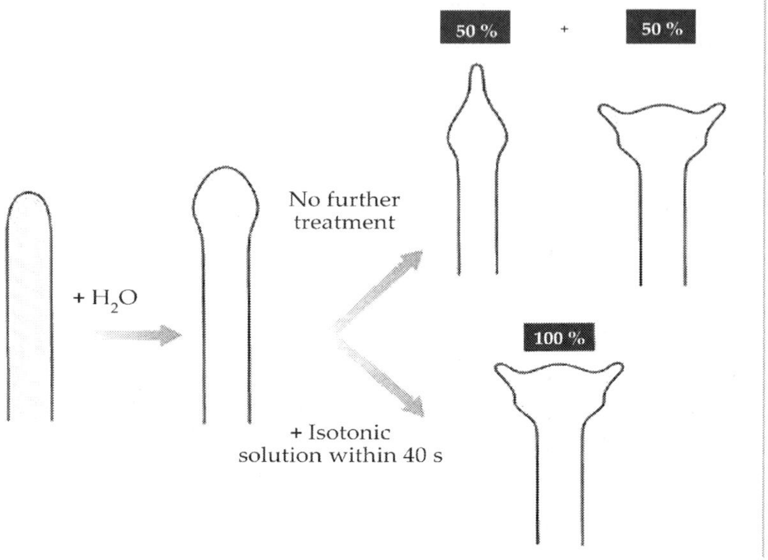

Figure- 4.2: Experiment of Robertson's on tip growth of hyphae.

To translate these discoveries, Robertson speculated that the ordinary example of apical development includes two free procedures: (i) ceaseless augmentation of a plastic, deformable tip and (ii) rigidification of the divider behind the expanding tip. [8] He imagined these two procedures as happening at a similar rate, yet with rigidification dependably somewhat behind the tip, similar to two autos going along two paths of a motorway at the very same speed however one is dependably marginally behind the other. At that point, if augmentation development is ended by an osmotic stun (including water) the procedure of rigidification will proceed and have a tendency to "surpass" the pinnacle. On the off chance that the tip straightens out to the new osmotic conditions in time it can develop on, however now from a more slender district of the pinnacle where the divider has not yet rigidified. [9]

Generally 50% of the tips in the first investigation appeared to have the capacity. Notwithstanding, if the tip can't change in time then the summit will be closed by rigidification, and development will just happen when new tips have been created, for this situation behind the first peak and by a procedure that takes a few minutes. [10] This would clarify why every one of the tips ceased for a few minutes when water was supplanted by the isotonic arrangement, in light of the fact that the tips would need to make two separate osmotic modifications and couldn't do as such before the pinnacle had rigidified. [11]

We should see later that this clarification was basically right. It is presently upheld by many lines of confirmation from divider enzymology and ultrastructural ponders. In any case, at this stage we should take note of some further focuses. Some of the time the hyphal tips swell and barged in light of flooding with water, maybe on the grounds that the divider at the extraordinary summit is excessively delicate, making it impossible to acclimate to quick changes in osmotic potential or maybe on the grounds that the divider at the outrageous peak is persistently being corrupted by divider lytic chemicals. Now and then the tips develop on as normal in the wake of being overwhelmed with water, yet a branch grows later from the position where the pinnacle had come to at the season of flooding. [12] Regardless, developing hyphal tips are exceptionally touchy to many sorts of aggravation and they have a tendency to react similarly – by a "stop–swell–branch" succession. This reaction can be evoked by gentle warmth or cool stun, by introduction to an extraordinary light bar, or notwithstanding when hyphal tips experience physical boundaries. Some morphological mutants of Neurospora and Aspergillus even demonstrate this example routinely amid development on agar plates. We should see that the stop–swell–branch succession happens amid the generation of a few separated structures, including the pre-entrance structures of contagious parasitic fungi of plants and creatures. [13]

Amass of Peak of Hyphae

Divider blend at the hyphal peak is a perplexing procedure, the points of interest of which are as yet not completely known, but rather from different lines of proof we can build a composite picture of divider development and ensuing divider development. A portion of the principle segments of this framework are talked about underneath.

Chitin Synthase

Chitin synthase catalyzes the union of chitin chains, and is consequently one of the essential proteins required in contagious divider development. Chitin is known to be framed in situ at the peak, as opposed to touching base in film bound vesicles. At the point when hyphal homogenates are tried for compound action in vitro, chitin synthase is found in no less than two structures: as an idle zymogen in chitosomes, and as a necessary film protein. [14] Chitosomes look like a portion of the microvesicles in the hyphal zenith. Be that as it may, the "shell" around a chitosome is not a phospholipid film, so chitosomes may be bundled inside phospholipid layers for transport to the tip, maybe in the multivesicular bodies that are some of the time found in electron micrographs. The zymogen type of chitin synthase, when embedded into the layer, must be enacted by a protease which most likely touches base at the pinnacle in different vesicles. [15]

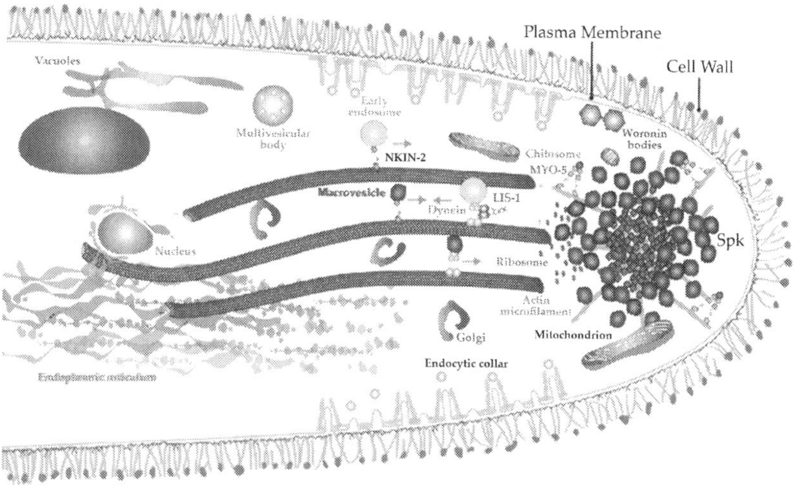

Figure- 4.3: Possible hyphal apex wall growth.

At that point the substrate is conveyed from the cytosol to the inward face of the chitin synthase protein (tied down in the film), and chitin chains are orchestrated and expelled from the layer external face. The substrate for chitin combination is N-acetylglucosamine, yet it is provided as a sugar nucleotide, UDPN-acetylglucosamine (where UDP = uridine diphosphate), with a high-vitality security required for chitin union. Obviously, there must be systems for managing the movement of chitin synthase amid divider development. This could be accomplished in various ways, including chemical inhibitors, in light of the fact that the cytosol is known to contain a chitin synthase inhibitor, which may keep any "overflow." [16]

Glucan Synthase

Glucan synthase is the other real compound required in divider development. It catalyzes the amalgamation of β-1,3-glucan chains, which frequently include the main part of the contagious divider. Like chitin synthase, glucan synthase is thought to touch base in vesicles and progresses toward becoming embedded in the plasma layer at the zenith. [17] The substrate for this compound is a sugar nucleotide (UDP-glucose), provided from the cytosol. Be that as it may, the movement of glucan synthase is managed uniquely in contrast to chitin synthase. The catalyst is made out of two subunits, one of which (on the film external face) contains the reactant site, and the other is a guanosine triphosphate (GTP) restricting protein. So the protein is believed to be enacted when GTP lands at the cytoplasmic face, at that point glucan chains are integrated and expelled into the divider. [18] The glucan anchors appear to experience advance change inside the divider. Specifically, short β-1,6-connected side chains create and interface the β-1,3-glucan chains. The quantity of these spread linkages increments dynamically as the divider develops behind the pinnacle.

Mannoproteins

Mannoproteins and different glycoproteins frame a generally little extent of the aggregate divider piece of parasitic hyphae, yet are more typical in yeasts and in the yeast-like period of dimorphic growths. These glycosylated proteins are among the few divider segments that are pre-shaped in the endoplasmic reticulum–Golgi complex and are conveyed to the summit in vesicles. [19]

Development and Cross-Connecting of Hyphal Divider

Different sorts of cross-linkage happen between the real divider polymers after these have been embedded in the divider, and this appears to happen continuously once again from the hyphal tip. For instance, basically immaculate glucans can be extricated from recently shaped parasitic dividers by utilizing hot antacid, yet in the more established divider areas an expanding extent of the glucan is soluble base insoluble, obviously on the grounds that it is complexed with chitin. In help of this view, the glucans can be removed in the wake of treating dividers with chitinase to debase the chitin. [20] The chitin and glucans are connected by covalent bonds. Little is thought about the procedure aside from that amino acids might be included, in light of the fact that the amino corrosive lysine is related with up to half of the glucan–chitin linkages in dividers of Schizophyllum cooperative (Basidiomycota). Notwithstanding these intermolecular bonds, the chitin chains connect with each other by hydrogen holding, to frame microfibrils. The glucans likewise connect with each other. These extra bondings behind the developing peak could serve to change over the first plastic divider into a dynamically more cross-connected and rigidified structure. [21]

Lytic Enzymes of Wall

There are restricting perspectives on whether divider lytic catalysts are essential for apical development. From one viewpoint, it has been recommended that the current divider must be mollified all together for new divider segments to be embedded, in which case divider development would include an adjust of divider lysis and divider amalgamation. Reliable with this, chitinase, cellulase (in Oomycota), and β-1,3-glucanase exercises can be found in hyphal divider portions, in spite of the fact that these proteins may exist for the most part in a dormant shape. [22] Then again, it has been contended that the generous cytoskeleton of tubulins and actin could fortify the hyphal tip, blocking the requirement for an inflexible divider and hence blocking the requirement for divider debasing proteins. Notwithstanding, there is doubtlessly divider lytic catalysts would be required for the generation of new tips (new hyphal branches) that rise up out of the already inflexible divider additionally again from the hyphal zenith. [23]

Steady Model of Fungal Wall Growth

Wessels (1990) proposed a relentless state model of contagious tip development that could make the part of tip-found divider lytic proteins superfluous, and furthermore could clarify a few different components of tip development. As per this model, the recently shaped divider at the outrageous tip is recommended to be viscoelastic, so that the divider polymers stream outwards and in reverse as new segments are consistently included at the extraordinary tip.

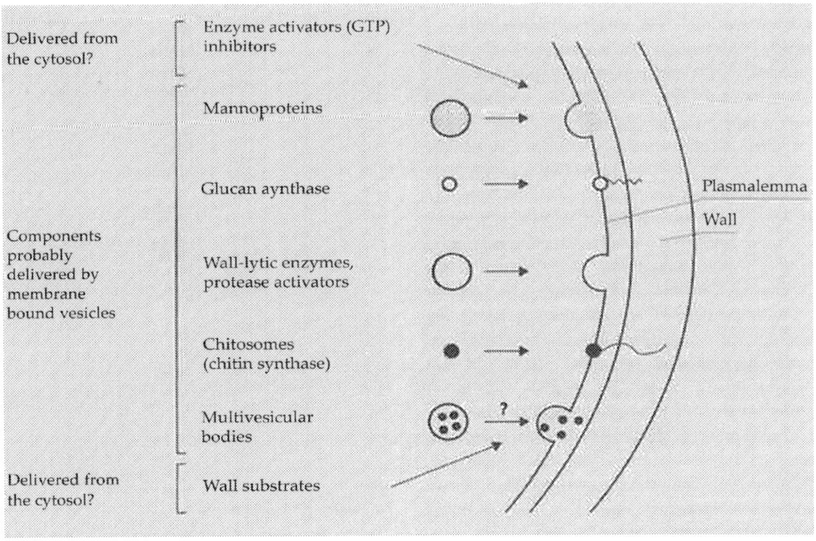

Figure- 4.4: Synthesis of fungal wall at hyphae tip.

At that point the divider rigidifies logically by the arrangement of additional bonds behind the tip. This is reminiscent of Robertson's unique thought of a plastic, deformable divider at the outrageous tip, with consequent rigidification happening behind the tip. In this manner, for instance, if the supply of vesicles to the tip is moderated or ended (by an osmotic stun, and so on.) the rate of hyphal expansion would moderate or stop, yet divider cross-connecting may be unaffected if holding happens suddenly inside the divider. We could accordingly have in any event a portion of the components of a binding together hypothesis of contagious tip development, in view of much biochemical and ultrastructural prove. It is as yet important to clarify how hyphae with a basically liquid (viscoelastic) divider could oppose turgor weight, however the response to this could lie in the basic help given by the huge actin cytoskeleton at the hyphal tip. Jackson and Heath in 1990 examined this for Saprolegnia ferax (Oomycota). [24] They demonstrated that treatment of hyphae with *cytochalasin E* (one of a few cytochalasins, which upset cell flow by authoritative to actin microfilaments) caused disturbance of the actin top and drove at first to an expansion in the rate of tip augmentation, yet then the tips swelled and burst. The weakest area of the tip, most defenseless to blasting, was not the outrageous summit where the actin top is densest yet on the shoulders of the peak where the actin is less thick and where the divider apparently has not yet rigidified adequately to make up for the weaker cytoskeleton. It will be reviewed that the shoulder of the zenith is the place new tips begin when hyphae are overflowed with water. [25]

Another critical part of the unfaltering state display is that it could clarify how parasitic fungi discharge catalysts into the earth for breakdown of complex polymers. Catalysts are moderately expansive particles, ordinarily in the scope of 30–50 kDa (KiloDaltons), and there is no confirmation that parasitic dividers have constant pores of this size through which compounds could be discharged into the earth. The viscoelastic divider model could resolve this issue if the catalysts are discharged from vesicles that wire with the hyphal tip and these compounds at that point stream outwards through the creating divider. [26]

Main Thrust for Development of Apical

Having considered the flow of divider development and divider rigidification, the rest of the inquiry concerns the main thrust for apical expansion. The cytoskeletal parts have developed as the most grounded contender for this, steady with many examinations on creature cells where distensions, for example, pseudopodia are connected to the polymerization of actin. Concentrates on Saprolegnia (Oomycota) have demonstrated that the summit can expand notwithstanding when hyphae have irrelevant turgor weight, probably in light of the fact that actin polymerization drives this procedure. Actin is copious in hyphal tips, and both tip expansion and cytoplasmic gushing can be stopped by treating parasitic fungi with cytochalasins ("cell-relaxers") which tie to actin. In S. cerevisiae there is solid confirmation that F-actin is included in the restriction of bud arrangement and that it connects with the engine protein myosin to transport vesicles to the bud site The subject of whether microtubules are straightforwardly required in parasitic tip development is more problematical.

Hyphal expansion can be stopped by the benzimidazole fungicides, the related azole medications, and griseofulvin, all of which meddle with microtubule work. Corresponding with this stoppage of development, there is a dynamic consumption of vesicles in the hyphal tip. Along these lines, microtubules should somehow be included in tip development, maybe by giving tramlines to vesicle cargoes. Calcium likewise is by all accounts personally included in tip development on the grounds that the tips of a few organisms, including Neurospora, and furthermore Saprolegnia (Oomycota), require outside calcium for proceeded with tip development. [27] In addition, the plasmalemma at the outrageous tip is accounted for to have a high centralization of extend enacted calcium channels, permitting the entrance of calcium when the layer is extended. The importance of this is the intracellular levels of free calcium are dependably firmly directed, cells keep up a low calcium level by sequestering abundance calcium in intracellular stores, for example, the endoplasmic reticulum, mitochondria, and vacuoles. So any confined entrance of calcium through the plasma layer would cause an annoyance, incorporating a communication with the cytoskeleton, since calcium is known to cause the constriction of F-actin.

There is bounteous confirmation for a part of calciummediated motioning in numerous parasitic fungi and different creatures, however the subtle elements of calcium flagging and how it identifies with tip development stay misty. Bartnicki-Garcia in 2002 gives an incredible and provocative audit of this and other extraordinary inquiries in hyphal tip development.

Germination of Spores and Orientation of Growth of Hyphal Tip

Parasitic fungi react too many sorts of natural flag, including signals that trigger spore germination (i.e. the creation of a hyphal tip where none existed before) and signs that change the introduction of hyphal tip development. Beneath we consider a few cases of these procedures.

Concentrations on Developing Pores

Some parasitic spores, for example, the uredospores of rust organisms (Basidiomycota), have a settled purpose of germination named the germ pore, where the divider is obviously more slender than somewhere else. So also, the zoospores (motile, flog cells) of plasmodiophorids, Chytridiomycota, and Oomycota, have a settled purpose of germination, and they settle and cling to open surfaces so that their future purpose of germ-tube outgrowth is situated by that surface. [28]

Be that as it may, numerous spores appear to have the capacity to sprout from any point on the cell fringe. The germination procedure frequently takes after a typical example. At first, the spore swells by hydration, at that point it swells advance by a dynamic metabolic process and new divider materials are joined over most or the greater part of the cell surface, the stage named nonpolar development. At long last a germ-tube (a youthful hypha) rises up out of a restricted point on the cell surface, and all consequent divider development is limited to this district. The main sign that a pinnacle will rise is the limited advancement of an apical vesicle group.

In the conidia of *Aspergillus niger* the move from nonpolar to enraptured development is temperature-subordinate. At an ordinary temperature of around 30°C, the spore at first fuses new divider material over the entire surface and afterward a peak is shaped. However when the spores are hatched at 44°C they keep on swelling for 24–48 hours, delivering mammoth gathered cells together to 20–25 μm breadth (a 175-crease increment in cell volume) with dividers up to 2 μm thick. At this stage the cells quit developing. Be that as it may, if these “mammoth cells” are moved down to 30°C preceding they quit developing they will react by creating a hyphal peak, and this carries on in an uncommon route: rather than shaping an ordinary hypha it delivers a little spore-bearing head. These perceptions recommend two things. In the first place, that the move from nonpolar to polar development in *A. niger* is temperature-subordinate – it is obstructed at the prohibitive temperature (44°C). Second, that the organism can even now “develop” at the prohibitive temperature: it achieves a formative stage at which it is resolved to sporulate, and it does as such when the temperature is brought down.

The creation of spores from developing spores with at least mediating development is named microcycle sporulation. It happens actually in a few parasitic fungi, particularly on the off chance that they develop in water movies in nutrientlimited conditions. For instance, microcycle sporulation has been accounted for some saprotrophs on leaf surfaces (e.g. Cladosporium, Alternaria spp), some leaf-tainting pathogens (e.g. Septoria nodorum), a few vascular shrink pathogens that colonize xylem vessels (e.g. Fusarium oxysporum), and the rhizosphere organism Idriella bolleyi which is a natural control specialist of root pathogens. [29] Every one of these organisms will develop to frame typical hyphae in supplement rich conditions, so their microcycling conduct in supplement poor conditions may be a methods for spreading to new and conceivably more ideal situations.

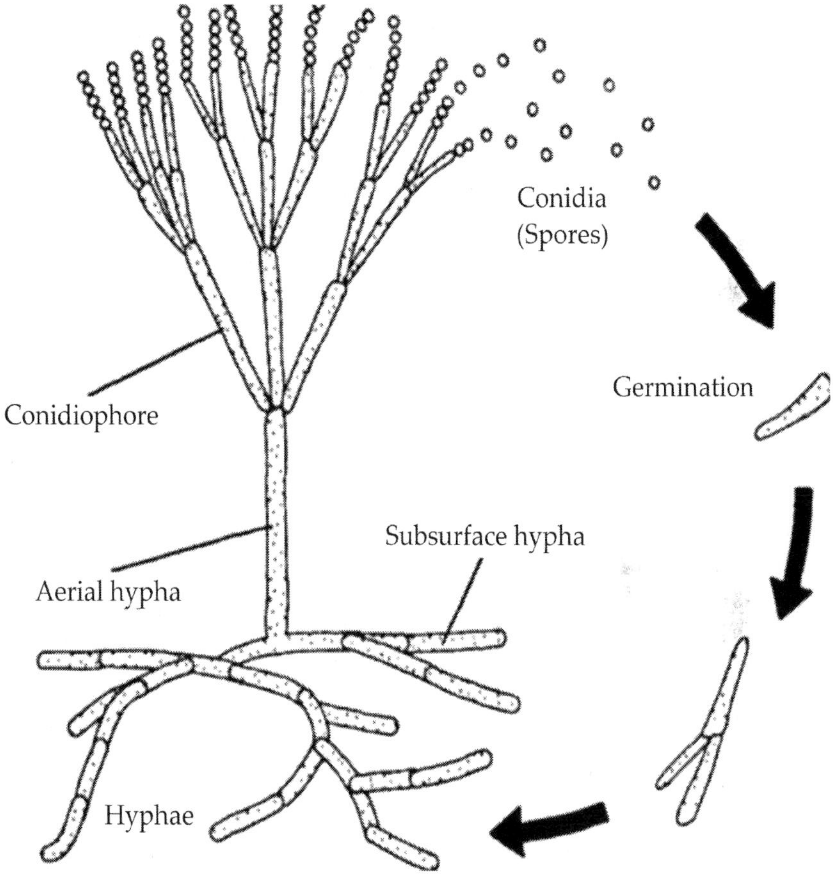

Figure- 4.5: Germination stages of Aspergillus niger spores.

SPORE GERMINATION TROPISMS

A tropism is characterized as a directional development reaction of a living being to an outside jolt. The spores of a few growths demonstrate this uniquely, a great illustration being the yeast-like organism Geotrichum candidum which is a typical reason for decay of dairy items. The tube shaped spores of this parasite develop commonly from one or other shaft, yet the site of germ-tube rise is affected emphatically by the nearness of neighboring spores when the spores are seeded thickly on agar and secured with a coverslip. In these conditions the germ-tubes dependably rise up out of the end farthest far from a touching spore, a wonder named negative autotropism. [30]

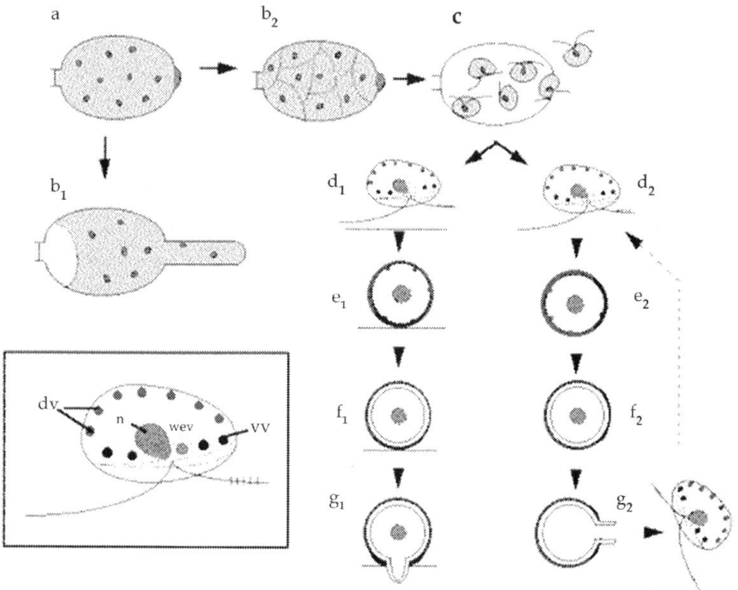

Figure- 4.6: Two way germination of phytophthora infestans sporangium.

The reasons for this conduct are as yet misty. From one perspective, it has been proposed to include the arrival of auto-inhibitors, which would collect maximally in the zone of contact of two spores yet could diffuse far from the "free" closures, prompting germination there. Then again, oxygen consumption in the zone of spore contact could be a basic variable for *G. candidum* in light of the fact that the spores dependably sprout towards an oxygen source (a little opening in a plastic coverslip put over the spore layer) and this positive tropism to oxygen could beat the negative autotropism of touching spore sets.

The spores of Idriella bolleyi (a mitosporic organism) additionally demonstrate negative autotropism, yet they demonstrate a considerably more terrific reaction when set in contact with grain root hairs. The spores of Idriella dependably develop far from living root hairs however towards dead root hairs and quickly infiltrate them. This conduct is by all accounts environmentally important in light of the fact that *I. holleyi* is a powerless fungal parasite of oat and grass roots. It misuses the root cortical cells as they begin actually behind the developing root tip, and in doing as such it rivals forceful root fungal pathogens that generally would utilize the dead cells as a nourishment base for disease. In this manner, the spore germination tropisms of I. bolleyi help to clarify its part as a natural control specialist of oat root pathogens, like the part of non-pathogenic strains of the take-all growth. The tropic signs for I. bolleyi spores appear to be very particular, since G. candidum and some other organisms tried in similar conditions demonstrated very unique reactions, for instance, G. candidum developed towards both living and dead root hairs.

Contagious spores can likewise demonstrate introduction reactions to electrical fields of adequately high quality (5– 20 V cm−1). For instance, in one investigation the spores of Neurospora crassa and Mucor mucedo were found to develop towards the anode, though spores of Emericella nidulans demonstrated no critical introduction reaction. [31] The substantial (more established) hyphae of these and other parasitic fungi demonstrated a variety of introduction reactions:

Neurospora hyphae developed towards the anode and shaped branches towards the anode, yet hyphae of *E. nidulans* and *M. mucedo* developed and expanded towards the cathode. In a later report the galvanotropic reactions of substantial hyphae were observed to be pH-and calcium-subordinate. Neurospora hyphae even changed from being unequivocally cathodotropic at pH 4.0 to emphatically anodotropic at pH 7. Given the scope of various reactions to electrical fields it is hard to compress this theme, but to state that contagious hyphae can be receptive to electrical/ ionic fields.

Hyphal Tropisms

In every one of the cases above we have utilized the term tropism freely to portray the position where a hyphal tip rises. In any case, entirely, a tropism is a bowing reaction that orientates a hypha in a specific bearing, for nearest to the amino corrosive source. Manavathu and Thomas in 1985 examined this for a strain of Achlya ambisexualis and found that, of all the single amino acids tried, just the sulfur-containing amino corrosive methionine could inspire hyphal tropism. Notwithstanding, angles of numerous other single amino acids would evoke tropism if the medium contained a uniform foundation of cysteine. As a clarification of this it was proposed that cysteine, when taken up by cells, can give one of its sulphydryl (- SH) gatherings to other amino acids and in this manner create methionine. In microbes the fascination of cells to a few sorts of compound is interceded by chemoreceptor edifices, and this includes a part for methionyl subordinates which give methyl gatherings to the inside areas of the receptor buildings.

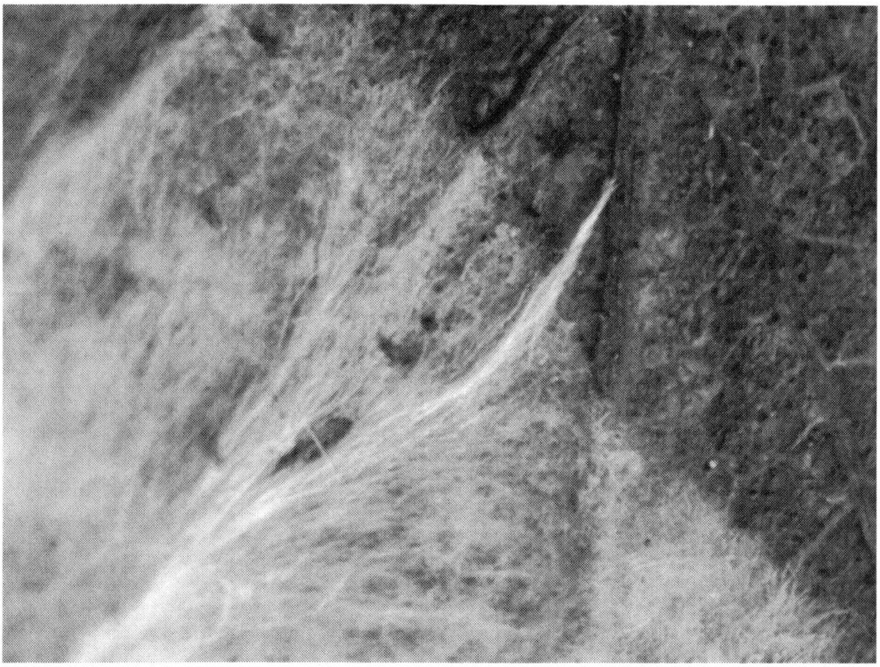

Figure- 4.7: Orientation of fungal hyphae towards nutrient source.

A comparative framework may be included in chemotropism by the Oomycota, on the grounds that Manavathu and Thomas found that few methyl-giver mixes could evoke a tropic reaction. In work with a strain of Achlya bisexualis, Schreurs *et al.* in 1989 found that the hyphal tips orientate towards the tips of micropipettes containing either phenylalanine or methionine. Likewise, when micropipettes containing attractant amino acids were put behind the hyphal tips the branches risen up out of the hyphae and developed towards the attractants. Subsequently it appears that the start of spreading and the tropism of hyphal tips are firmly related reactions to natural signs, and in some Oomycota these reactions may be intervened by plasma film receptors for particular amino acids. [32]

The hyphae of numerous parasitic fungi demonstrate tropic reactions to non-supplement elements of potential biological pertinence. For instance, germ-tubes emerging from spores of the arbuscular mycorrhizal growths (Glomeromycota) can develop towards unpredictable metabolites (maybe aldehydes) from roots, some wood-spoiling parasitic fungi (e.g. Ascomycota, Chaetomium globosum) orientate towards unpredictable mixes from crisply cut wood squares, and hyphae of the seedling pathogen Athelia (Sclerotium), rolfsii (Basidiomycota) orientate towards methanol and other short-chain alcohols from newly deteriorating natural issue. Sexual pheromones additionally evoke introduction reactions.

YEAST CELL CYCLE

As opposed to parasitic hyphae, which develop constantly from a hyphal tip, yeasts normally develop by a rehashed maturing procedure to deliver settlements of single cells. As appeared for Saccharomyces cerevisiae, in each "turn" of the yeast cell cycle a youthful bud rises up out of an anticipated point on the mother cell. The bud develops apically by directing of divider parts to the bud tip, yet at a later stage the method of development switches and divider segments are embedded consistently finished the phone, so that the bud turns out to be more swollen. In the interim, the cell has experienced mitosis, and one of the two little girl cores enters the bud. The last phase of cytokinesis (cell division) happens by the improvement of an essential septum, made out of chitin, between the mother and little girl cells. This plate of chitin is then overlaid by β-glucan and mannan (the significant divider segments) to frame the auxiliary septum, and the cells at last independent, leaving the chitin plate on the mother cell.

The yeast cell cycle has been examined seriously as a model of the direction of cell development and division. Four phases in this cycle are perceived: G1 (first hole), S (DNA combination), G2 (second hole), and M (mitosis). In each turn of the cycle a bud rises, develops to about full size, gets one of the girl cores from atomic division and afterward isolates from the parent cell. At its speediest the cycle takes around 1.5

hour in S. cerevisiae, however the time can differ inside wide breaking points, and contingent upon the accessibility of supplements. All of this variety happens in G1, in light of the fact that S, G2, and M together possess a pretty much consistent time. [33]

The most imperative checkpoint in this cycle is named begin. It happens amid G1 in *S. cerevisiae* and it is the phase where the phone coordinates all the data from intracellular and ecological signs to decide if the cell cycle will proceed, or enter stationary stage, or the cell will experience sexual proliferation. Numerous cell division cycle (CDC) qualities have been distinguished in the maturing yeast *S. cerevisiae*, and homologous qualities that direct improvement have additionally been found in the indirectly related parting yeast, Schizosaccharomyces pombe.

Investigation of the qualities and quality results of these two life forms has helped us to see how cells build up their extremity of development. In sprouting yeast the main bud creates at one of the shafts of the cell. At the point when this bud has created and isolated from the mother cell, the following bud emerges at a direct contiguous toward the bud scar. It develops at first by polar development (arrange 4) however then by divider development over the vast majority of the bud surface (organize 5). At the season of bud rise, a "tag" or "historic point" is set down at the site where another bud will shape, and a ring of septin proteins is saved now. [34]

YEAST CELL CYCLE

At a later phase of progress the cell apparatus, including actin microfilaments, will guide vesicles and divider forerunners to this site, bringing about the restricted advancement of a septum to isolate the girl cell. The parting yeast, Schizosaccharomyces pombe, develops uniquely in contrast to Saccharomyces, since it produces tube shaped cells that stretch out at both closures, with labels at the shafts of the cell. At the point when the cell has achieved a basic volume the core partitions, and labels at that point coordinate the creation of a septum at the site where the cell will separate (stages 5–7). Both Saccharomyces and Schizosaccharomyces can experience mating affected by pheromones delivered by cells of inverse mating sorts, so this speaks to yet another formative stage including the foundation of polar locales.

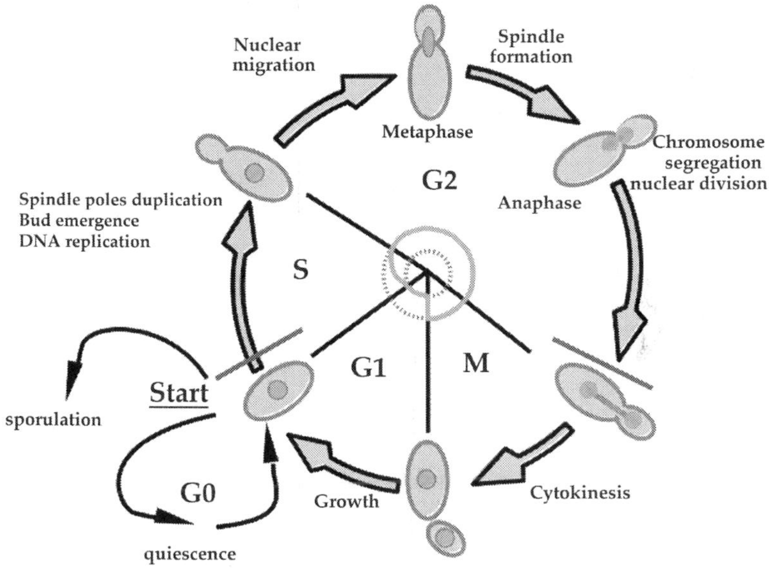

Figure- 4.8: Cell cycle of yeast.

The CDC42 quality is a key part in the foundation of spellbound development of Saccharomyces. It encodes a GTP (guanosine triphosphate)- restricting protein, which acts working together with the protein results of different qualities (CDC24 and BEM1) to enroll septins and actin microfilaments to the destinations where confined divider development will happen. Less is thought about the foundation of extremity in splitting yeast, however homologs of a portion of the key sprouting yeast qualities have been distinguished in Schizosaccharomyces – for instance, the parting yeast qualities ral1/sdc1 (homologous to CDC24) and ral3/scd2 (homologous to BEM1). Some human qualities likewise demonstrate base succession homology with CDC qualities of Saccharomyces, showing a level of quality protection through developmental time. Along these lines, the yeast cell cycle can be utilized to research cell occasions of significantly more extensive criticalness, including the direction (or deregulation) of typical cell division. [35]

Cell Cycle in Mycelial Fungi

The embodiment of the yeast cell cycle depicted above is that, amid a specific timeframe (which shifts relying upon supplement accessibility and other natural elements) one cell produces two cells, while the biomass duplicates, and the quantity of cores additionally pairs. This brings up the issue of whether an identical cycle happens in mycelial parasitic fungi. The principal clear confirmation that there is such a cycle in mycelial parasitic fungi was acquired with a surprising growth, Basidiobolus ranarum (beforehand doled out to the Zygomycota, yet now exchanged to Chytridiomycota in light of its little subunit rRNA quality succession). This parasitic fungi normally develops as single vast cells in the rear gut of frogs or different amphibiams, where it causes next to zero harm.

Be that as it may, it will develop as hyphae with finish, unperforated septa on agar plates. *B. ranarum* likewise has to a great degree huge cores (around 20 μm distance across) which are effortlessly observed by light microscopy and which are masterminded frequently, one for every cell. The hyphae reach out by tip development on agar, orchestrating protoplasm and drawing it advances as the tip develops. At the point when a basic volume of protoplasm has been integrated, the huge focal core isolates and a septum is set down at the purpose of atomic division, making two cells, each with one core. [36] The new apical cell at that point develops on, and it rehashes the entire procedure when enough protoplasm has been combined. The penultimate cell, which has been separated by the entire septum, creates another branch summit and its protoplasm streams into this. Basically, consequently, two hyphal tips (each with a given protoplasmic volume) are shaped from the first single tip cell, and there is a reasonable connection between cytoplasmic volume, atomic division, and stretching, precisely like the phone cycle of yeasts. In any case, it is named a duplication cycle as opposed to a cell cycle, and the cells stay joined to each other. [37]

A comparative duplication cycle has been appeared to happen in numerous parasitic fungi, even those with ordinary, punctured septa. This was finished by enabling the spores of different parasitic fungi to sprout on agar and the creating states were shot at interims. From the photos, the quantity of hyphal tips and the aggregate hyphal length were recorded at various circumstances and used to figure a hyphal development unit (G) where:

G = Total length of mycelium

After introductory vacillations in the early phases of development, the estimation of G ended up noticeably steady and normal for each parasitic species or strain. For instance, a G estimation of 48 μm (identical to a hyphal volume of 217 μm3) was ascertained for hyphae of Candida albicans, and estimations of 32 μm (629 μm3), 130 μm (4504 μm3), and 402 μm (11,986 μm3) were found for a wild-sort strain and two "spreading" mutants of Neurospora crassa. [38]

The steadiness of these qualities for singular strains exhibits that, as a province develops, the quantity of hyphal tips is straightforwardly identified with the cytoplasmic volume. For instance, when a settlement of C..albicans had delivered an extra 48 μm length of hypha (or a hyphal volume of 217 μm3) it had sufficiently integrated protoplasm to create another tip. We can in this manner consider a parasitic province as being made out of various "units" (the hyphal development units), each of which speaks to a hyphal tip in addition to a normal length of hypha (or volume of cytoplasm) related with it. They are not seen as independent units since they are consolidated, however in a few regards they are comparable to the different cells delivered in the yeast cell cycle. Truth be told, the duplication cycle of an ordinary mycelial growth, Emericella nidulans, has been appeared to be intently connected with an atomic division cycle. The apical compartment develops to about twice its unique length, at that point the few cores in this compartment isolate pretty much synchronously and a septum is set down close to the center of the apical compartment. After this, a progression of septa are framed in the new subapical compartment to partition it into littler compartments, each with only maybe a couple cores, while the multinucleate tip develops on and will rehash the procedure at the appropriate time. Prior in this section we said the fringe development zone of a parasitic settlement – the length of hypha expected to help the ordinary augmentation rate of tips at a state edge. It can be assessed by cutting the hyphae at various separations behind the tips, and it can be as extensive as 5–7 mm. This is very not quite the same as the hyphal development unit which ranges from around 30 to 400 μm. [39] The distinction is clarified by the way that the hyphal development unit is measured in supplement rich conditions and is a genuine impression of development (increment in biomass, or quantities of tips) though the fringe development zone is an impression of the rate of expansion of a province edge, and it applies to more seasoned settlements, where a portion of the hyphae end up noticeably driving hyphae, which are significantly more extensive and have substantially speedier augmentation rates than the rest.

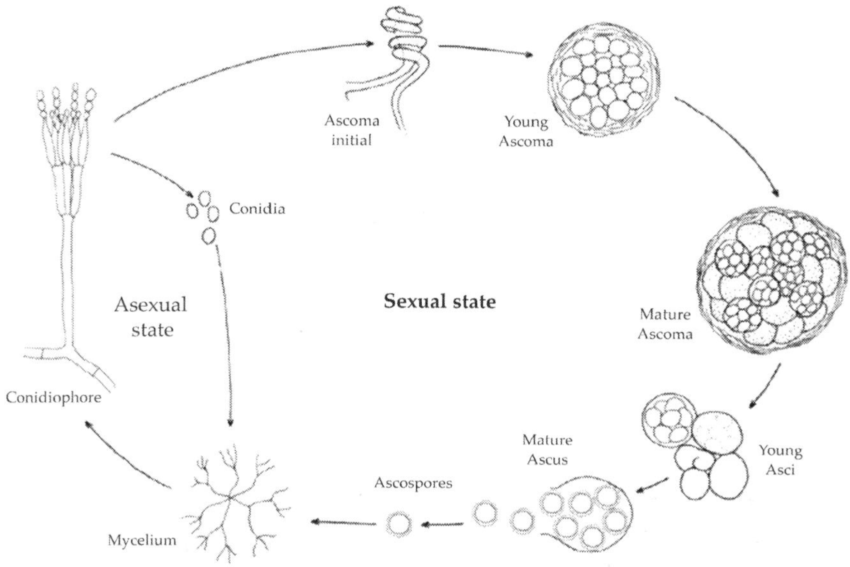

Figure- 4.9: Life cycle of mycelial fungi.

Fungal Mycelium as a System Capturing Nutrients

The parasitic mycelium is a profoundly productive and versatile gadget for catching supplements. Over an extensive variety of supplement fixations, a parasitic province will stretch out over an agar plate at a similar rate, whether on water agar or a standard supplement rich medium. In any case, on water agar the state is inadequately spread, while on supplement rich agar the expanding design is thick. This high level of versatility is a key component of parasitic fungi, particularly in soil and waterways where supplements are probably going to happen in restricted pockets. [40] A youthful larch seedling was immunized with a mycorrhizal fungal parasite and developed in peat against the substance of a straightforward perspex box. The root framework itself is very restricted: it comprises of the district set apart by two-fold sharpened stones (<<) where the roots are wrapped by a mycorrhizal sheath. The greater part of the stretching system that we see is an arrangement of totaled contagious hyphae, named mycelial ropes which investigate the soil for supplements. When they locate a restricted pocket of natural supplements they deliver a mass of hyphae to misuse the supplement rich zone.

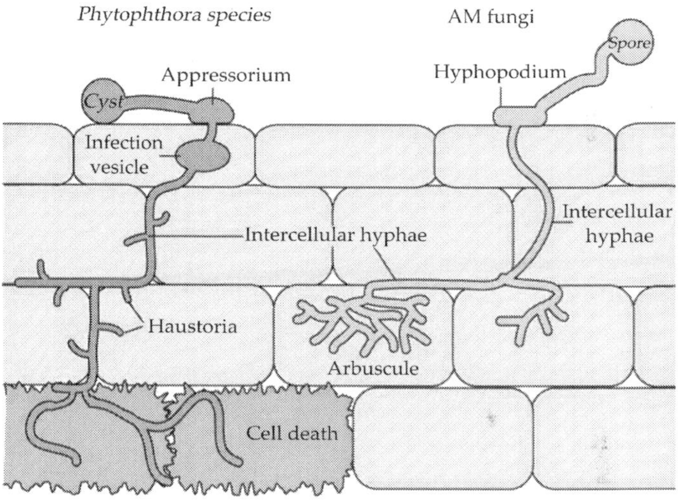

Figure- 4.10: Fungal nutrient capturing system.

Ceaseless Culture Framework versus Cluster Culture Framework

Clump culture frameworks are utilized normally in industry in light of the fact that valuable essential metabolites, for example, natural acids, and optional metabolites, for example, anti-infection agents, are delivered in the deceleration and early stationary stages. Bunch societies are additionally utilized for blending and wine-production, on the grounds that the way of life stock is the attractive item. [41] Constant culture frameworks are another option to group culture. In these frameworks, culture medium is included at a consistent moderate rate, and a relating volume of the old culture medium together with a portion of the contagious biomass is evacuated by a flood gadget. Such culture frameworks are checked naturally so that variables, for example, pH, temperature, and broke up oxygen focus are kept up at the coveted levels. They are mixed energetically to keep the living being in suspension and to encourage dissemination of supplements and metabolic results. [42]

There are different sorts of constant culture framework however the most well-known sort is the chemostat. In this framework the groupings of supplements are balanced purposely so that one basic supplement is at a moderately low fixation while every other part are available in overabundance. At the point when the quantity of cells in the way of life begins to build, the rate of exponential development ends up noticeably restricted by the development constraining supplement. At this stage the rate of development of the way of life can be controlled exactly by changing the rate at which crisp culture medium is provided, this is named the weakening rate of the way of life. [49] In any case, take note of that the organism is continually developing exponentially, just the rate of exponential development is represented by the weakening rate. In principle, by modifying the weakening rate the way of life development rate can be changed in accordance with any coveted level, up to μmax (any further increment in weakening rate would cause "wash-out" in light of the fact that the cells would be expelled by flood quicker than they can develop). By and by, notwithstanding, these societies wind up plainly insecure as they approach μmax on the grounds that at that point even a minor, impermanent change in development rate can cause wash-out.

Chemostats are valuable for some exploratory purposes, on the grounds that the physiology of organisms can change at various development rates or because of various development constraining supplements. These progressions can be considered in detail in chemostats though they happen transitorily in clump societies. For instance, when *Saccharomyces cerevisiae* is developed at low weakening rates (moderate development) in glucose-constrained culture it utilizes a significant extent of the substrate for generation of biomass. By differentiate, *S. cerevisiae* produces ethanol to the detriment of biomass at higher weakening rates. It changes from cell creation to liquor generation in conditions favoring fast digestion, despite the fact that glucose is the development restricting supplement in both cases. Chemostats additionally are valuable for mechanical procedures, since cells or cell items (e.g. anti-infection agents) can be recovered consistently from the flood medium. [43]

By and by, be that as it may, the majority of the customary modern procedures depend on cluster societies, either on the grounds that the cost of changing over to persistent culture frameworks does not legitimize the expanded proficiency or in light of the fact that it is hard to outline and work full generation scale chemostat frameworks. [48] The group societies utilized mechanically are regularly "bolstered cluster" frameworks, in which supplements or different substrates are added intermittently to maintain the creation of valuable metabolites. By keeping cells interminably development restricted in stationary stage and afterward "nourishing" the way of life with chose metabolic forerunners, bunch groups can be utilized to create generous amounts of anti-infection agents or other business items.

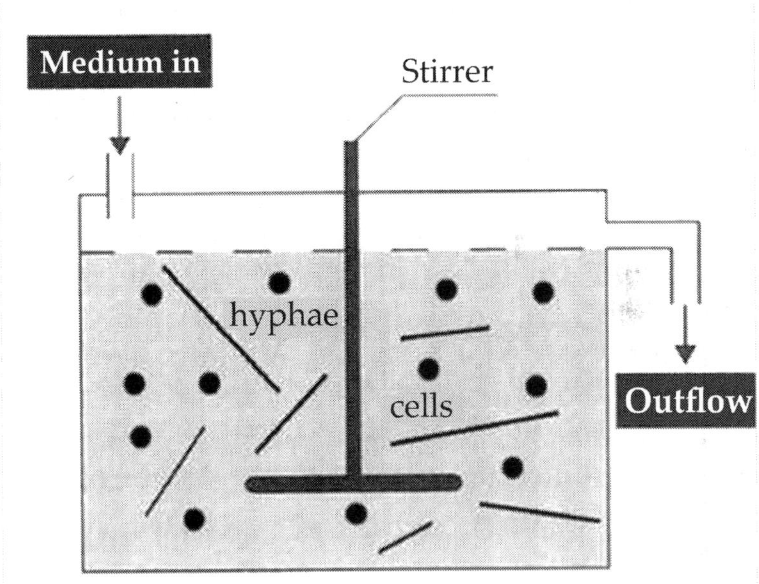

Figure- 4.12: Fungal biomass formation from continues culture.

Commercial Biomass Production of Fungi

The most intriguing late utilization of constant culture frameworks has been the improvement of a totally novel sustenance item, named Quorn mycoprotein. Truth be told, this is the sole survivor of the tremendously proclaimed "singlecell protein" upheaval of the late 1900s, when researchers and worldwide guide offices endeavored to create protein-rich nourishments from microbial biomass, to meet the approaching protein lack in the creating scene.

The improvement of Quorn is a noteworthy mechanical achievement, which assumed control 20 years to achieve fulfillment. Presently Quorn items are accessible in numerous stores and are generally utilized as contrasting options to meat items on account of their nutritious profile. Quorn mycoprotein is created monetarily from chemostat societies of the growth Fusarium venenatum, which is developed at 30°C of every a medium made out of glucose (the carbon source), ammonium (the nitrogen source), and other mineral salts. [44] The parasitic mycelium is recovered consistently from the way of life surge, at that point adjusted to hold the sinewy surface, and vacuumdried on a channel bed before being constituted into meat-like pieces. For this situation a persistent culture framework was regarded essential to guarantee a high level of reproducibility of the item, and furthermore for monetary reasons in light of the fact that the yield of mycelium over some stretch of time was around five times higher than if a progression of group societies were utilized. [47] Glucose is utilized as the development restricting supplement in the generation framework, and the weakening rate is set to give a multiplying time (μ) of 0.17–0.20 h−1, which is underneath the μmax of 0.28 h−1.

The rate of substrate change to protein is to a great degree high, around 136 g protein being created from each 1000 g sugar provided. For examination with this, the proportional generation of protein by chickens, pigs, and steers would be around 49, 41, and 14 g separately. Trinci in 1992 has portrayed the many stages in the improvement of this maturation innovation.

At first, numerous potential growths were screened to locate an appropriate creature for business utilize. At that point a reasonable substantial scale fermenter framework must be planned in light of the fact that parasitic mycelia have thick properties in arrangements, so the way of life are hard to blend to accomplish sufficient oxygenation – the organism utilizes 0.78 g oxygen for each 1 g biomass created. The framework utilized financially is a 40 m3 transport fermenter about the tallness of Nelson's segment! Packed air is utilized to circulate air through and flow the way of life. This maintains a strategic distance from the warming that would be caused by a mechanical stirrer, saving money on cooling costs. There is a potential issue with the high nucleic corrosive substance of a microbial biomass when utilized as a human sustenance source, in light of the fact that our bodies process RNA to uric corrosive which can cause gout-like side effects. Along these lines, the nucleic corrosive substance of the gathered mycelium should have been decreased while holding however much of the protein as could reasonably be expected. This was accomplished by abusing the higher warmth resilience of RNAases than of proteases. The way of life outpouring containing the biomass is gathered in a vessel and its temperature is raised quickly to 65°C and kept up at this level for 20–30 minutes. The development of the growth is halted at this temperature and the proteases are pulverized so that generally little protein is lost, however the ribosomes separate and the contagious RNAases corrupt a significant part of the RNA to nucleotides which are discharged into the spent culture filtrate. At that point the mycelium can be collected for drying. A last issue has still not been conquered totally and it constrains the productivity of business generation. [45]

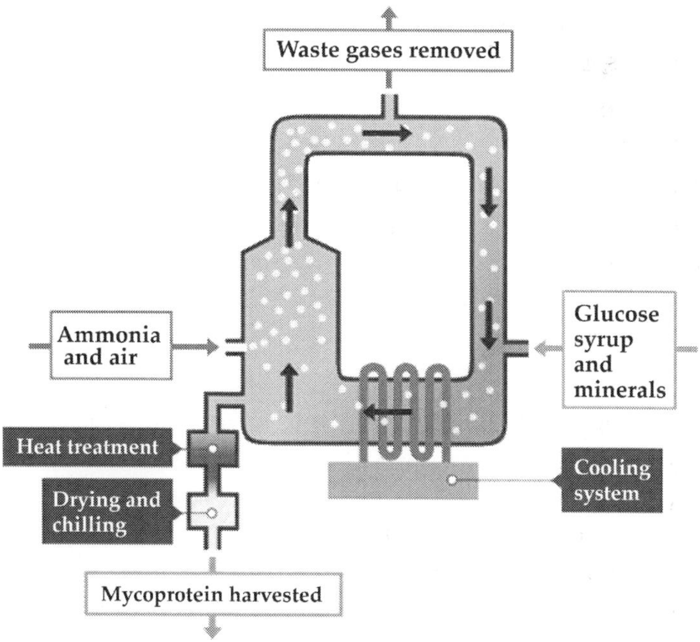

Figure- 4.13: Production of myco-protein from fungal biomass.

Amid delayed culture in fermenter vessels the parasite is subjected to determination weight and it changes to "provincial structures" with a high expanding thickness however moderately low augmentation rate of the primary, driving hyphae. These structures prevail over the wild-sort after around 500–1000 hours of constant culture. Their hyphal development unit lengths go from 14 to 174 μm, contrasted and 232 μm for the wild-sort, so they give an essentially less stringy biomass, which is undesirable at last item. The generation runs must be ended rashly to keep away from this issue. By the by, the advancement of Quorn is a huge mechanical accomplishment and a business achievement. [46]

SUMMARY

- Fungus grow with the help of hyphae. Branches of hyphae orientate and arise themselves for extreme effectiveness of nutrient catching. We describe methods of fungal growth in continuous culture and batch culture systems, and connecting this to industrial-scale processes, comprising the commercial creation of mycoprotein. Apical growth is major adaptation of fungi. Fungi adapt it during process of convergent evolution.
- Certain experiments were done by different scientists. Colonies of *N. crassa* and *Fusarium oxysporum* were grown on nutrient rich agar, water was also added into it and then perceived their tips activities. Tips stop growth and swelled into diamond shape, then grow into new tip behind the original apex. Then hyphal assembling occurs by chitin synthase, glucan synthase, manno-proteins.
- After that cross-linking and maturation of hyphal wall occurred. Steady model of fungal wall growth, orientation of hyphal growth tips and germination of spores, cell cycle of various fungus, commercial biomass production, and driving force behind growth of apical also has been discussed in this chapter.

REFERENCES

1. Youngner, V. B., & McKell, C. M. (1972). *The Biology and utilization of grasses*. Oxford: Elsevier Science.
2. Smith, J. H. (1923). *On the apical growth of fungal hyphae.*
3. Bartnicki-Garcia, S, Bracker, C E, Gierz, G, López-Franco, R, & Lu, H. (n.d.). *Mapping the growth of fungal hyphae: orthogonal cell wall expansion during tip growth and the role of turgor.*
4. Riquelme, M. (2000). *The role of the spitzenkörper in fungal growth and morphogenesis.*
5. Swinburne University of Technology. Faculty of Engineering and Industrial Sciences, Filipponi, Luisa, Hanson, Kristi L., Lee, Abraham P., & Nicolau, Dan V. (2005). *Fungal growth in confined microfabricated networks*. SPIE.
6. Reynaga-Peña, Cristina G., Gierz, Gerhard, & Bartnicki-Garcia, Salomon. (n.d.). *Analysis of the role of the Spitzenkörper in fungal morphogenesis by computer simulation of apical branching in Aspergillus niger*. The National Academy of Sciences of the USA.
7. Thomson, D. D., University of Aberdeen., & Biotechnology and Biological Sciences Research Council (Great Britain). (2014). *The exploratory behaviour of Candida albicans hyphae.*
8. Spohr, Anders Bendsen, Dam Mikkelsen, C., Carlsen, Morten, Nielsen, Jens Bredal, & Villadsen, John. (1998). *On-line study of fungal morphology during submerged growth in a small flow-through cell*. John/Wiley & Sons, Inc. John/Wiley & Sons Ltd.
9. Wang, Allen, Lane, Shelley, Tian, Zhen, Sharon, Amir, Hazan, Idit, & Liu, Haoping. (n.d.). *Temporal and Spatial Control of HGC1 Expression Results in Hgc1 Localization to the Apical Cells of Hyphae in Candida albicans* ▿ . American Society for Microbiology.
10. Wu, Qindong, Sandrock, Tanya M., Turgeon, B. Gillian, Yoder, Olen C., Wirsel, Stefan G., & Aist, James R. (1998). *A Fungal Kinesin Required for Organelle Motility, Hyphal Growth, and Morphogenesis*. The American Society for Cell Biology.
11. Spohr, Anders Bendsen, Dam Mikkelsen, C., Carlsen, Morten, Nielsen, Jens Bredal, & Villadsen, John. (1998). *On-line study of fungal morphology during submerged growth in a small flow-*

through cell. John/Wiley & Sons, Inc. John/Wiley & Sons Ltd.

12. Piippo, S. (Sari). (2010). *Grazing tolerance of biennial meadow plants in relation to resource availability*. University of Oulu.
13. Allsup, C., Ken, P., Augspurger, C., Lankau, R., & Miller, A. (2014). *Overcompensation of Ipomopsis aggregata following ungulate herbivory: getting to the root of it*.
14. Webster, J. (1989). *Introduction to fungi*. Cambridge: Cambridge University Press.
15. Inoue, S. (1997). *The role of cytoplasmic dynein in hyphae of Nectria haematococca*.
16. Martinez, R. F. W. (2008). *Overexpression, biochemical characterization and crystallization of chitin synthase 2 from Sacharomyces cerevisiae*.
17. Wang, Z. (1998). *Cloning and characterization of the chitin synthase 3 (WdCHS3) and chitin synthase 4 (WdCHS4) genes of Wangiella dermatitidis and expression studies of WdCHS3*.
18. Jollès, P., & Muzzarelli, R. A. A. (1999). *Chitin and Chitinases*. Basel: Birkhäuser Verlag.
19. Adams, D. J. (1980). *Chitin synthase of coprinus cinereus*. University of Aberdeen.
20. Beth, D. A. (1998). *Chittin biosynthesis in Neurospora crassa: Structure and function of chitin synthase genes*. Rehovot: publisher not identified.
21. Yeager, A. R. (2003). *Design and synthesis of mechanistic probes for chitin synthase*.
22. Nino-Vega, G. A. (1996). *Molecular analysis of chitin synthase genes of Paracoccidioides brasiliensis*.
23. Kafka, M. D. (2002). *Design and synthesis of 2-N-Acetylglucosamine derivatives as potential chitin synthase inhibitors*.
24. Starr, T. L., & University of California, Berkeley,. (2008). *Genetic and biochemical analysis of adaptor protein-mediated transport of chitin synthase III in Saccharomyces cerevisiae*.
25. Culp, D. W. (2000). *Functional analysis of chitin synthase genes from Aspergillus nidulans*.

26. Antelo, L. (2002). *Glucan synthase of Phytophthora sojae: Characterization, purification and attempts of molecular cloning.*
27. Töller, A. (2011). *Studies of plant innate immunity provide new functional insights on class IIa WRKY transcription factors and reveals a role for two glucan synthase-like genes in gametophyte development.*
28. Garcia, E. O., Deising, H. N., & Vögele, R. T. (2013). *Infection structure-specificity of [beta]-1,3-glucan synthase is essential for pathogenicity of Colletotrichum graminicola and evasion of glucan-triggered immunity.* Halle, Saale: Universitäts- und Landesbibliothek Sachsen-Anhalt.
29. Purdue University., United States., United States., United States., & Nicholas, C. C. (2009). *Identifying the catalytic components of cellulose synthase and the maize mixed-linkage beta-glucan synthase.* Schomburg, D., & Stephan, D. (1996). *Enzyme handbook: 12.* Berlin: Springer.
30. Sutcliffe, J. A., & Georgopapadakou, N. H. (1992). *Emerging Targets in Antibacterial and Antifungal Chemotherapy.* Boston, MA: Springer US.
31. Bossche, H. ., Odds, F. C., Kerridge, D., & Symposium on Topics in Mycology on Fungal Dimorphism. (1993). *Dimorphic fungi in biology and medicine.* New York: Plenum.
32. Dan, J. M. (2008). *Exploiting pattern recognition receptors to create a Cryptococcus neoformans vaccine.*
33. Mansour, M. K. (2005). *Host response to mannosylated proteins from the pathogenic yeast, Cryptococcus neoformans.*
34. Rayes, E. (1996). *Characteristics of mannoprotein generated emulsions.*
35. In Jordão, A. M., & In Cosme, F. (2016). *Recent advances in wine stabilization and conservation technologies.*
36. Diako, C., Ross, C. F., & Washington State University. (2016). *Influence of wine components on the chemical and sensory quality of wines.*
37. Ryan, C. (1998). *Development of a method for the enzymatic determination of mannoprotein in wine.*

38. Lallemand (Firm). (2006). *New trends in the use of specialty inactivated yeast during winemaking: Proceedings of Les XVIIIes Entretiens scientifiques Lallemand*. Blagnac, France: Lallemand.
39. Tanner, W., & Loewus, F. A. (1981). *Plant Carbohydrates II: Extracellular Carbohydrates*. Berlin, Heidelberg: Springer Berlin Heidelberg.
40. Washington, D.C: United States. Dept. of Energy. Office of Science.
41. Moore, D., Robson, G. D., & Trinci, A. P. J. (2011). *21st century guidebook to fungi*. Cambridge: Cambridge University Press.
42. Bennett, J. W., & Lasure, L. L. (1991). *More Gene Manipulations in Fungi*. Oxford: Elsevier Science.
43. Roussos, S., Lonsane, B. K., Raimbault, M., & Viniegra-Gonzalez, G. (1997). *Advances in Solid State Fermentation*. Dordrecht: Springer Netherlands.
44. Chávez, C. L. R., & Aguilar, J. O. A. (2015). *An integrated process for the in situ product recovery of prodigiosin by micellar aqueous two phase system from a culture of Serratia marcescens*.
45. In Vogel, H. C., & In Todaro, C. L. (2014). *Fermentation and biochemical engineering handbook: Principles, process design, and equipment*.
46. Buckles, W. R. (2006). *Dietary considerations for increasing conjugated linoleic acid in ruminant animals*.
47. Kamm, B. (2014). *Microorganisms in Biorefineries*. (Microorganisms in biorefineries.) Berlin, Heidelberg: Springer Berlin Heidelberg.
48. In Buckeridge, M. S., & In De, S. A. P. (2017). *Advances of basic science for second generation bioethanol from sugarcane*.
49. Roussos, S., Lonsane, B. K., Raimbault, M., & Viniegra-Gonzalez, G. (1997). *Advances in Solid State Fermentation*. Dordrecht: Springer Netherlands.

CHAPTER

5

FUNGAL PRODUCTS AND METABOLISM

In this section we talk about the fungus fundamental metabolic pathways, as a reason for seeing how they develop on various sorts of substrate and in various ecological circumstances. We additionally cover a portion of the fungus unusual and unique parts of digestion, comprising the generation of an extensive variety of auxiliary metabolites of ecological and commercial criticalness, for example, the penicillin anti-infection agents, and vital mycotoxins like the exceptionally cancer-causing aflatoxins and the harmful ergot alkaloids. A portion of the material in this section will be well-known, fundamental organic chemistry, however it is displayed in the particular setting of biological fungus.

HOW FUNGI GET VITALITY IN VARIOUS CIRCUMSTANCES

The spine to the focal vitality yielding pathway, in which sugars are separated by means of the Embden– Meyerhof (EM) pathway (glycolysis) and the tricarboxylic corrosive (TCA) cycle. A number of the intermediates can be attracted off to create other fundamental metabolites or for blend of an extensive variety of specific auxiliary metabolites. [1]

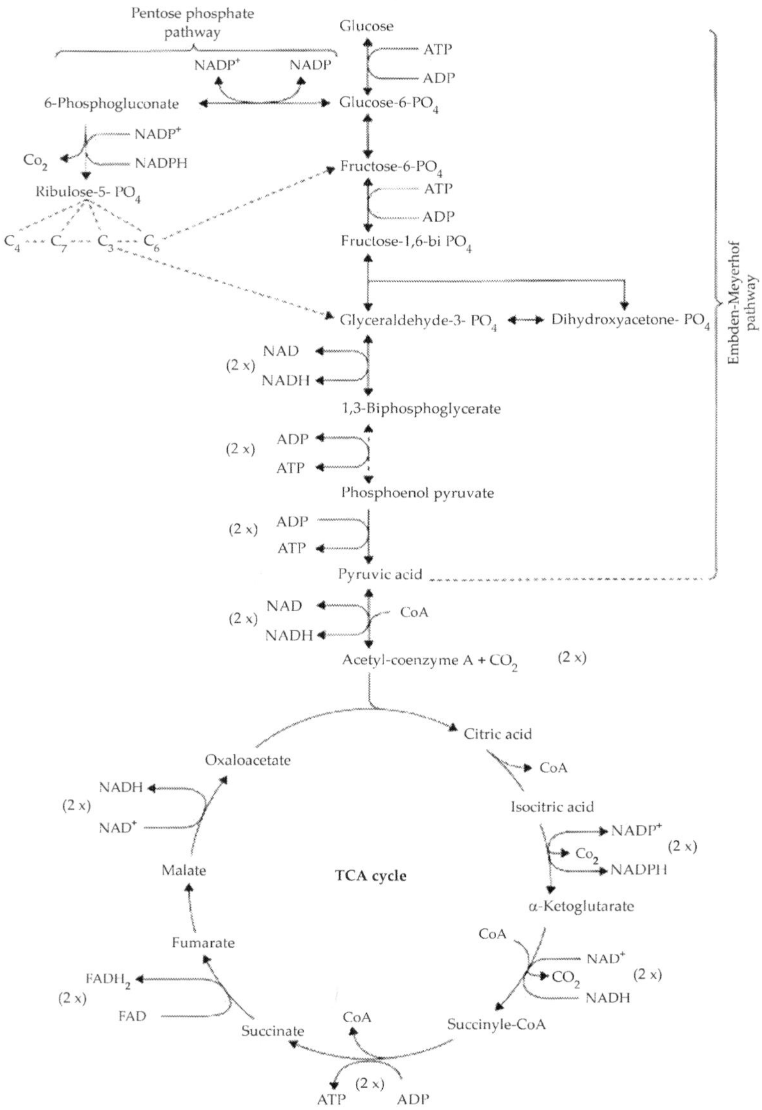

Figure- 5.1: Tricarboxylic acid cycle showing central metabolic pathways of fungi.

In spite of the fact that parasitic fungi can acquire vitality by oxidizing an extensive variety of mixes, it is helpful to start by considering a growth developing on a straightforward sugar, for example, glucose. In the arrangement of enzymic ventures of the Embden–Meyerhof pathway, glucose is changed over to glucose-6-phosphate, at that point to fructose-6-phosphate, lastly to fructose-1,6-biphosphate. These phosphorylation responses happen to the detriment of vitality given by two atoms of the vitality rich compound, adenosine triphosphate (ATP). At that point fructose-1,6-biphosphate is part into two 3-carbon mixes, glyceraldehyde-3-phosphate and dihydroxyacetone phosphate. These two mixes are interconvertible, and in a further arrangement of enzymic steps they are changed over to pyruvic corrosive. Pyruvic corrosive is one of the key intermediates of focal digestion, since it speaks to a branch-point: the response steps that take after will rely upon whether the organism is developing in the nearness or non-attendance of oxygen.

Within the sight of oxygen, pyruvic corrosive (a 3-carbon compound) is transported to the mitochondria, where it is changed over to the 2-carbon compound acetyl-coenzyme-An (acetyl-CoA) by partner with an atom of coenzyme-An discharging a particle of CO_2. [2] At that point acetyl-CoA consolidates with oxaloacetate (a 4-carbon compound) to create citrus extract (6-carbon). At last, in the responses of the TCA cycle, citrus extract is changed over back to oxaloacate, with the loss of two particles of CO_2.

The net consequence of this entire arrangement is that one atom of glucose is totally oxidized to six particles of CO_2 (and six atoms of water), as per the exact condition for high-impact breath:

$$C_6H_{12}O_6 + 6O_2 \rightarrow 6CO_2 + 6H_2O$$

Recollecting that each progression after the cleavage of fructose-1,6-biphosphate into two 3-carbon mixes happens twice, as appeared in. The oxidation of any substance should dependably be combined with a relating decrease of another substance, and this part is served by three nucleotides:

NAD^+ (nicotinamide adenine dinucleotide), $NADP^+$, and FAD (flavin adenine dinucleotide). These mixes acknowledge electrons and are correspondingly diminished to NADH, NADPH and $FADH_2$. [3] These nucleotides at that point should be reoxidized for the entire procedure to proceed, and this is accomplished by passing their electrons along an electron transport chain, where oxygen is the terminal electron acceptor.

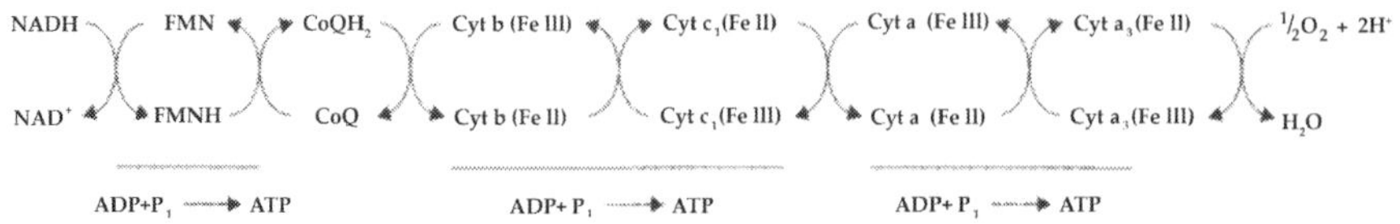

Figure- 5.2: Respiratory electron transport chain.

The electron transport chain is appeared in streamlined. It comprises of a progression of electron transporters which are situated in the mitochondrial film, and in a few cases traverse this layer. They are adjusted in a particular succession.

At first, NADH [or NAD(P)H] is reoxidized to NAD^+ [or $NAD(P)^+$] by exchanging its electrons to the transporter particle, flavin mononucleotide (FMN), which winds up plainly diminished to FMNH. (Note that, for our motivations, we can respect FMN/FMNH as proportional to FAD/ $FADH_2$, the basic point is that a flavin atom goes about as a transporter in the electron-transport chain.) The flavin nucleotide, thusly, exchanges the electrons to coenzyme Q, et cetera down the chain, until the point when the last stride when oxygen acknowledges the electrons and is, itself, diminished to water. [4] The bearers in the electron transport chain create a proton intention constrain, on the grounds that protons (H^+ particles) are expelled to the outside of the mitochondrial film while OH^- particles amass inside. This polarization of the layer is utilized to drive the union of ATP (from ADP + inorganic phosphate) when protons re-enter through a film found ATPase (additionally named ATP synthase).

The electron transport chain is, basically, an electrochemical slope, and at three phases along this inclination adequate vitality is discharged to incorporate a particle of ATP from the oxidation of NADH/NAD (P) H. Be that as it may, just two atoms of ATP are created from the oxidation of flavin nucleotides (e.g. FMN).

Vitality Yield from Vigorous Breath

We can figure the hypothetical vitality yield from glucose amid oxygen consuming breath by computing the quantity of ATP atoms that could be orchestrated:

2 ATP from the EM pathway down to pyruvic corrosive (4 ATP delivered however 2 ATP utilized at first to phosphorylate glucose).

2 ATP from the TCA cycle (one in each turn of the cycle, however 2 particles of pyruvate must be prepared through this cycle).

30 ATP from the re-oxidation of 10 pyridine nucleotides (NADH/NADPH).

4 ATP from the re-oxidation of 2 flavin nucleotides.

Total ATPs are 38

In any case, the genuine ATP yield will be considerably less than this, for no less than two reasons:

- Intermediates are persistently drawn from the pathways for biosynthetic responses, so this speaks to lost potential ATP.
- The noteworthy part of NADP/NADPH (instead of NAD^+/NADH) is to give diminishing energy to biosynthetic responses, as opposed to produce ATP.

At this stage we should specify the exceptional part of the pentose-phosphate pathway. It can be utilized as another option to the EM pathway for producing vitality from sugars (giving 1 ATP rather than the 2 ATP from the EM course). [5] In any case, its significant part is in biosynthesis, it produces some imperative intermediates, for example, ribose-5-phosphate for the amalgamation of nucleic acids and erythrose-4-phosphate for the blend of fragrant amino acids.

WHAT OCCURS WHEN OXYGEN IS CONSTRAINING?

Without a terminal electron acceptor (generally oxygen), the electron transport chain can't work, so the pool of diminished nucleotides (NADH, and so on.) collects and digestion would quickly stop. [6] In any case, growths, and numerous different life forms, can at present acquire some vitality without oxygen by one of the accompanying responses:

$$CH_3.CO.COOH \xrightarrow{\text{Pyruvic dehydrogenase}} \text{acetaldehyde} + CO_2$$

$$CH_3.CO.COOH + NADH \xrightarrow{\text{Alcohol dehydrogenase}} CH_3CH_2OH + NAD+$$

$$CH_3.CO.COOH + NADH \xrightarrow{\text{Lactic dehydrogenase}} CH_3.CHOH.COOH + NAD+$$

In both cases the final result (ethanol or lactic corrosive) is more lessened than pyruvic corrosive, so the responses are combined with the reoxidation of NADH to NAD+. This permits the Embden–Myerhof pathway to proceed, and the cells discharge either lactic corrosive or ethanol into the encompassing medium. Most yeasts and mycelial parasitic fungi create ethanol, this is the premise of the mixed beverages industry. In any case, a few Chytridiomycota create lactic corrosive (e.g. Blastocladiella, Allomyces), as do people when our tissue oxygen level is exhausted. [7]

Vitality yielding responses of this sort, where an interior inorganic compound is the terminal electron acceptor, are characterized by the biochemical term maturation. In the principal condition over the terminal electron acceptor is acetaldehyde, since this acknowledges electrons from NADH and is, itself, diminished to ethanol. Thus, in the second condition, pyruvic corrosive acknowledges electrons and is itself decreased to lactic corrosive.

As far as vitality yield, the transformation of pyruvic corrosive to either ethanol or lactic corrosive is extremely wasteful as just 2 moles of ATP are delivered from each mole of sugar used (contrasted and the potential 38 ATP from vigorous breath). [8] Parasitic fungi in this way require a plenteous supply of sugars for development in anaerobic conditions. Fungal parasitic fungi additionally should be provided with an extensive variety of different supplements in anaerobic conditions, on the grounds that the TCA cycle and a few different responses don't work to give the antecedents to biosynthesis.

Elective Terminal Electron Acceptors

At any rate a few fungal parasitic fungi (e.g. Emericella nidulans, Neurospora crassa) have an option methods for adapting to anaerobic conditions: they utilize nitrate set up of oxygen as the terminal electron acceptor in the electron transport chain, so that the full TCA cycle works. This gives a hypothetical yield of 26 ATP from every particle of glucose used. [9] The yield is lower than when oxygen is the terminal electron acceptor, in light of the fact that the vitality differential along the electrontransport chain from NADH/NADPH to nitrate is adequate to produce just 2 ATP (and just a single ATP from flavin nucleotides to nitrate).

All vitality yielding pathways which (i) include an electron-transport chain and (ii) utilize outside inorganic substances as the terminal electron acceptor are portrayed by the term breath. Be that as it may, a qualification is made between high-impact breath (when oxygen is the terminal electron acceptor) and anaerobic breath (when nitrate is the terminal electron acceptor).

Focal Part of the Vitality Yielding Pathways

It can be seen that many sorts of substrate can be utilized as potential vitality sources, gave that they can be nourished into one of the vitality yielding pathways. For instance, pentose sugars, for example, xylose (a noteworthy part of hemicelluloses) can be encouraged into the pentose-phosphate pathway and used to give vitality. [10] An amino corrosive like glutamic corrosive can be deaminated to a natural corrosive (α-ketoglutaric corrosive for this situation) which will encourage into the TCA cycle and yield a hypothetical 9 ATP amid its transformation to oxaloacetate. Also, if acetic acid derivation is provided as a substrate it can be connected to coenzyme-An, and acetyl-CoA would then be able to consolidate with oxaloacetate (giving citrate) and be handled through the TCA cycle. Unsaturated fats can be corrupted to acetyl-CoA by a procedure named β-oxidation and would then be able to be used similarly. The main impediment in every one of these cases is that oxygen (or maybe nitrate) is required as a terminal electron acceptor. The main substrates that can supply vitality through aging are sugars and sugar subordinates, through the Embden–Meyerhof pathway or the pentose-phosphate pathway. [11]

Coordination of Digestion: Pathways Adjusting

We noted before that few intermediates of the fundamental vitality yielding pathways fill in as antecedent metabolites for biosynthesis. For instance:

- Dihydroxyacetone phosphate (in the Embden– Meyerhof pathway) can be changed over to glycerol for lipid blend.
- Acetyl-CoA is utilized for blend of unsaturated fats and sterols.
- α-ketoglutaric corrosive is utilized to deliver amino acids of the essential glutamate "family" (arginine, proline, glutamate).
- Oxaloacetate is utilized for the similarly imperative aspartate family (aspartate, lysine, threonine, methionine, and isoleucine).

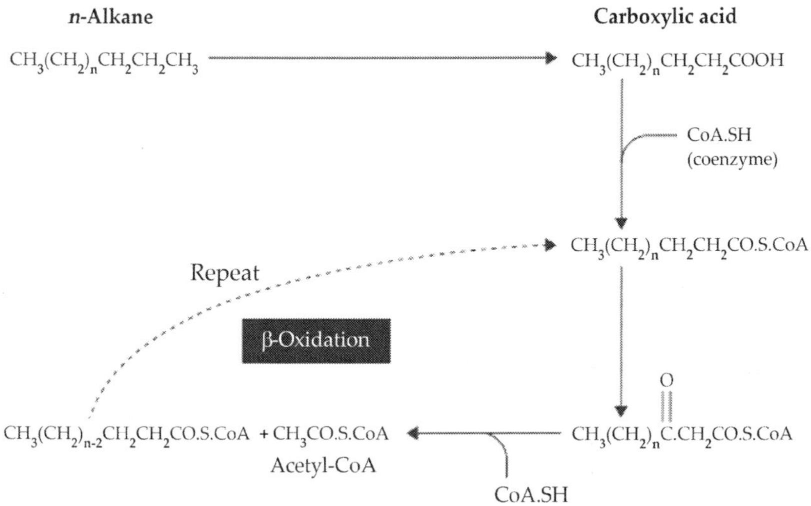

Figure- 5.3: Reactions in β-oxidation.

Along these lines, the inquiry emerges, in what manner can the pathways for vitality generation proceed when intermediates are evacuated? The primary issue would emerge when intermediates of the TCA cycle are evacuated, consequently breaking the cycle. [12] This issue is overwhelmed by unique anaplerotic responses (truly "topping off" responses) which renew the missing intermediates. One such response succession is the glyoxylate cycle, portrayed later in an alternate setting. Another kind of anaplerotic response includes the coupling of CO_2 to pyruvic corrosive, to give oxaloacetate, as takes after:

$$\text{Pyruvate} + \text{ATP} + HCO_3^- \xrightarrow{\text{Pyruvate carboxylase}} \text{oxaloacetate} + \text{ADP} + \text{Pi}$$

The vitamin biotin serves a pivotal part as the cofactor of pyruvate carboxylase, going about as a benefactor or acceptor of CO_2. Biotin likewise is the cofactor in other carboxylation responses. This clarifies why a few organisms should be provided with biotin in the development medium if a parasite can't incorporate it.

HOW SUGARS GENERATED FROM NON-SUGARS

Sugars are constantly required for the amalgamation of contagious dividers, nucleic acids, and capacity mixes, so how are they created when a parasite is developing on non-sugar substrates? This is finished by a procedure named gluconeogenesis (the stage of sugars another). [13] A large portion of the means are essentially an inversion of the Embden–Myerhof pathway, however the progression amongst phosphoenolpyruvate and pyruvate in this pathway is irreversible thus should be skirted.

Consider the instance of a parasite developing on acetic acid derivation (2-carbon) as the sole carbon source. After take-up, acetic acid derivation is changed over to acetyl-CoA and is utilized to create oxaloacetate by the glyoxylate cycle which is a shortcircuited type of the TCA cycle. [14] The principal response step is the cleavage of isocitrate (a 6-carbon transitional of the TCA cycle) to yield succinate (4-carbon) and glyoxylate (2-carbon). At that point glyoxylate is consolidated with acetyl-CoA to deliver malate and thus oxaloacetate. Starting here, oxaloacetate is decarboxylated to yield phospoenolpyruvate (PEP), demonstrated as follows, and sugars are framed from PEP by inversion of whatever remains of the Embden–Myerhof pathway:

$$\text{Oxaloacetate} + \text{ATP} \xrightarrow{\textit{Phosphoenolpyruvate carboxykinase}} \text{phosphoenolpyruvate} + CO_2 + \text{ADP} + \text{Pi}$$

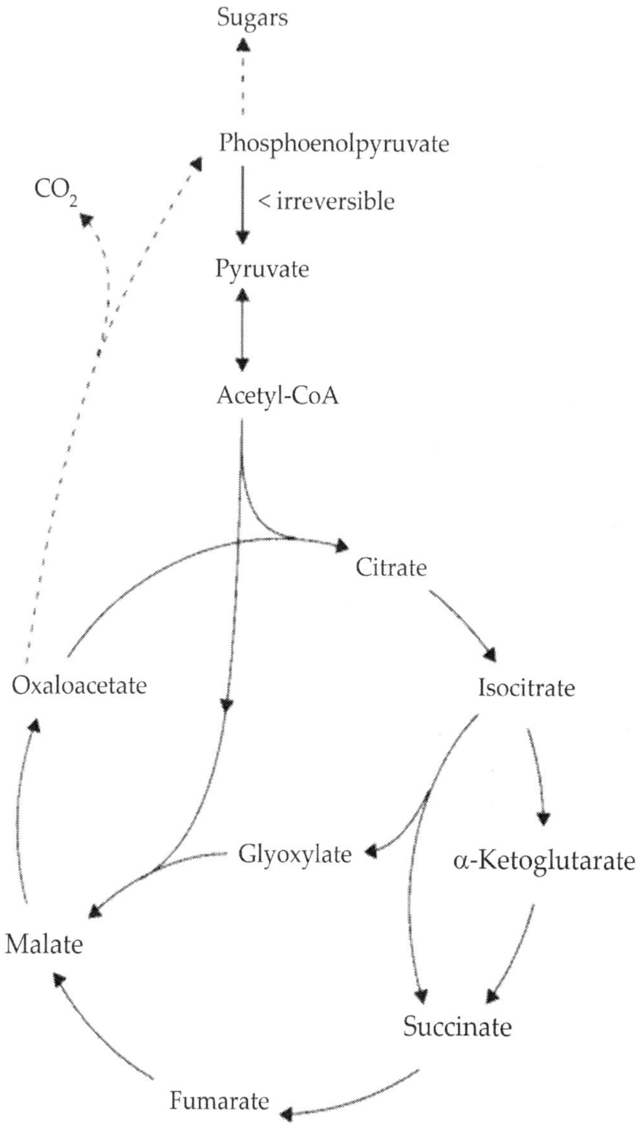

Figure- 5.4: When fungi are grown on non-sugar substrates, role of the glyoxylate cycle.

The essential pretended by the glyoxylate cycle is prove by the way that its compound levels increment more than 20-overlap when acetic acid derivation is provided as the sole carbon source, and mutant strains of growths that can't develop on acetic acid derivation regularly are disturbed in one of the means of this pathway. An indistinguishable procedures from depicted above could be utilized to produce sugars from different substrates like unsaturated fats, natural acids, amino acids, ethanol, and so forth. [15]

Discharges of Natural Acids as Business Items

Fungal growths are critical business wellsprings of natural acids. For instance, if *Aspergillus niger* is developed at high glucose levels (15–20%) and low pH (around 2.0) it will change over the vast majority of the sugar to citrus extract and discharge this into the way of life medium. A lot of oxaloacetate must be created for this, by the carboxylation of pyruvic corrosive examined before.

Additionally Rhizopus nigricans (Zygomycota) delivers a lot of fumaric corrosive, another TCA cycle middle. In both cases the development of the organism is extremely (and deliberately) limited by low pH or some other component, yet the fundamental metabolic pathways keep on operating. [16] The organism carries on like an auto with its motor ticking over: the fuel (substrate) is not utilized for development, so a helpful metabolic halfway is discharged as a sort of fumes item. We should see later that optional metabolites, for example, anti-infection agents are created comparatively. These sorts of process are named vitality slippage, the way that they happen at all shows that they are fundamental, maybe on the grounds that parasitic fungi need to keep their ordinary metabolic procedures working amid periods when development is incidentally stopped. [17]

Mobilizable and Vitality Stockpiling Mixes of Fungi

Fungal growths have a trademark scope of mobilizable and vitality stockpiling mixes, very not quite the same as those of plants, yet strikingly like those of creepy crawlies. The fundamental vitality stockpiling mixes of growths incorporate lipids, glycogen (a α-connected polymer of glucose), and trehalose (a nonreducing disaccharide). The fundamental mobilizable starches are trehalose and straight-chain sugar alcohols (polyols, for example, mannitol and arabitol, in spite of the fact that ribitol is basic in the Zygomycota. The Oomycota have none of these regular "parasitic starches." Instead their capacity mixes are lipids and solvent mycolaminarins (β-connected polymers of glucose), and they translocate glucose or comparable sugars.

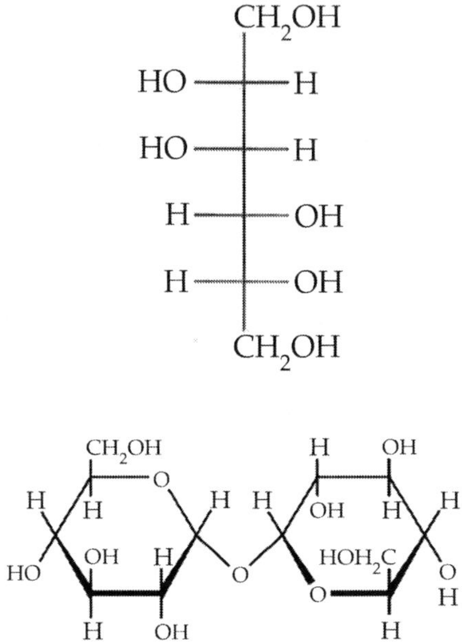

Figure- 5.5: Structure of trehalose and mannitol.

All these fungal parasitic starches can be gotten from the more natural sugars. For instance, trehalose is a disaccharide of glucose, and mannitol is gotten from fructose by a one-stage response including the catalyst polyol dehydrogenase. Be that as it may, a state of uncommon intrigue is that these mixes are interconvertible, and appear to have changed parts in various parts of a mycelial arrange. Brownlee and Jennings in 1981 researched this for the dry-decay parasite Serpula lacrymans (Basidiomycota). [18] They formulated a basic test framework in which little pieces of wood were colonized by *S. lacrymans* and were then put on a sheet of Perspex, so that the parasite developed over the Perspex, bolstered by supplements translocated from the wood squares. [19]

The growth developed at first as a wide mycelial state, however then framed mycelial lines behind the settlement edge. Diverse areas of the settlement could then be examined to decide the distinctive extents of solvent sugars that they contained. The mycelial strings contained elevated amounts of trehalose, however low levels of different sugars or sugar alcohols.

The mid zone of the states contained abnormal amounts of trehalose and arabitol (a sugar liquor), and these abnormal states were kept up in the submarginal and peripheral zones of the provinces. In additionally analyzes, 14C-marked glucose was added to the wood squares, and the name was finished diverse zones of the mycelial states. [20] This again affirmed arabitol and particularly trehalose were the principle frames in which sugars and sugar alcohols are translocated inside the mycelial framework, gathering in the mid zone and submarginal zone. The transformation of trehalose (a disaccharide) to arabitol close to the edge of the settlement is probably going to expand the osmotic potential, drawing water towards the state edge so that the parasite can develop crosswise over supplement free zones.

Serpula lacrymans is outstanding to colonize the auxiliary timbers in structures. It needs water to colonize the timbers at first, yet then its fast rate of cellulose breakdown and the consequent breakdown of glucose (to CO_2 and H_2O) creates "metabolic" water, which is some of the time oozed as beads on the hyphae. Along these lines, the organism can spread a few meters crosswise over brickwork or mortar, drawing the water and supplements advances to colonize different timbers. [21] Comparative mycelial lines and fanatics of hyphae are seen in numerous ectomycorrhizal growths of timberland trees, where they are mindful both for the spread of mycorrhizal improvement and for catching mineral supplements and translocating them back to the root framework. [58]

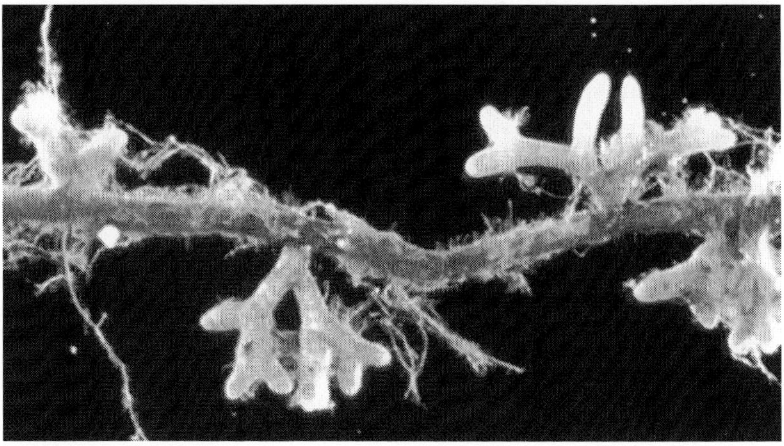

Figure- 5.6: Basidiomycota mycorrhizas on a tree root.

The "contagious sugars" are likewise involved in many plant-parasitic and cooperative affiliations. The underlying investigations on this were made by Harley with the ectomycorrhizal fungal parasitic fungi of beech trees. These fungal parasitic fungi create a generous sheath of tissue around the root tips, and this sheath can be analyzed away for particular synthetic investigation. At the point when the leaves of tree seedlings were presented to $14CO_2$, the name was discovered basically as sucrose in the leaf, stem, and root tissues, yet was then found as mannitol and trehalose in the contagious sheaths. [22] Comparable discoveries have been made for plants contaminated by rust or fine buildup organisms (the biotrophic plant fungal pathogens). [59] At the point when leaves are presented to $14CO_2$, the mark is discovered at first in the regular plant sugars (glucose, sucrose) however spores created by the biotrophic growths are found to contain the name as contagious starches. Most plants don't utilize these mixes (with the exception of in seeds and organic products), so the transformation of plant sugars to the ordinary contagious starches speaks to a "metabolic valve", a restricted stream of supplements from the host to the parasite. This may add to the achievement of biotrophic plant pathogens. [23]

The real pathways of supplement translocation in organisms are misty, however supplements can move both advances and in reverse in hyphae, from areas of relative wealth to relative lack. The tubular vacuolar arrangement of fungal growths, might be huge in this regard since it can transport fluorescent colors against the general stream of cytoplasm. The run of the mill contagious sugars may have a few other critical parts in parasitic physiology. For instance, mannitol is a typical constituent of contagious vacuoles, where it has a noteworthy part in directing cell pH. [24]

Formation of Chitin

Chitin is a trademark part of contagious dividers. The blend of this polysaccharide takes after the example of combination of numerous polysaccharides and can be spoken to by the accompanying general condition:

$$\text{Donor-sugar unit} + \text{Acceptor} \longrightarrow \text{Donor} + \text{Acceptor-sugar unit}$$

For instance, in the union of chitin, fructose-6-phosphate (from the Embden–Myerhofpathway) is at first changed over to N-acetylglucosamine (GlcNAc) by progressive increases of an amino gathering (from the amino corrosive glutamine) and an acetyl gathering (from acetyl-CoA). [25] At that point GlcNAc responds with uridine triphosphate (UTP) to shape UDP-GlcNAc in addition to inorganic phosphate. The high vitality (initiated) sugar unit is then added to the lengthening chitin chain, by the protein chitin synthase.

UTP + GlcNAc ⟶ UDP-GlcNAc + Pi

Then:

UDP-GLcNAc + n [GlcNAc] ⟶ UDP + n+1[GlcNAc]

An ineffectively or nonacetylated type of chitin, named chitosan, is found in Zygomycota, for example, Mucor spp. It is integrated in an indistinguishable route from chitin yet is then deacetylated by the compound chitin deacetylase. [26]

CHITIN

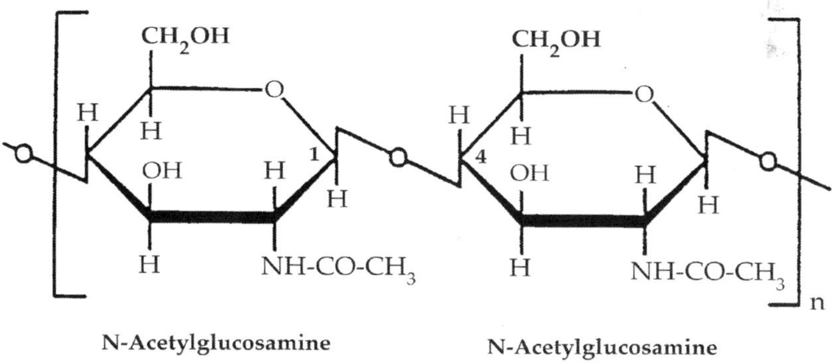

Figure- 5.7: Structure of chitin.

Biosynthesis of Lysine

Lysine is a basic amino corrosive that must be provided as a dietary supplement for people and many homestead creatures, since they can't integrate it. Lysine is created monetarily by vast scale maturation utilizing the bacterium Brevibacterium flavum.

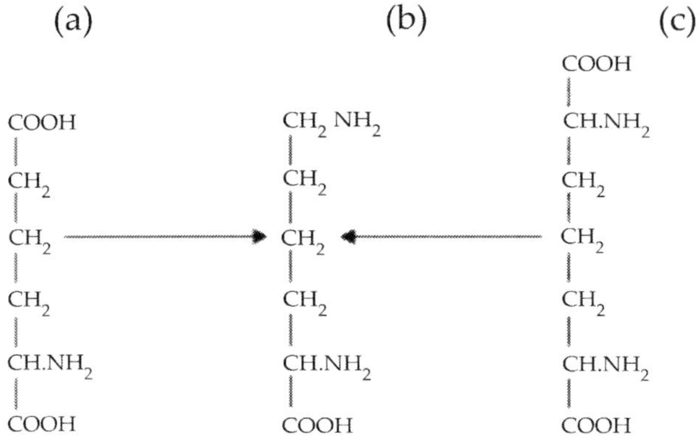

Figure- 5.8: (a) α-Aminoadipic acid (b) Lysine (c) Diamiopimelic acid, showing intermidiates of lysine formation pathways.

The intriguing component of this amino corrosive is that it is integrated by two particular pathways that are totally unique in relation to each other. [27] These pathways are named DAP and AAA, after their trademark intermediates, α-diaminopimelic corrosive and α-aminoadipic corrosive. The AAA pathway is discovered just in the chitin-containing growths and a few euglenids. Every single other living being that orchestrate lysine, the plants, microorganisms and Oomycota utilize the DAP pathway. Such a noteworthy contrast amongst parasitic fungi and different life forms shows a reasonable, and old, developmental dissimilarity.

Secondary Metabolism

The term optional digestion alludes to an extensive variety of metabolic responses whose items are no specifically or clearly required in typical development. [28] In this regard, auxiliary digestion varies from delegate digestion (the typical metabolic pathways talked about before in this section). A large number of auxiliary metabolites have been depicted from growths, and the main components that they have in like manner are:

- They have a tendency to be delivered toward the finish of the exponential development stage in bunch culture or when development is substrate-constrained in consistent culture.

- They are delivered from regular metabolic intermediates yet by unique enzymatic pathways encoded by particular qualities.
- They are not fundamental for development or ordinary digestion.
- Their generation has a tendency to be variety, species-or even strain-particular.

Enthusiasm for this various scope of mixes stems predominantly from their business or natural essentialness. For instance, the penicillins (from Penicillium chrysogenum), the fundamentally comparative cephalosporins (from Acremonium, or Cephalosporium species), and griseofulvin (from *P. griseofulvum*) are anti-infection agents delivered financially from organisms. The obscurely pigmented melanins in some parasitic dividers are additionally optional metabolites, just like the carotenoid colors in the conidia of growths, for example, *Neurospora crassa*. [29] These mixes help to shield cells from harm caused by responsive oxygen species, for example, superoxide and hydrogen peroxide. Some optional metabolites are plant hormones, for example, the gibberellins utilized industrially in cultivation. The underlying revelation of gibberellins was made amid investigations of a plant infection which is the bakanae sickness of rice, caused by the growth Gibberella fujikuroi. In this ailment the shoots stretch uniquely inferable from the generation of gibberellins by the parasitic organism. Some optional metabolites are included in separation, cases being the contagious sex hormones. Others are promoted as pharmaceuticals, including ciclosporin A which smothers organ dismissal in transplant surgery. [30]

Facilitate cases incorporate the mycotoxins, for example, the aflatoxins delivered by Aspergillus flavus and *A. parasiticus* and the ergot alkaloids created by Claviceps purpurea and related species. This rundown of cases could go on, yet numerous optional metabolites have no undeniable part in the life of the maker living being, and changed strains that don't create these mixes regularly develop and additionally the wild-sort strains in culture. This brings up the issue of why optional metabolites are delivered in such differing qualities and wealth, particularly since they are encoded by tapes of qualities that could be required to be lost on the off chance that they give no specific preferred standpoint. [57]

One conceivable recommendation is that the procedure of optional digestion is vital, paying little respect to the final results. As indicated by this view, auxiliary digestion goes about as an overspill or escape valve, to expel intermediates from the essential metabolic pathways when development is briefly confined. This is like the clarification we noted before for the overproduction of natural acids: a creature needs to keep up the fundamental metabolic pathways amid periods when development is limited, however the regular metabolic intermediates can't be permitted to gather since they would disturb typical digestion. [31] Thus, it is contended that these intermediates are shunted into auxiliary metabolic pathways, whose items are either sent out from the cell or amass as (overwhelmingly) idle mixes. In reality, at one phase optional digestion was named "shunt digestion." It could then be contended that the qualities encoding the auxiliary metabolic pathways are allowed to transform (surely more so than those encoding essential digestion) and determination weight would support transformations that prompt items that advantage a fungal growth.

For instance, anti-infection agents could be helpful with regards to an area, mycotoxins as creature antifeedants, melanin for security against UV harm, sex hormones for drawing in accomplices, and flavor or smell segments of toadstools for pulling in creepy crawlies for spore dispersal. In the event that this perspective of optional digestion is right, at that point it appears that numerous growths have still to discover helpful parts for their auxiliary metabolites, or we presently can't seem to discover them! [32]

Regardless, auxiliary digestion is under tight administrative control. As noted above, optional metabolites regularly are created towards the finish of the exponential development stage in bunch societies, when development is constrained by a basic deficiency of a specific supplement however when different supplements are as yet accessible. In constant culture frameworks, optional metabolites can be delivered all through the exponential development stage. [33] The basic component is that the qualities encoding optional digestion are curbed by abnormal amounts of specific supplements, yet in chemostats the way of life medium can be composed so that these repressor substrates are the development restricting supplements, constantly display at low fixation since they are used when they enter the chemostat. [56]

Pathways and Antecedents of Auxiliary Digestion

We see that a couple of key intermediates of the essential metabolic pathways give the beginning stages to the pathways of auxiliary digestion. The absolute most critical optional metabolic pathway is the polyketide pathway, which appears to have no other part aside from in auxiliary digestion.

Table- 5.1: Secondary metabolites from various precursors and pathways

Precursor	Pathway	Metabolites
Sugars	_	Muscarine, Kojic acid
Aromatic amino acids	Shikimic acid	Few lichen acids
Aliphatic amino acids	Several, comprising peptide formation	Penicillins
		Fusaric acid
		Ergot alkaloids
		Lysergic acid
		Sporidesmin
		Beauvericin
		Dextruxins
Organic acids	TCA cycle	Rubratoxin
		Itaconic acid
Fatty acids	Lipid metabolism	Polyacetylenes
Acetyl-CoA	Polyketide	Patulin
		Usnic acid
		Ochratoxins
		Griseofulvin
		Aflatoxins
Acetyl-CoA	Isoprenoid	Trichothecenes
		Fusicoccin
		Sex hormones: antheridol, oogonial, etc.
		Cephalosporins
		Viridin

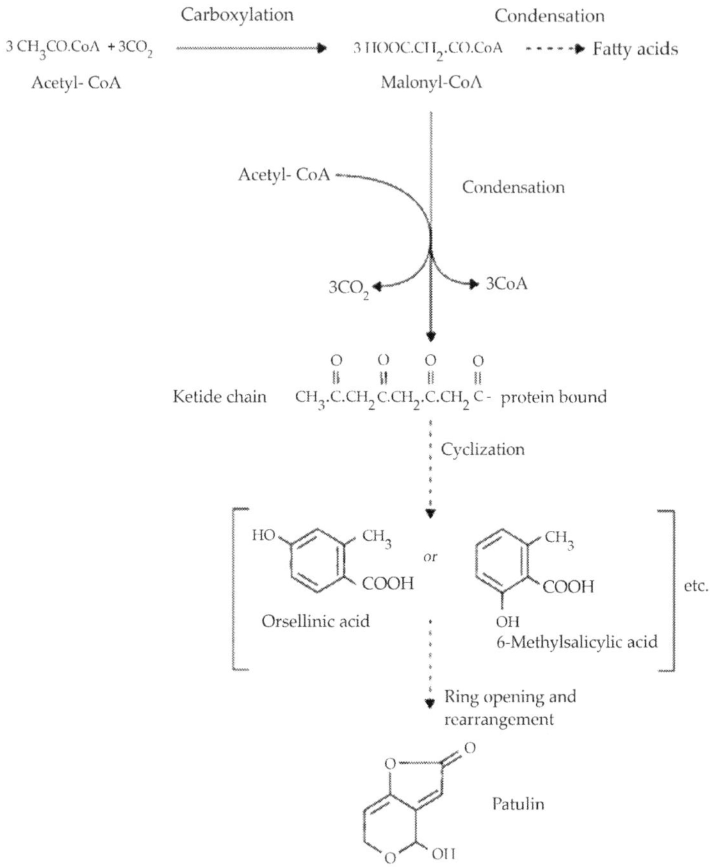

Figure- 5.9: Life cycle of polyketide pathway.

For this situation the forerunner is acetyl-CoA, which is carboxylated to frame malonyl-CoA (a typical occasion in the union of unsaturated fats), at that point at least three particles of malonylCoA consolidate with acetyl-CoA to shape a chain. This chain experiences cyclization, at that point the ring frameworks are altered to give an extensive variety of items. [34] These incorporate the anti-toxin griseofulvin (from Penicillium griseofulvum) used to treat dermatophyte contaminations of people, the strong aflatoxins (from *A. parasiticus* and *Aspergillus flavus*) talked about later, the ochratoxins from different Penicillium and Aspergillus spp., and the anti-toxin or mycotoxin patulin from Penicillium patulum. [35]

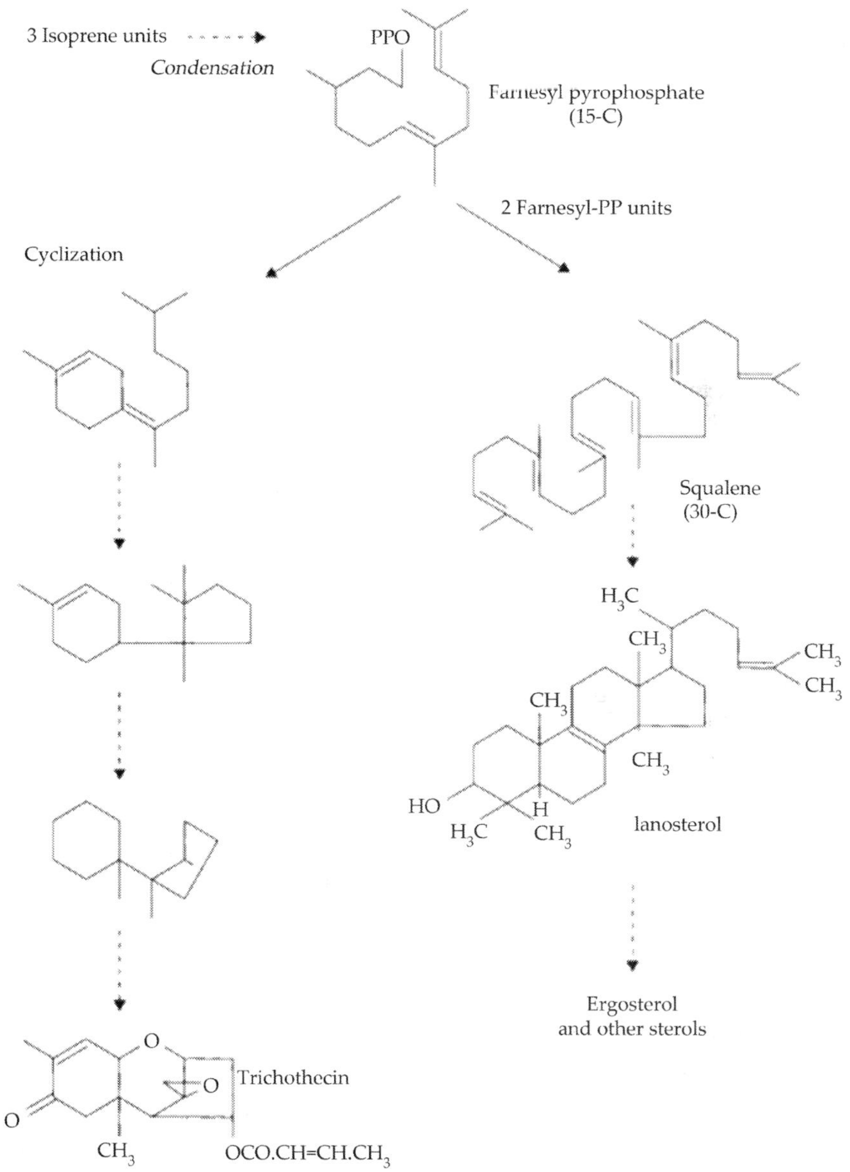

Figure- 5.10: Pathway of isoprenoid for synthesis of numerous fungal pheromones, cephalosporin antibiotics mycotoxins, and synthesis of sterols.

Another vital auxiliary metabolic pathway of growths is the isoprenoid pathway for the combination of sterols. Once more, acetyl-

CoA is the forerunner, however three atoms of this gather to frame mevalonic corrosive (a 6-carbon compound) which is changed over to a 5-carbon isoprene unit. At that point the isoprene units gather make a beeline for tail to frame chains of different lengths, and the chains experience cyclization and further changes. [37] The results of this pathway incorporate the mycotoxins of Fusarium spp. that develop on sodden grain, for example, T-2 poison and the trichothecenes.

The shikimic corrosive pathway, utilized typically for the generation of sweet-smelling amino acids, gives the antecedents to the stimulating auxiliary metabolites, lysergic corrosive and psilocybin in toadstools of Psilocybe, and the poison muscarine in toadstools of Amanita muscaria. Different pathways lead from aliphatic amino acids to the penicillins (see beneath), amatoxins and phallotoxins (in toadstools of the "passing top," Amanita phalloides), from unsaturated fats to the unstable polyacetylenes delivered by mycelia and natural product collections of a few Basidiomycota, and from intermediates of the TCA cycle to the rubratoxins of Penicillium rubrum. Against this foundation, we now consider a couple of cases of auxiliary metabolites of extraordinary intrigue. [36]

PENICILLINS

Penicillin was found by Alexander Fleming in 1929 as a metabolite of Penicillium chrysogenum (initially misidentified as *P. notatum*) that hindered the development of Staphylococcus. Be that as it may, Fleming couldn't cleanse the compound in stable shape, and this was just accomplished in the mid 1940s by two British researchers, Florey and Chain, working in the USA amid the Second World War. For this work, they imparted to Fleming the Nobel Prize for Medicine. [55]

Penicillin is most dynamic against Gram-positive microorganisms. Its method of activity is to keep the cross-connecting of peptides amid the blend of peptidoglycan in bacterial cell dividers, so that the dividers are debilitated and the cells are powerless to osmotic lysis. Astoundingly, penicillin is as yet one of the cutting edge anti-infection agents after over 60 years of use. Likewise of intrigue is Structure of penicillins. The way that the early maturation frameworks that made penicillin creation conceivable were at first produced for citrus extract generation. Be that as it may, the cutting edge penicillins, similar to the related cephalosporins (peptide anti-infection agents), are semisynthetic items. [38]

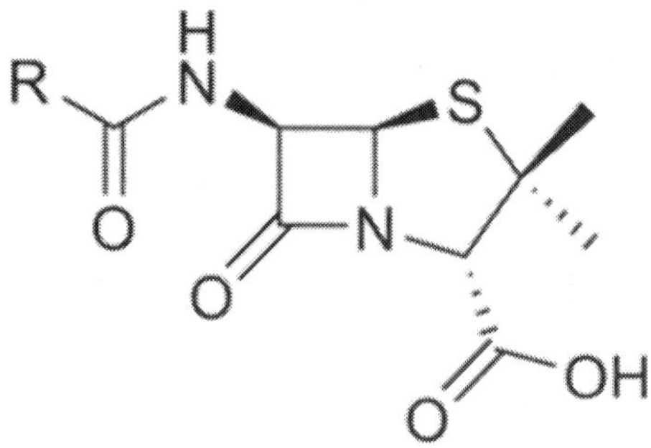

Figure- 5.11: Structure of Penicillin.

The fundamental structure of penicillin is a ring framework gotten from two amino acids, d-valine, and l-cysteine, yet is incorporated from a tripeptide antecedent (α-aminoadipic acid–cysteine–valine) by the supplanting of α-aminoadipic corrosive with an acyl gathering (appeared as "R"). This progression is catalyzed by the chemical acyl transferase. [39] In the early years of business generation, it was found that adjustment of the way of life medium would create penicillins with various acyl gatherings, giving diverse properties on the penicillin particle. So a scope of penicillins were created economically via painstakingly controlling the supply of acyl antecedents in the way of life vessels. In any case, it was then discovered that few microscopic organisms deliver the catalyst penicillin acylase, which can be utilized to evacuate the acyl side chain and leave the essential atom, 6-aminopenicillanic corrosive (6-APA). [40] Scientific experts could then join any coveted side fasten to this atom, with a high level of exactness. So the current generation strategy for semisynthetic penicillins includes three phases:

- Culture of the growth to create most extreme measures of a penicillin, as a rule *penicillin G* which was the sort initially found. This is finished by bolstered cluster culture in which glucose is included stages to anticipate concealment of the optional metabolic qualities, while the antecedent amino acids are provided in overabundance. Additionally, the pH and air circulation are painstakingly controlled, in light of the fact that the penicillin atom separates above pH 7.5 is an issue that kept

Fleming from delivering penicillin reliably, however which Florey and Chain illuminated by thoroughly controlling the aging conditions.

- Recovery of penicillin from the way of life filtrate and treatment with penicillin acylase to create 6-APA.
- Chemical expansion of particular acyl gatherings to create a scope of "semisynthetic" penicillins with various properties. For instance, oxacillin and firmly related mixes are impervious to some bacterial β-lactamase that are the catalysts that sever the ring of *penicillin G* and along these lines inactivate the anti-microbial. Ampicillin has noteworthy movement against some Gram-negative microorganisms, while the characteristic penicillins act mostly against Gram-positives. *Penicillin V* has improved imperviousness to debasement by stomach corrosive, so it is one of the best penicillins for oral organization, while *penicillin G* is defenseless to corrosive. [41]

Notwithstanding these advances, all penicillins are defenseless to breakdown by the plasmid-encoded β-lactamases of some enteric microbes, and penicillins cause hypersensitive responses in a few patients. These issues have been drawn nearer by building up a basically related gathering of anti-infection agents, the cephalosporins, initially found as results of an organism called "Cephalosporium acremonium", however now delivered economically from strains of microscopic organisms (Streptomyces spp).

MYCOTOXINS

Mycotoxins are an assorted scope of mixes from various antecedents and pathways. They ordinarily cause poisonous quality when people or higher creatures ingest them over a moderately drawn out stretch of time, from low fixations in despicably put away sustenance or creature feedstuffs. [54] In any case, we can't reject the likelihood that even short exposures to these poisons are dangerous. The issue in setting up this lies in the way that numerous years may slip by before the impacts of presentation wind up plainly obvious.

A hefty portion of the mycotoxin issues result from the uncalled for capacity of sustenance and sustain items. In any case, some Fusarium species develop on the grains of standing oat trims in wet field conditions and can deliver mycotoxins before the grain is reaped. [42] The Fusarium poisons in this way posture specific issues. A few toadstools additionally contain fatal poisons, the exemplary illustrations being the "demise top" Amanita phalloides, and the "crushing blessed messenger" Amanita virosa. The danger of these is presented by the phallotoxins and amatoxins, yet these are intensely poisonous mixes and are not regularly classed as mycotoxins.

Table- 5.2: Representative mycotoxins in foodstuffs

Toxin	Representative fungi	Foodstuff	Effects
Aflatoxins	A. Parasiticus, Aspergillus flavus	Peanuts, oilseeds	Hepatocacinomas, nephrotoxic
Ergot Alkaloids	Claviceps purpurea	Cereals, grasses	Neurotoxic
Fuminosins	Fusarium moniliforme	Maize	Esophageal cancer
Ochratoxin A	Penicillium spp. And Aspergillus	Grain crops	Kidney carcinoma, nephrotoxic
Patulin	Penicilliumexpansum, Aspergillus clavatus	Apples	Hemorrhage, contact edema
Sporidesmin	Pithomyces chartarum	Grass	Facial eczema of cattle, sheep
Sterigmatocystin	Aspergillus spp.	Grain, Oil seeds	Hepatocarcinogen
Trichothecenes	Stachybotrys chartarum	Cereals	Abortive, Blistering, estrogenic
Zearalenone	Fusarium	Cereals	Vulvovaginitis

Ergot Alkaloids and Related Poisons

The ergot growth, Claviceps purpurea produces sclerotia that create set up of the grain in tainted oats and grasses, at that point the sclerotia (named ergots) tumble to the ground and overwinter close to the dirt surface. In the accompanying summer they deliver minute fruiting bodies, taking after drumsticks, which contain numerous perithecia that discharge ascospores. [43] The planning of arrival of these ascospores matches with the time when the grass blossoms open, so the parasite taints the creating ovary and after that forms into another sclerotium, to rehash the life cycle. The gathering and processing of ergot-defiled oats has caused various passings throughout the hundreds of years. In one type of the ailment, named c onvulsive ergotism, the sensory system is influenced and causes fierce writhings.

Figure- 5.11: Ergots of Claviceps purpurea.

In another frame, gangrenous ergotism, the blood vessels contract, causing oxygen starvation and genuine harm in the tissues. The ergot alkaloids have some vital restorative uses, to alleviate a few sorts of headache and to control draining after labor. They can likewise be mishandled, the indole core of one of the ergot alkaloids, ergotamine, is lysergic corrosive, and this can be adjusted artificially to create the stimulating medication, LSD (lysergic corrosive diethylamide). [44]

As opposed to the ergot alkaloids, the wellbeing peril postured by the numerous organisms that develop subtly on put away nourishment and feedstuffs was just perceived in moderately late circumstances. In the 1950s there were episodes of canine hepatitis (related with specific sorts of puppy sustenance) and of liver growth in cultivated rainbow trout, related with cottonseed dinner in the fish nourishment. In 1959, 100,000 youthful turkeys comes to death in Britain subsequent to being encouraged on fungal parasite sullied cottonseed dinner. The cause was followed to another sort of poison, named aflatoxins, some of which are among the most intense known cancer-causing agents.

AFLATOXINS

Aflatoxins are created mostly by Aspergillus flavus (subsequently the name aflatoxin) and *A. parasiticus*, both of which are normal in tropical and subtropical conditions. They develop on the root frameworks and harvest garbage of groundnuts (peanuts), giving inoculum to colonization of the underground groundnut organic products. *A. flavus* can be disengaged from the shells of any peanuts purchased from a market, it is viewed as dim or dark fixes inside the shells. In any case, most strains don't deliver aflatoxin, and even the strains that do create these mixes require particular conditions. Oil-rich yields, for example, shelled nut foods grown from the ground are particularly positive for aflatoxin creation, reliable with the finding that aflatoxin generation in research center culture is empowered by lipids. [45]

After breakdown of the lipids by lipases, the unsaturated fats are processed to acetyl-CoA through β-oxidation, at that point the aflatoxins can be combined from acetyl-CoA by the polyketide pathway. The Hicks *et al.* in 2002 gives subtle elements of the hereditary qualities and biosynthesis of these mixes. Aflatoxins can be distinguished in concentrates of debased sustenances by thin-layer chromatography, since they demonstrate regular green or blue fluorescence under UV illumination.

This shading distinction is identified with the ring structure of the particle and has prompted the refinement between aflatoxins of the G (green-fluorescing) and B (blue) sorts. This is additionally connected to harmfulness, since aflatoxin B1 is more lethal than aflatoxin G1. The harmfulness of every one of these mixes relies upon the nearness of a twofold bond on the end furan ring. Aflatoxins B1 and G1 have this twofold bond and they are cancer-causing agents, though alfatoxins B2 and G2 do not have the twofold bond and are just feebly dangerous. A comparative contrast is found in the drain of dairy animals that are encouraged on aflatoxin-polluted nourish.

Around 5% of the aflatoxin allow by a cow can be found as **aflatoxin M** in the drain, and it happens as either aflatoxin M1 or M2 relying upon whether the dairy animals have ingested aflatoxin B1 or B2. At the point when aflatoxins are ingested from the gut they are passed to the liver, where the aflatoxins with a twofold bond (at position *) are used to an exceedingly receptive however shaky shape with an epoxide connect at the position of the twofold bond on the end furan ring. [53] This epoxide bunch empowers the poison to tie to DNA and depurinate it, making harm the genome which can prompt hepatomas. Then again, the epoxide can offer ascent to dihydrodihydroxyaflatoxin, which is intensely poisonous. All created nations now force strict points of confinement on the levels of aflatoxins and other normal mycotoxins in nourishments expected for human utilization.

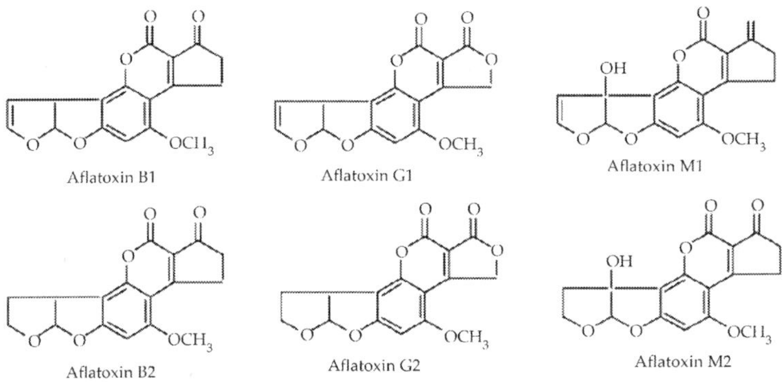

Figure- 5.12: Few common Aflatoxins.

SPORIDESMIN

The mitosporic organism Pithomyces chartarum is a saprotroph that develops on the amassed dead leaf sheaths at the bases of field grasses. It is basic in parts of South Africa, New Zealand, and Australia where it causes a condition called facial skin inflammation in sheep and cows. The prominent side effects are viewed as rankling injuries on parts of the body, for example, the face and udders that are presented to daylight. However this is an optional manifestation, and the essential driver is a mycotoxin, sporidesmin, which is available just in the spores of Pithomyces, not in the mycelium. [52] The poison is ingested when the creatures brush on pastures where the organism is sporulating, prompting corruption of the liver, and scarring and incomplete blockage of the bile conduit, so that the fractional breakdown results of chlorophyll amass in the blood. They are photoactive mixes so they cause photosensitization of the skin where it is not ensured by a covering of hair.

P. chartarum requires very particular conditions for sporulation, a mix of moderately high temperature and high moistness over a time of days. Along these lines, in nations where facial skin inflammation is basic there is a determining framework with the goal that ranchers can bring the creatures into fenced in areas when the danger of introduction to spores is high. Fungicides can likewise been utilized to control this issue.

Some More Poisons

More than 300 mycotoxins have been distinguished to date and, albeit a large number of them won't not be a seriouscause for concern, they could display a potential risk through long haul, low-dose introduction. We can't consider these mixes, yet to end this section we will quickly consider three further cases of poisons and conceivably lethal conditions.

PATULIN

Patulin was initially found as a widespectrum antibacterial anti-microbial yet its improvement was surrendered due to its mammalian danger. It is a little, straightforward metabolite created by numerous types of Penicillium and Aspergillus, including the regular apple-spoil parasite Penicillium expansum which causes a delicate, watery decay when spores enter the apple skin through injuries. Patulin has been distinguished in industrially created squeezed apple and other organic product juices, particularly if these are produced using halfway rotting apples. It can cause edema and draining when ingested, and furthermore is by all accounts cancer-causing in test creatures. It is imprudent to eat any piece of a decayed apple.

Roquefort Cheddar

Roquefort cheddar and other blue-veined cheeses are delivered from goats' drain and vaccinated with the parasite *Penicillium roqueforti*. [47,48] Roquefort cheddar contains low levels of the mycotoxin roquefortine, yet these levels are not thought to be risky. You settle on your decision! [46,48,49]

Wiped out Building Disorder

This is a condition related with clamminess and buildup, which energizes the development of form fungal parasitic fungi. The manifestations incorporate migraines, weariness, and general discomfort, yet the exact causes are difficult to decide. Stachybotrys chartarum is one of the dimly pigmented growths that generally happen in these ecological conditions. [50] It is unequivocally cellulolytic and develops on substrates like the paper sponsorship of plasterboard. This parasite was ensnared in the demise of thousands of stallions in the Soviet Union, when the creatures were bolstered on Stachybotrys-defiled feed in the 1930s. Stachybotrys is one of a few growths (counting Fusarium spp.) that create the trichothecene poisons. There has been theory that Stachybotrys is embroiled in wiped out building disorder, yet cautious research has reliably neglected to give any proof that Stachybotrys or its poisons are included in debilitated building disorder. [51]

SUMMARY

- This chapter comprises knowledge regarding fungal basic metabolic pathways, which explains about how fungi grow in different environmental conditions and on different types of substrates. Mechanisms through which fungi attain their energy in different conditions. It cover topics of vitality yielding from aerobic respiration, electron transport chain, and roles of energy yielding pathways.
- Organic acids are important products made by fungus. For example, *Aspergillus niger* when grown at low pH of nearly 2.0 and high glucose medium of about 15 to 20 %, then it will make citric acid from most of the sugars and discharge it into cultural medium. Fungus also have ability to store energy and mobilizable compounds of fungi. Certain types of formations also done by fungus, for example, chitin and lysine biosynthesis.

- This section told about the precursors and pathways of secondary metabolism, discovery of penicillin, effects of mycotoxins, toxins related to ergot alkaloids, aflatoxins, sporidesmin, patulin, Roquefort cheese, and sick building syndrome. Nearly more than 300 mycotoxins have been discovered now, that might not creating serious problems, but could make possible difficulties by low-dosage and long term exposure.

REFERENCES

1. Fowler, S., Roush, R., Wise, J., OpenStax College., & Rice University. (2016). *Concepts of biology*.
2. Edmunds, Peter J. (2012). *Growth, respiration and photophysiology of coral massive Porites spp. in the experiment of Moorea*. PANGAEA.
3. Tan, V. C. (2013). *Developmenty of aqueous two-phase system (ATPS) to separate crude bioactive phenolics from Phanerochaete Chrysosporium biomass*.
4. Konno, R. (2009). *D-amino acids: Practical methods and protocols: Free D-amino acids*. New York: Nova Biomedical Books.
5. Starr, C., In Taggart, R., In Evers, C. A., & In Starr, L. (2016). *Biology: The unity and diversity of life*.
6. Goodenough, J., & McGuire, B. (2014). *Biology of humans: Concepts, applications, and issues*.
7. In Ramawat, K. G., & In Merillon, J. M. (2015). *Polysaccharides: Bioactivity and biotechnology*.
8. Nebel, B. J. (2016). *Elementary science education: Building foundations of scientific understanding*.
9. Campbell, N. A. (2003). *Biology: Concepts & connections*. San Francisco: Benjamin Cummings.
10. Schofield, D. A. (1994). *Regulation of chitin synthesis in Candida albicans*.
11. Saldivar-Guerra, E., & Vivaldo-Lima, E. (2013). *Handbook of Polymer Synthesis, Characterization, and Processing*. Chicester: Wiley.

12. Johnson, B. R. (1980). *Morphological and physiological responses of wood-decay fungi to polyoxin inhibitors of chitin synthesis.*
13. Jollès, P., & Muzzarelli, R. A. A. (1999). *Chitin and Chitinases.* Basel: Birkhäuser Verlag.
14. Johnson, B. R., & Forest Products Laboratory (U.S.). (1980). *Responses of wood decay fungi to polyoxin D, an inhibitor of chitin synthesis*. Madison, Wis: Forest Products Laboratory.
15. Wendisch, V. F. (2007). *Amino acid biosynthesis: Pathways, regulation and metabolic engineering*. Berlin: Springer.
16. Gebhardt, H. (2005). *Impact of trehalose and mycolate biosynthesis on the cell envelope of a Corynebacterium glutamicum L-lysine production strain.*
17. Ehmann, D. E. (2000). *Enzymology of domains of non-ribosomal peptide synthetases involved in lysine and enterobactin biosynthesis.*
18. Henrich, G. W. (1970). *N-acyl lysine derivatives and Neurospora crassa growth*. Nashville.
19. Campbell, R. A. (1998). *Lysine biosynthesis: Synthesis of enzyme inhibitors and substrates.*
20. Dvořáčková, I. (1990). *Aflatoxins and human health*. Boca Raton, Fla: CRC Pr.
21. Eaton, D. L., & Groopman, J. D. (1994). *The Toxicology of aflatoxins: Human health, veterinary, and agricultural significance.* San Diego: Academic Press.
22. United States. (2008). *Loss adjustment procedures for aflatoxins.* Washington, D.C.: U.S. Dept. of Agriculture, Risk Management Agency.
23. Basappa, S. C. (2009). *Aflatoxins: Formation, analysis and control.* Oxford: Alpha Science International.
24. Briggs, L. R. (1997). *Development of immunoaffinity columns and immunoassays for the collection, partial purification and measurement of sporidesmin A and its metabolites.*
25. Dalefield, R. (2017). *Veterinary toxicology for Australia and New Zealand.*
26. In Blood, F. R. (1970). *Essays in toxicology: Volume 2.*
27. Ronaldson, J. W. (1978). *Fungal metabolite chemistry: Sporidesmin,*

Crepidotine.

28. Cordiner, S. J. (1984). *Effects of sporidesmin on isolated liver cells.*
29. Whyte, B. E. (1994). *Mechanisms of sporidesmin toxicity in Saccharomyces cerevisiae.*
30. Woodcock, J. C. (1998). *Metal interactions with the mycotoxins sporidesmin and gliotoxin.*
31. Cordiner, S J, & Jordan, T W. (1983). *Inhibition by sporidesmin of hepatocyte bile acid transport.*
32. Yu, F. L. (1993). *Protein variation related to sporidesmin induced liver damage and to genetic .ion for or against sporidesmin induced damage.*
33. Flåøyen, A. (1993). *Studies on the aetiology and pathology of alveld with some comparisons to sporidesmin intoxication: Arne Flåøyen.* Oslo: Norwegian College of Vetetrinary Medicine.
34. Boyd, E. M. (n.d.). *Patulin.*
35. Lindroth, S. (1980). *Occurence, formation and detoxification of patulin mycotoxin.* Espoo: Technical Research Centre of Finland.
36. Singh, J. (1966). *Mechanism of antifungal action of Patulin.*
37. Singh, G. (1950). *Studies on patulin.*
38. Walker, C., Central Science Laboratory (Great Britain), & Great Britain. (2006). *Patulin in apple purée: November - December 2005 : report.* York: Central Science Laboratory.
39. Lam, K. S. (1983). *Regulation of antibiotic biosynthesis: A study of induction of patulin pathway enzymes.* Ottawa: National Library of Canada.
40. Olberg, I. H. (1985). *Mykotoksiner i næringsmidler og fôr: Patulin, aflatoksin, ochratoksin, zearalenone.* Oslo: Norges Veterinærhøgskole.
41. Stroka, J., Anklam, E., & European Commission. (1998). *Validation of analytical methods to determine the content of alfatoxin, ochratoxin and patulin in foodstuffs of vegetable origin.* Luxembourg: EUR-OP.
42. Burdaspal, P. A., Ma, L. T., & European Commission. (2001). *Validation of analytical methods to determine the content of*

aflatoxin, ochratoxin and patulin in foodstruffs of vegetable origin: Report on workpackage 7.4. Luxembourg: EUR-OP.

43. Harris, K. L. (2007). *Patulin, surveillance in Michigan apple cider and juice, and factors influencing its production and concentration in apple products*.
44. Roquefort Association. (1964). *44 famous recipes featuring Roquefort cheese: The king of cheese and cheese of kings*. New York, N.Y: Roquefort Association.
45. Bryant, H. W., & Hammer, B. W. (1940). *Bacteriology of cheese: V*. Ames, Iowa: Agricultural Experiment Station, Iowa State College of Agriculture and Mechanic Arts.
46. Studd, W., & Landmark Media, Inc. (2007). *The legend of Roquefort - France*. Falls Church, Va: Landmark Media.
47. Roquefort Association. (1959). *50 exclusive recipes featuring Roquefort: The cheese that makes eating an elegant experience*. New York, N.Y: The Association.
48. Evans, A. C. (1918). *Bacterial flora of Roquefort cheese*. Washington: Dept. of Agriculture.
49. Roquefort Association. (1970). *Roquefort chefmanship recipes: Make the good things in life taste even better!*. New York, N.Y: Roquefort Association.
50. Goss, E. F., Nielsen, V., & Mortensen, M. (1935). *Iowa blue cheese*. Ames, Iowa: Agricultural Experiment Station, Iowa State College of Agriculture and Mechanical Arts.Thom, C., Currie, J. N., & Matheson, K. J. (1914). *Studies relating to Roquefort and Camembert type of cheese*. Storrs, Conn: Storrs Agricultural Experiment Station.
51. Baechler, M. C. (1991). *Sick building syndrome: Sources, health effects, mitigation*. Park Ridge, N.J., U.S.A: Noyes Data Corp.
52. Berglund, B., Raw, G. J., & International Council for Building Research, Studies and Documentation. (1997). *Sick building syndrome: The design of intervention studies*. Rotterdam, Netherlands: CIB.
53. Clinton, P. (1994). *Sick building syndrome: A bibliography*. London: Joseph Clarke.

54. Goldstein, W. E. (2011). *Sick building syndrome and related illness: Prevention and remediation of mold contamination*. Boca Raton, FL: CRC Press.
55. Tate, N. (1994). *The sick building syndrome: How indoor air pollution is poisoning your life, and what you can do*. Far Hills, N.J: New Horizon Press.
56. Teeuw, K. B. (1993). *Sick building syndrome: The role of airborne microorganisms and endotoxin*.
57. Matthews, B. L. (2008). *Chemical sensitivity: A guide to coping with hypersensitivity syndrome, sick building syndrome and other environmental illnesses*. Jefferson, N.C: McFarland.
58. Michelle, C. (2006). *Sick building syndrome and the problem of uncertainty: Environmental politics, technoscience, and women workers*. Durham: Duke University Press.
59. Lawson, L. (2000). *Staying well in a toxic world: Understanding environmental illness, multiple chemical sensitivities, chemical injuries, and sick building syndrome*. Evanston, IL: Lynnword Press.

CHAPTER

6

FUNGI NUTRITION

Fungus have very basic necessities for nutrition. They require a source of natural supplements to supply their vitality and to source for skeleton of carbon for combination of cell. In any case, given a straightforward vitality source, for example, glucose, numerous fungus can blend all their other cell parts from inorganic sources such as phosphate or ammonium particles, nitrate particles and follow levels of different minerals, for example, iron, magnesium, potassium, and calcium. Parasitic fungus that typically develop in a host domain or in other supplement rich substrates may require extra parts, yet at the same time the supplement necessities of most organisms are very basic. So, organisms need to catch supplements from their environment. The cell divider keeps parasitic fungus from overwhelming nourishment particles, so organisms retain straightforward, dissolvable supplements through the divider and plasma layer. Much of the time this is accomplished by discharging proteins to decompose complex polymers and after that engrossing the supplements discharged by these depolymerase catalysts. Parasitic fungus create an enormous scope of these proteins, to decomposet diverse sorts of polymer.

Actually, there is not really any normally happening natural aggravate that can't be used as a supplement source by some parasitic fungus. In this section we consider the numerous adjustments that parasitic fungus have developed for supplement catch.

NUTRIENT NECESSITIES OF FUNGI

For routine research center culture, most parasitic fungi are developed on media with common segments, for example, potatodextrose agar (concentrate of bubbled potatoes, with 2% w/v glucose), malt extricate agar (as a rule containing 2% malt concentrate), or cornmeal (or cereal) agar, a culture medium regularly used to advance sexual improvement. [1] Contagious development media are normally somewhat acidic (pH 5–6) and rich in sugars. This appears differently in relation to bacterial culture media, for example, "supplement agar" (meat concentrate and peptone) which are generally impartial or basic (pH 7–8) and rich in natural nitrogen. This distinction mirrors the higher gross protein substance of microorganisms contrasted and growths. Different sorts of agar are utilized to culture particular organisms or to advance particular formative stages. [2]

In any case, keeping in mind the end goal to decide the base supplement necessities, an organism must be developed in synthetically characterized fluid culture, since agar contains polluting influences. A run of the mill negligible medium, containing just mineral salts and glucose. A few normal organisms will develop on this medium. The principle special cases are noted underneath. [3]

- Some growths should be provided with at least one vitamins, the most well-known necessities being for little measures of thiamine, biotin, or both.
- Some growths can't use nitrate or ammonium as their sole nitrogen source, rather they require a wellspring of natural nitrogen. In these cases the prerequisite typically can be met by providing a solitary amino corrosive, for example, asparagine. A few Basidiomycota fall in this class.
- Some parasitic fungi have more particular individual necessities, which will be found in their characteristic living spaces. For instance, some Oomycota (e.g. Phytophthora

infestans) should be provided with sterols for development, some other Oomycota (e.g. Leptomitis lacteus) require sulfur-containing amino acids, for example, cysteine, since they can't utilize inorganic sulfur. A portion of the parasitic fungi from creature manure should be provided with the iron-containing haem gathering, a part of cytochromes in the respiratory pathway. [4]

Table- 6.1: Usual supplements need by fungus comprising the thiamine or vitamins biotin

Chemical	Quantity
$NaNO_3$ or NH_4NO_3	2g
L-asparagine at equivalent nitrogen content	
KH_2PO_4 (Possibility in form of buffered mixture with K_2HPO_4)	1 g
$MgSO_4$	0.5 g
KCI	0.5 g
$CaCl_2$	0.5 g
$FeSO_4$, $ZnSO_4$, $CuSO_4$	0.005-0.01 g each
Sucrose of glucose	20 g
Distilled water	1 liter

It ought to be stressed that every one of these remarks allude to the base prerequisites for development, not to the ideal necessities. Likewise, they identify with substantial (vegetative) development, and extra supplements might be required to finish the life cycle. [5] For instance, a few Pythium and Phytophthora spp. try not to require sterols for hyphal development however they require them for both sexual and agamic generation. Numerous parasitic fungi have additional supplement prerequisites for development in imperfect ecological conditions. For instance, Saccharomyces cerevisiae has generally straightforward prerequisites for development in oxygen consuming society yet should be provided with an extensive variety of vitamins and other development considers anaerobic conditions, since a portion of the fundamental metabolic pathways that regularly supply these necessities don't work in anaerobic conditions.

VITALITY SOURCE AND CARBON IN FUNGI

A colossal scope of natural mixes can be used by some growth. Supplements are organized in rough request of basic many-sided quality from left to right, and in inexact request of their level of use (vertical hub). At one extraordinary, the least difficult natural compound, methane (CH_4), can be utilized by a couple of yeasts. They change over methane to methanol by the compound methane monooxygenase. [6] At that point this is changed over to formaldehyde by the protein methanol dehydrogenase. At that point formaldehyde joins with ribulose-5-phosphate, and two further response steps produce fructose-6-phosphate.

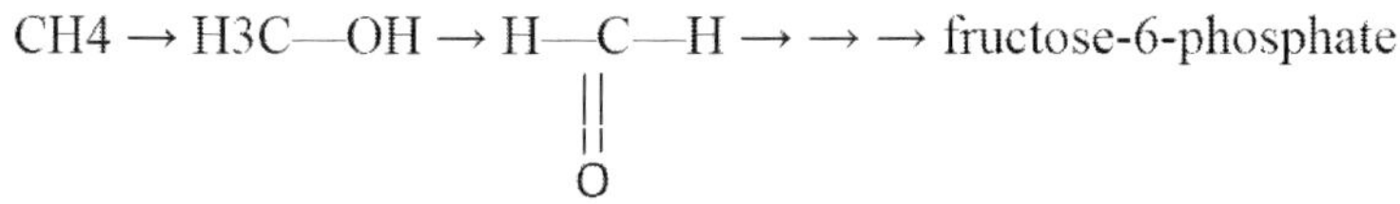

A few more yeasts (e.g. Candida spp.) and a couple of mycelial parasitic fungi can develop on the more extended chain hydrocarbons (C9 or bigger) in oil based commodities. The fundamental confinement with these mixes, as with methane, is that they are not miscible with water thus development is limited to the water–hydrocarbon interface. Two mycelial organisms are infamous for their development on the long-chain hydrocarbons in flying lamp fuel: Paecilomyces varioti (a mitosporic parasite), and Amorphotheca resinae (Ascomycota). *A. resinae* has the questionable qualification of having caused no less than one noteworthy aircrash, by blocking channels and consuming the dividers of fuel stockpiling tanks, yet this issue is effectively dodged.

A bigger number of organisms can use the normal alcohols, for example, methanol and ethanol. Truth be told, ethanol is a magnificent carbon hotspot for Armillaria mellea, Emericella nidulans, and Candida utilis, it can even be their favored carbon substrate. Glycerol and unsaturated fats will bolster the development of a few growths, and can be the favored substrates for a couple of organisms, for example, the basic "sewage parasite," Leptomitus lacteus (Oomycota). [7] Amino acids additionally can be used, however they contain overabundance nitrogen in connection to their carbon content, so ammonium is discharged amid their digestion and this can regularly bring down the pH to development inhibitory levels unless the way of life medium is firmly cushioned.

Most by far of organisms can use glucose and different monosaccharides or disaccharides. Some additionally can utilize sugar subordinates, for example, aminosugars (e.g. glucosamine), sugar acids (e.g. galacturonic corrosive), or sugar alcohols (e.g. mannitol). Be that as it may, the usage of any sugar or sugar subordinate relies upon the nearness of proper film transport proteins, and these have diverse substrate-specificities. [8] Growths for the most part have a constitutive transport protein for glucose. It transports glucose in inclination to different sugars in a blend, however it has a moderately low restricting specificity thus without glucose, it will transport some different sugars. The take-up of these sugars prompts the instigated amalgamation of their particular transporter proteins.

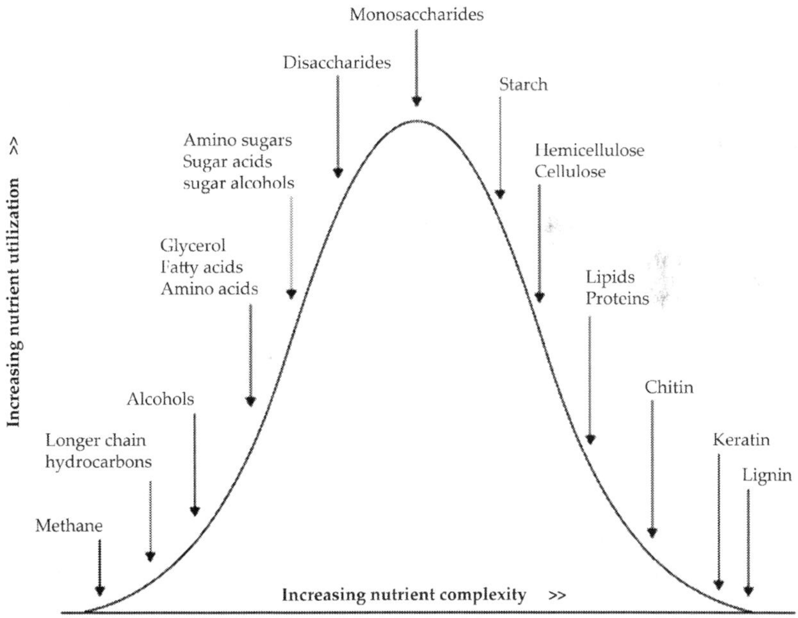

Figure- 6.1: Fungal main carbon substrate.

This can be shown effectively when Saccharomyces cerevisiae is developed in a shaken fluid culture medium containing a disaccharide, for example, lactose as the sole carbon source. The organism debases lactose to glucose and galactose by methods for a divider bound compound, β-galactosidase, at that point the parasite develops quickly by taking up the glucose specially, while galactose stays in the way of life medium. [9] At the point when the glucose is exhausted the development rate moderates for around 30 minutes while a galactose bearer protein is orchestrated, and after that the development rate increments once more, utilizing galactose as the substrate. This sort of biphasic development bend is named a diauxic development bend. Sucrose is utilized as a part of a comparative way: it is severed to glucose and fructose by the divider bound catalyst invertase, at that point the glucose is taken up before fructose. Be that as it may, a couple of organisms can't utilize sucrose as a substrate (e.g. Ascomycota, Rhizopus nigricans, Sordaria fimicola, Zygomycota) since they don't create invertase. [10]

The usage of polymers requires the arrival of particular proteins. For instance, numerous growths can use starch (by delivering amylase), some can use lipids (by creating lipases) or proteins (by creating proteases), and some can develop on cellulose, hemicelluloses, pectic mixes, or chitin. Regularly, these chemicals happen as buildings, comprising of at least two catalysts that assault the polymer at various destinations, acting synergistically. [11] We will see this on account of cellulose breakdown, later in this part. At the extraordinary end of the range, a couple of exceedingly specific organisms corrupt the most basically intricate, cross-connected polymers. For instance, lignin is debased by chemicals of a couple of wood-spoiling Basidiomycota named "white decay parasitic fungi". Essentially, "hard keratin" in the horns and feet of creatures is debased by a couple of keratinophilic parasitic fungi, including species identified with the dermatophytic pathogens of people.[12]

To outline this area, growths abuse an extensive variety of natural supplement sources, however in all cases they rely upon the take-up of basic, dissolvable supplements which diffuse through the divider and enter the cells by means of particular transport proteins. These exclusive permit the section of little atoms, for example, amino acids, monosaccharides, and little peptides of a few amino acids. Indeed, even disaccharides, for example, sucrose and cellobiose (a breakdown result of cellulose) may should be corrupted to monosaccharides before they are taken up by generally organisms. Any bigger atoms must be separated by extracellular chemicals (depolymerases) which are discharged by the organism.[13]

ADJUSTMENTS FOR SUPPLEMENT CATCHING IN ORGANISM

Electrical part in Hyphal Development

The developing tips of fungal organisms produce an electrical field around them. It can be mapped by setting a small vibrating cathode in the fluid film around a hypha. Regularly, the outside of the hypha is more electronegative at the summit than additionally back, demonstrating that present (which is sure by tradition) enters at the tip and exits in the subapical areas. In most organisms that have been inspected in adequate detail, for instance Achlya (Oomycota) and Neurospora (Ascomycota), the current is by all accounts conveyed by protons (H^+) in light of the fact that it compares with a slope of pH along the hyphal surface and the present still happens when other applicant particles are decreased or wiped out from the development medium. [14] In any case, in marine parasitic fungi the current may be conveyed by K^+. For quite a while it was theorized that the electrical field may be the main thrust for apical development, maybe by influencing the exercises of film proteins or the cytoskeleton. Be that as it may, this now appears to be improbable in light of the fact that the course of current stream can be switched without influencing tip development. Rather, the electrical field is by all accounts personally included in supplement take-up. [15]The take-up of natural supplements is an energydependent procedure. Particles (e.g. H^+) are pumped to the outside of the cell by film found particle pumps, utilizing vitality gotten from the separation of ATP.

The particles then re-enter the cell through particular sorts of film proteins named symporters, which all the while transport a natural atom into the cell. This appears to apply for the most part in, in light of the fact that their plasma films have been appeared to have H^+-coupled transport frameworks for sugars, amino acids, nitrate, ammonium, phosphate, and sulfate, and also other particle channels, for example, extend actuated channels for calcium, and nonselective channels for both cations and anions. [16] Be that as it may, growths are thought to vary from most different creatures in one noteworthy regard: in most cell sorts the particle pumps and the symport proteins likely happen in nearness to each other, while in tip-developing organisms the particle pumps are believed to be most dynamic behind the peak, while the symport proteins are dynamic near the tip. This spatial division would clarify the outside electrical field and it bodes well naturally. The supplement take-up symporters would be of most an incentive at the hyphal tip, which broadens constantly into new zones of supplements, while promptly behind the tip there is a thick zone of mitochondria that could supply the ATP required for the proton pumps.[17]

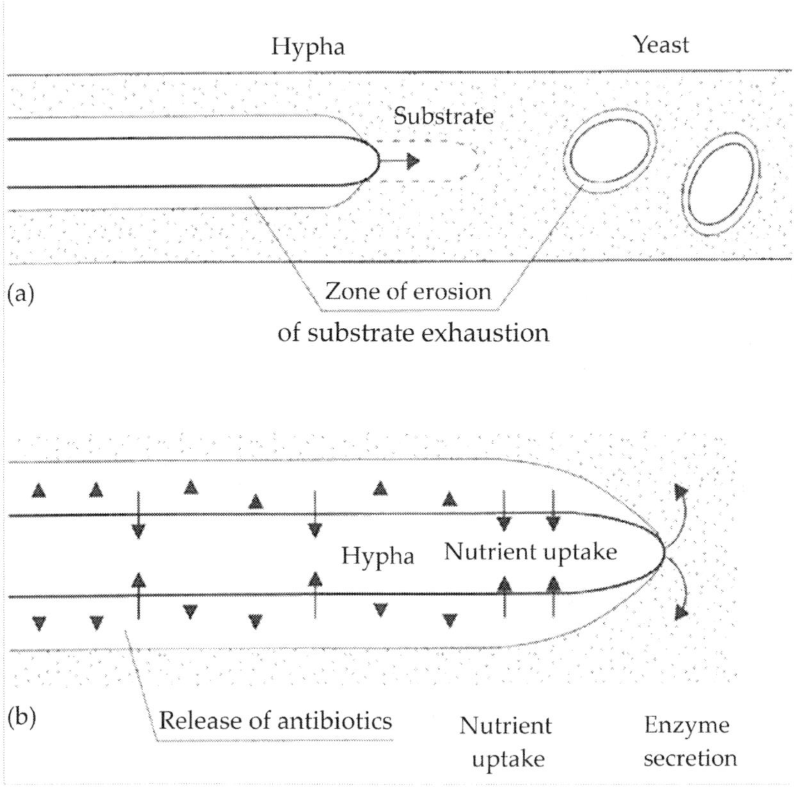

Figure- 6.2: Fungal strategies for growth (a) shows elongation of hyphae (b) defense of substrate through the hyphae of fungi.

Catalyst Discharges

The need to acquire supplements has, truly, formed the way that parasitic fungi develop. The rate of dispersion of dissolvable natural supplements to a stationary cell would constantly have a tendency to be development restricting, and the need to discharge degradative chemicals forces significantly more prominent limitation[18]s. Proteins are substantial particles, around 20,000–60,000 Daltons on account of parasitic cellulases, so they don't diffuse a long way from the hyphal surface. Subsequently, growths make confined zones of substrate disintegration (supplement exhaustion) when developing in substrates, for example, cellulose, and the hyphae must expand persistently into crisp zones. This clarifies why yeast cells never deliver depolymerase catalysts, since they would have no chance to get of getting away from the disintegration zones they had made. Rather, yeasts happen in situations rich in straightforward solvent supplements as on leaf, products of the soil surfaces, or on mucosal layers on account of Candida albicans.

They depend either on a consistent supply of supplements from the basic substratum, or on water streams to bring new supplies of supplements.[19] The huge size of compounds makes a potential issue for their discharge through the hyphal divider, since this would require the nearness of constant pores of adequately expansive size. Assessments of divider porosity can be gotten by concentrate the porousness of dividers to atoms of referred to sub-atomic mass, for example, industrially accessible dextrans, polyethylene, and glycols.

These investigations propose a cut-off of divider porosity at in the vicinity of 700 and 5000 Da, which is much lower than the extent of generally compounds. [20]These discoveries apply for the most part to yeasts, and may be distinctive for mycelial parasitic fungi. Notwithstanding, late investigations demonstrate that compounds are discharged principally and maybe solely in the areas of new divider development. This was first demonstrated when a cellulase quality was cloned into yeast (*S. cerevisiae*) and the chemical was observed to be discharged from the development locales of the bud. Additionally, a glucoamylase has been appeared to be discharged only at the tips of *Aspergillus niger*. So it appears that compounds bound for discharge through the divider are transported to the hyphal tip, where they are discharged by exocytosis and might stream outwards with the recently shaped divider parts. [21]Some of these chemicals would achieve the cell surface and be discharged into the outer condition, while others may progress toward becoming secured in the divider and fill in as divider bound proteins, for instance, debasing disaccharides and other little atoms (oligomers) to monosaccharides, amino acids, and so on.

Defense of Territory

The arrival of compounds to debase a polymer speaks to a speculation of assets, so we can anticipate that it will be combined with barrier of the domain, keeping different living beings from sharing the breakdown items. [22]There are a couple of known cases of noncellulolytic parasitic fungi that develop in close relationship with cellulose-degraders, using a portion of the protein breakdown items. However, these parasitic fungi may develop in a mutualistic affiliation that advantages the cellulolytic growth. By and large, three elements may cause polymerdegrading parasitic fungi to shield their domain.

- The union of depolymerases is firmly managed by criticism components, talked about later, so the rate of catalyst creation is coordinated to the rate at which the breakdown items can be used. [23]
- The last phases of polymer breakdown are accomplished by divider bound compounds, so that the most promptly utilizable monomers are not accessible to different life forms, we will see this later on account of cellulose.
- A polymer-corrupting growth may deliver anti-infection agents or other suppressive metabolites. This is hard to show at the size of individual hyphae, however Burton and Coley-Smith in 1993 revealed that antibacterial mixes were discharged by hyphae of Rhizoctonia species, individuals from the Basidiomycota that are known to corrupt cellulose.[24]

It might be noteworthy that anti-toxins are created for the most part by polymer-corrupting parasitic fungi and their generation in vitro is related with nutrientlimiting development conditions. These are the conditions that polymer-debasing growths would encounter more often than not, on the grounds that depolymerases are delivered just when all the more promptly accessible supplements are hard to come by. In this manner, antibiosis may have advanced not as a forceful technique but rather for the protection of region.[25]

Breakdown of Cellulose as Investigation of Extracellular Catalysts

Cellulose is the most copious characteristic polymer on earth, speaking to around 40% of all the plant biomass that is created and reused on a yearly premise. Growths assume the central part in debasing cellulose and furthermore in using the cellulose breakdown items. In this way, we will consider the structure of cellulose and the part of cellulase catalysts in some detail as a model of how proteins work in indigenous habitats. [26]

Science of Cellulose

Regarding its substance structure, cellulose is a moderately straightforward polymer. It comprises fundamentally of since quite a while ago, unbranched, chains of in the vicinity of 2000 and 14,000 glucose deposits, connected by β-1,4 securities. The β setup implies that every glucose unit is turned 180 degrees to the following, and the entire tie is comprised of these rehashing sets. Such a particle ought to be effortlessly debased by the extracellular catalysts of organisms, however the physical structure of cellulose presents issues of chemical access to the substrate. [27]The individual cellulose chains stack firmly together in a close crystalline way to frame micelles, which are strengthened by hydrogenbonding, and the micelles themselves adjust to shape microfibrils of around 10 nm distance across, unmistakable in electron micrographs. These unbending, insoluble microfibrils display a very considerable impediment to enzymatic corruption, however a few sorts of organism have advanced methods for debasing them. [28]

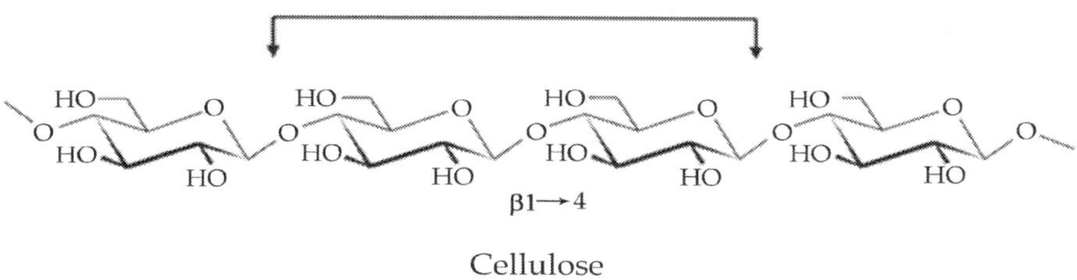

Figure- 6.3: Structure of cellulose chain.

The Cellulase Compound

No less than three sorts of compound are included in the entire breakdown of a cellulose anchor to its constituent glucose units, and some of these catalysts exist in numerous types of various sub-atomic mass. All things considered they are named "cellulase" or, all the more exactly, the cellulase catalyst complex. The three principle proteins in this complex are:

- An endoglucanase (endo-β-1,4-glucanase), which acts aimlessly focuses inside a cellulose chain, breaking the atom into progressively littler pieces. This chemical is found in various structures, running from around 11,000 to 65,000 Da. [29]
- An exo-acting protein, named cellobiohydrolase, which acts just on the finishes of cellulose chains, separating off progressive disaccharide units (cellobiose). This catalyst is more uniform than the endoglucanase, extending from around 50,000 to 60,000 Da. [30]
- β-glucosidase (or "cellobiase"), which severs the disaccharide cellobiose to glucose, for take-up by the growth. This is a divider bound protein, in the same way as other catalysts required in the last phases of debasement of polymers.[31]

The consolidated activities of these three sorts of catalyst are demonstrated diagrammatically. Every one of the three proteins act synergistically and are firmly directed, to guarantee that a cellulose-corrupting fungal parasite does not discharge sugars at a speedier rate than it can utilize them. The endoglucanases, by assaulting the cellulose chains indiscriminately, logically make more finishes on which cellobiohydrolase can act. [32] In any case, the subsequent cellobiose can tie to the dynamic site of cellobiohydrolase, and aggressively restrain the catalyst activity. In this manner, if cellobiose aggregates then the rate of cellulose breakdown will consequently be hindered, and this control interfaces the rate of breakdown to the rate at which the parasite requires glucose for development and digestion. Likewise, the direction of cellulose breakdown is accomplished by the regular criticism framework named catabolite suppression, whereby qualities encoding the compounds are subdued when all the more promptly utilizable substrates like glucose are accessible. [33]

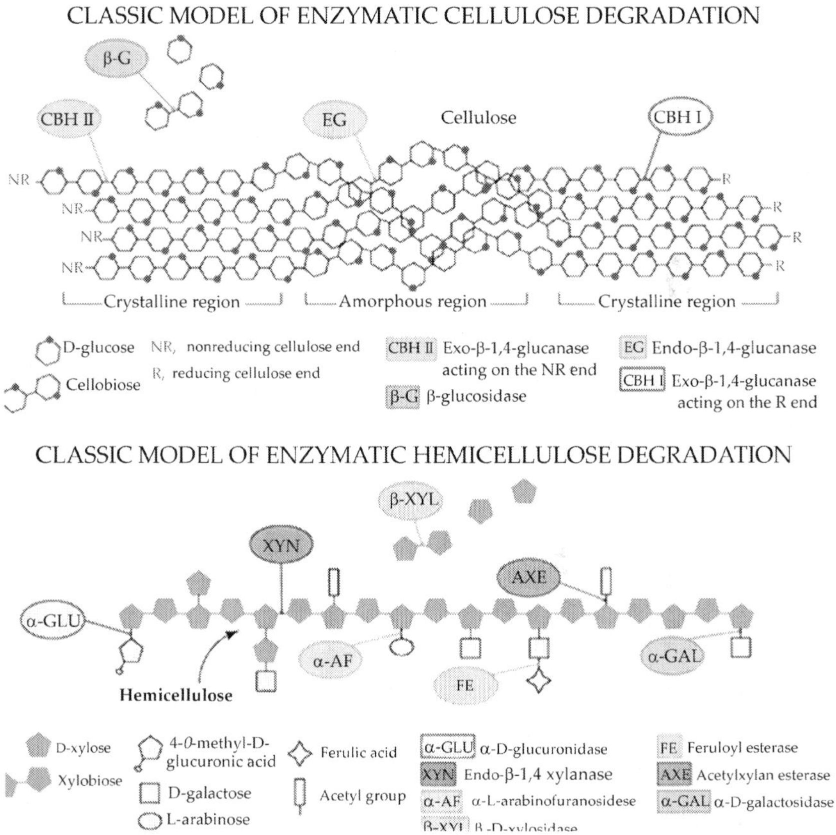

Figure- 6.4: Enzymatic and structural breakdown of cellulose and hemicellulose.

The nearness of cellulose is known to initiate the combination of cellulase proteins. It is hard to envision that cellulose itself could be the inducer since it is insoluble. Rather, cellulose-corrupting organisms are thought to deliver low (constitutive) levels of cellulase proteins so that the breakdown items, for example, cellobiose can go about as signs for catalyst enlistment when cellulose is available in nature. Cellobiose has been found to go about as a feeble inducer of cellulases in a few parasitic fungi, and a subordinate of cellobiose, named sophorose, is a considerably more grounded inducer for Trichoderma reesei, the growth utilized ordinarily as a model living being for investigations of cellulose activity. [34]

In rundown, by a blend of quality constraint (by abnormal amounts of glucose or cellobiose), aggressive restraint of catalyst activity (by the halfway breakdown results of cellulose, for example, cellobiose), and quality enlistment (without glucose however nearness of low levels of cellobiose or its subordinates), the rate of cellulose breakdown is coordinated intently to the rate at which an organism can utilize the sugars discharged from the substrate. [35]

Natural and Commercial parts of Cellulose

The capacity to corrupt "crystalline" cellulose (cotton textures, and so forth.) is found in generally couple of parasitic fungi, yet a few growths can debase the "solvent substituted celluloses, for example, carboxymethyl cellulose in which around 30% of the glucose buildups bear a substituent methyl gathering. These dissolvable celluloses frame gels in water and are created financially for some reasons, for example, backdrop glues and paint thickeners.[36] The parasitic fungi that corrupt these items discharge an endoglucanase yet not cellobiohydrolase, and they incorporate a few basic leaf-surface saprotrophs, for example, Cladosporium spp. also, Sydowia polyspora.

This brings up the issue of why a few parasitic fungi create just piece of the cellulase chemical complex on the off chance that they can't corrupt common types of cellulose. An imaginable clarification was found over 40 years back by Taylor and Marsh (1963) yet has been to a great extent disregarded.[37] These specialists found that few Pythium spp. (which are not for the most part thought to be cellulolytic organisms) could corrupt cotton filaments taken from unopened cotton bolls on cotton plants, yet couldn't debase the cotton strands once the cotton bolls had opened and the strands had air-dried. Basic rewetting of the dried filaments did not render them degradable. In any case, they could be corrupted on the off chance that they were "swollen" by treatment with KOH. The suggestion is that the cellulose in regular, wet plant cell dividers might be vulnerable to assault, giving a substrate that few parasitic fungi can utilize. Be that as it may, the cellulose chains bond together firmly when the tissues have dried, and after that lone a confined gathering of parasitic fungi can debase them. [38]

Among the regular cellulose-degraders is Trichoderma reesei, initially separated from spoiling cotton texture. By normal, rehashed choice in lab conditions, a few strains of this parasite have been acquired that can discharge as much as 30 g dry weight of cellulose chemicals per liter of culture soup. The quality for a cellobiohydrolase of this parasite has been cloned into Saccharomyces cerevisiae under the control of a constitutive promoter. The yeast at that point secretes a lot of a practical cellulase, even without the substrate. Such catalysts, regardless of whether delivered by traditional or hereditarily designed strains, have critical business parts, for instance in rubbing denim items.[39]

MINERAL SUPPLEMENT REQUIREMENTS: PHOSPHORUS, IRON, AND NITROGEN

Fungus require numerous mineral supplements in any event follow sums (Table 6.1) yet nitrogen, phosphorus, and iron legitimacy unique specify in view of their hugeness for contagious exercises and collaborations in nature.[40]

Nitrogen

Of all the mineral supplements, nitrogen is required in the biggest sums and can regularly be the restricting element for contagious development in common environments. Parasitic fungi don't settle air N2 yet they can utilize many joined types of nitrogen. We will talk about this in connection to the chemical intervened grouping underneath, which speaks to the ordinary pathway for absorption of nitrogen by parasitic fungi and numerous different creatures.[41]

$$NO_3^- \xrightarrow{\text{Nitrate reductase}} NO_2^- \longrightarrow NH_4^+ \xrightarrow{\text{Glutamine dehydrogenase}} \text{Glutamate} \xrightarrow{\text{Glutamine synthase}} \text{Glutamine}$$

All fungus can utilize amino acids as a nitrogen source. Frequently they should be provided with just a single amino corrosive, for example, glutamic corrosive or glutamine, and from this they can create the various fundamental amino acids by transamination responses. The standard type of these responses is demonstrated as follows, utilizing for instance, the generation of alanine from glutamic corrosive, or the other way around:

HOOC.CH_2.CH. (NH_2).COOH + CH_3.CO.COOH ⟷ HOOC.CH_2.CO.COOH +CH_3.CH (NH_2).COOH

Mostly fungus can utilize alkali or ammonium (NH_4) as a nitrogen source. After take-up, smelling salts/ammonium is joined with natural acids, for the most part to create either glutamic corrosive (from α-ketoglutaric corrosive) or aspartic corrosive, at that point the other amino acids can be framed by transamination responses, as noted previously. [42]The moderately couple of parasitic fungi that can't utilize ammonium and along these lines rely upon natural nitrogen sources incorporate some water molds (Saprolegnia and Achlya spp.), a few Basidiomycota, and the mycoparasitic Pythium spp. for example, P. oligandrum. Be that as it may, smelling salts/ammonium is frequently not a perfect nitrogen source in culture media, notwithstanding for a hefty portion of the organisms that can utilize it. The reason is that $NH4+$ is taken up in return for $H+$, and this can quickly bring down the pH of a culture medium to 4.0 or less, restraining the development of numerous growths. [43]

Numerous growths can utilize nitrate as their sole nitrogen source, changing over it to ammonium by the chemicals nitrate reductase and nitrite reductase. The growths that can't utilize nitrate incorporate Saccharomyces cerevisiae and numerous Basidiomycota, while some otherparasitic fungi, for example, Neurospora crassa are actuated to express the qualities for nitrate reductase and nitrite reductase just when a favored nitrogen source is inaccessible. [44]

By alluding to the nitrogen absorption pathway (above), plainly any parasite that can utilize NO_3 (or another nitrogen source towards the left of this pathway) should likewise have the capacity to utilize alternate types of nitrogen towards the privilege. Yet, the administrative controls on nitrogen take-up guarantee that nitrogen sources are not really utilized as a part of the ways we may anticipate. In the event that a growth is provided with a blend of nitrogen sources, at that point ammonium is taken up in inclination to either nitrate or amino acids. [45] The reason is that ammonium, or glutamine which is one of the primary amino acids framed from it, keeps the amalgamation of layer take-up proteins for other nitrogen sources, and furthermore keeps the combination of chemicals required in nitrate usage.

Phosphorus

All creatures require noteworthy measures of phosphorus, as phosphates, for generation of sugar phosphates, nucleic acids, ATP, layer phospholipids, and so on. In any case, phosphorus is regularly inadequately accessible in indigenous habitats, on the grounds that even solvent phosphate composts are soon rendered insoluble when they complex with natural issue or with calcium and magnesium particles in soil.[46] Plant roots, specifically, experience issues in separating phosphorus from soil, since they drain the little pool of solvent phosphate in their prompt region and afterward need to rely upon the moderate solubilization and dispersion of phosphate from assist away. By differentiate, organisms are very capable at getting phosphorus, and they accomplish this in a few ways:

- They react to basically low levels of accessible phosphorus by expanding the movement of their phosphorus-take-up frameworks.[47]
- They discharge phosphatase compounds that can cut phosphate from natural sources.
- They solubilize inorganic phosphates by discharging natural acids to bring down the outer Ph.
- Their hyphae, with a high surface range/volume proportion, broaden consistently into new zones of soil.[48]

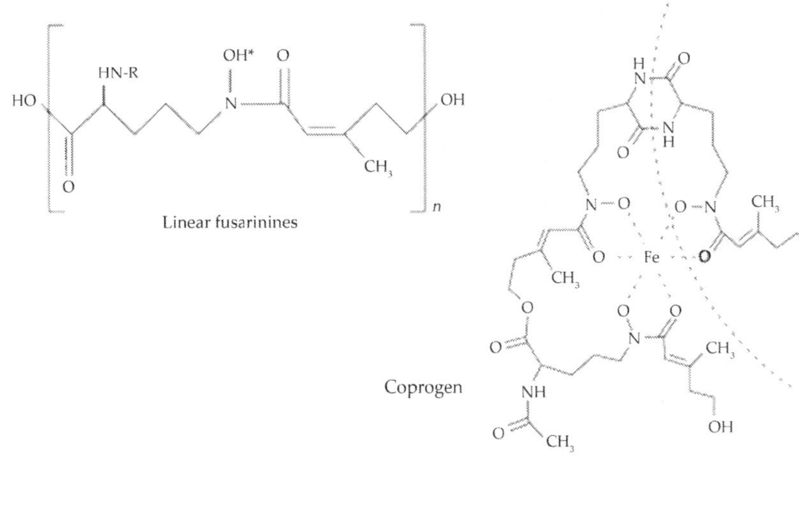

Figure- 6.5: Hydroxamate siderophores of fungi used to seizure ferric iron.

Notwithstanding these focuses, parasitic fungi collect and store phosphates in overabundance of their quick prerequisites, regularly as polyphosphates in the vacuoles. It is not astounding that plants have mycorrhizal relationship with organisms, since this is presumably the most effective route for the plant to acquire its phosphorus supply. The mycorrhizal parasitic hyphae have a high surface zone for take-up of phosphorus and other mineral supplements. The cost to the plant as far as providing sugars to the organism would be a great deal not as much as the cost of persistently creating new roots. [49]

Iron

It is required in moderately little sums however is basic as a giver and acceptor of electrons in cell forms, incorporating the cytochrome framework in oxygen consuming breath. Press ordinarily happens in the ferric (Fe_3^+) frame, insolubilized as ferric oxides or hydroxides at a pH over 5.5, and it is taken up by an alternate procedure contrasted and other mineral supplements. Press must be "caught" from nature by the arrival of iron-chelating mixes named siderophores. These mixes chelate a ferric particle (Fe_3^+) at that point they are reabsorbed through a particular film protein and Fe_3^+ is lessened to Fe_2^+ inside the cell, causing its discharge in light of the fact that the siderophore has a lower liking for Fe_2^+ than for Fe_3^+. At last, the siderophore is sent out again to catch a further ferric particle. Siderophores and their particular film proteins are delivered just in light of iron-restricting conditions. All the contagious siderophores that have been described to date are of the hydroxamate sort. Their structures, capacities, and applications are audited by Renshaw et al. In spite of their high proclivity for Fe_3^+, these contagious siderophores have much lower liking than do the siderophores (e.g. pseudobactin and pyoverdine) delivered by fluorescent pseudomonads which are regular on plant roots. This raises the likelihood that fluorescent pseudomonads could be utilized for the control of plant-pathogenic parasitic fungi in the root zone of yields.

For instance, pseudobactin-creating pseudomonads can stifle germination of the chlamydospores of Fusarium oxysporum on low-press media, while mutant pseudomonads, inadequate in siderophore generation, are ineffectual.

Efficiency of Substrate Utilization

Modern microbiologists are exceptionally worried about the proficiency of substrate change into microbial cells or cell items. Microbial biologists have been less worried about this, despite the fact that it has imperative ramifications for natural procedures. The two most generally utilized criteria of the productivity of substrate change are the financial coefficient and the Rubner coefficient:

$$\text{Economic coefficient (expressed as a percentage)} = \frac{\text{Dry weight of biomass produced}}{\text{Dry weight of substrate used up}}$$

$$\text{Rubner coefficient (expressed as a proportion)} = \frac{\text{Heat of combustion of biomass produced}}{\text{Heat of combustion of substrate used up}}$$

Run of the mill esteems for the monetary coefficient extend from 20 to 35 if a growth is developed in a weaken medium, yet the qualities fall notably if the medium is made logically wealthier – parasitic fungi appear to be "inefficient" if provided with abundance substrate. Qualities for the Rubner coefficient normally are higher than for the monetary coefficient. For instance Aspergillus niger had a Rubner coefficient of 0.55–0.61 (equal to 55–61%) over a scope of social conditions, contrasted and a financial coefficient of 35–46. The distinction is clarified essentially by the amalgamation of lipid stockpiling saves amid development, since lipids have higher calorific esteems than the starches utilized as substrates in culture media. In any case, the essential point uncovered by every one of these qualities is that a significant extent of the substrate provided to an organism is devoured in vitality generation as opposed to being changed over into biomass.

Substrate transformation efficiencies are hard to acquire in normal frameworks, yet Adams and Ayers (1985) did this in research facility conditions by gathering every one of the spores delivered by a mycoparasite (Sporidesmium sclerotivorum) when it was developed on sclerotia of its principle parasitic host, Sclerotinia minor. This exploratory framework impersonates the conditions in nature, since sclerotia are delivered as torpid survival structures on contaminated host plants and afterward overwinter in the dirt, where they can be assaulted by the mycoparasite. The detailed substrate change productivity was outstandingly high: a monetary coefficient of 51–60 and a Rubner coefficient of 0.65–0.75. This mycoparasite has a surprising method for parasitizing the host sclerotia: it infiltrates a portion of the sclerotial cells at first however then develops transcendently between the sclerotial cells, searching little measures of solvent supplements that hole from them, and subsequently making supplement stretch. The host cells react by changing over vitality stockpiling saves (basically glycogen) to sugars, which spill from the phones to help advance development of the mycoparasite. This basically noninvasive method of parasitism is additionally utilized by endophytic organisms in plants. They develop gradually and scantily, between or inside the plant cell dividers, misusing supplements that break from the host cells. As far as substrate proficiency we ought to likewise take note of that growths can develop by oligotrophy, utilizing to a great degree low levels of supplements (oligo = few) on silica gel or glass. They appear to develop by rummaging follow measures of unpredictable natural mixes from the air.

FUNGUS THAT CAN'T BE REFINED

To close this part we should record that few parasitic fungi still can't be developed in research center culture. They have been named commit parasitic fungi however now more generally are named biotrophic parasitic fungi. A considerable lot of them are critical in natural and financial terms, including the pervasive arbuscular mycorrhizal growths (Glomeromycota), rust parasitic fungi (Basidiomycota), fine buildup organisms (Ascomycota), and wool molds (Oomycota). These create supplement retaining haustoria or identical structures in have cells.

It stays to be checked whether some of these parasitic fungi will ever be developed in axenic culture (i.e. isolate from their hosts). Be that as it may, huge advance has been made in refined the rust organisms, beginning with Puccinia graminis (dark stem rust of wheat) and a few other rust animal varieties (Maclean 1982). P. graminis was found to develop gradually, and simply after a drawn out slack stage. Its straight augmentation rate on agar run from 30 to 300 μm day−1, contrasted with rates from 1 with 50 mm day−1 for organisms that are generally developed in research facility conditions. P. graminis tends to release crucial supplements, for example, cysteine, into the development medium, so a high spore thickness is required in the inoculum to limit the dissemination of supplements far from individual sporelings. Additionally, P. graminis produces self-inhibitors which can bring about "staling" of the way of life. This is valid for a few organisms on the off chance that they are developed on sugar-rich media, and it causes a dynamic diminishment and inevitable ending of development. It can be overwhelmed by utilizing moderately powerless media. Some other rust growths require moderately high groupings of CO_2 – for instance, Melampsora lini which causes flax rust. In the event that consideration is paid to every one of these focuses then a few rust growths can be kept up in research facility culture. Notwithstanding, during the time spent being refined (or of getting to be noticeably culturable) some of them change irreversibly to a "saprotrophic" shape that can't reinfect plants.

SUMMARY

- Nutritional needs of fungi are not very complex. They require source of organic nutrients in order to fulfill their need of cellular synthesis of carbon skeleton. Fungi can't engulf food particles due to presence of cell wall, thus, fungi absorb soluble nutrients by the cell membrane and cell wall. Fungi release depolymerase enzymes to degrade complex polymers, and then absorb the released nutrients. In labortary they are usually grow on oatmeal, malt extract agar, and potatodextrose agar.
- Carbon and energy sources are also being discussed in this section. Adjustments for capturing nutrients in fungus includes electrical role in hyphal growth, enzyme secretions, and defense of territory. Study of breakdown of cellulose by extracellular enzymes involves cellulose chemistry, cellulase enzyme, and commercial and ecological aspects of cellulose.
- In the last section topics are included regarding nutrient requirnments of minerals. Fungi need some elements in very small amounts called as trace elements, while some elements are required in sufficiently large amounts, such as, iron, phosphorus and nitrogen. These minerals are very important for interactions and activities of fungi in nature.

REFERENCES

1. Smith, A. J. H. (2013). *The role of ectomycorrhizal fungi on fertilized and unfertilized nursery grown white spruce.* (Masters Abstracts International, 52-2.)
2. Aristizabal, C., & University of Miami. (2008). *Arbuscular mycorrhizal fungi enhance the acquisition of mineral nutrients from leaf litter by morella cerifera.*
3. Addington, G. A. (2011). *Role of belowground fungal mutualisms and nutrient cycling alteration in invasion success of Polygonum cuspidatum Sieb [sic] & Zucc. (Polygonaceae) in the eastern United States.*
4. Cocklin, R. R., & IUPUI ScholarWorks. (2009). *Multiple, nutrient sensing kinases converge to phosphorylate an element of cdc34 that increases saccharomyces cerevisiae lifespan.*
5. Altman, P. L., Katz, D. D., Federation of American Societies for Experimental Biology., National Library of Medicine (U.S.), & American Institute of Biological Sciences. (1968). *Metabolism: Compiled and edited by Philip L. Altman and Dorothy S. Dittmer. Prepared under the auspices of the Committee on Biological Handbooks.* Bethesda, Md: Federation of American Societies for Experimental Biology.
6. Griffin, D. H. (1994). *Fungal physiology.* New York: Wiley-Liss.
7. Ryle, M., & Ørskov, E. R. (1990). *Energy Nutrition in Ruminants.* Dordrecht: Springer Netherlands.
8. Starr, C., In Taggart, R., In Evers, C. A., & In Starr, L. (2016). *Biology: The unity and diversity of life.*
9. Molles, M. C. (2013). *Ecology: Concepts and applications.* New York: McGraw-Hill.
10. Leadbetter, E. R., & Poindexter, J. S. (1985). *Bacteria in nature.* New York: Plenum Press.
11. In Benkeblia, N. (2014). *Polysaccharides: Natural fibers in food and nutrition.*
12. Stirling, G. R. (2014). *Biological control of plant-parasitic nematodes: Soil ecosystem management in sustainable agriculture.* Wallingford, Oxfordshire: CABI.

13. United States., United States., & United States. (2010). *The Role Fungi and Yeast in Monitored Natural Attenuation*. Washington, D.C: United States. Dept. of Energy.
14. Gadd, G. M., & Sariaslani, S. (2017). *Advances in Applied Microbiology*. Saint Louis: Elsevier Science.
15. Zeng, J., & Washington State University. (2012). *Degradation of wheat straw cell wall by white rot fungi Phanerochaete chrysosporium*.
16. Lawrence Berkeley National Laboratory., United States., & United States. (2012). *Transcription Factor Amr1 Induces Melanin Biosynthesis and Suppresses Virulence in Alternaria brassicicola*. Washington, D.C: United States. Dept. of Energy. Office of Science.
17. Flatt, P. R. (1992). *Nutrient regulation of insulin secretion*. London: Portland Press.
18. Renckens, B. A. M. (1980). *Regulation of the enzyme secretion by the pancreas*. S.l: s.n..
19. Kuijpers, G. A. J. (1984). *Transport mechanisms of pancreatic fluid and enzyme secretion*. Katholieke Universiteit te Nijmegen.
20. Schorger, A. W. (1926). *The chemistry of cellulose and wood*. New York: McGraw-Hill Book Co.
21. Marsh, J. T. (2008). *An introduction to the chemistry of cellulose*. Osler Press.
22. In Browning, B. L. (1975). *The chemistry of wood*. Huntington, N.Y: R.E. Krieger Pub. Co.
23. Rowell, R. M., & American Chemical Society. (1984). *The Chemistry of solid wood*. Washington, D.C: American Chemical Society.
24. Heuser, E., West, C. J., & Esselen, G. J. (1924). *Textbook of cellulose chemistry for students in technical schools and universities: As well as for cellulose experts*. New York: McGraw-Hill Book Co.
25. In Honeyman, J. (1959). *Recent advances in the chemistry of cellulose and starch*. Place of publication not identified: publisher not identified.
26. Miles, F. D., & Imperial Chemical Industries, ltd. (1955). *Cellulose nitrate: The physical chemistry of nitrocellulose : its formation and use*. London: Oliver and Boyd for Imperial Chemical Industries.

27. Burk, R. E., & Grummitt, O. (1947). *The chemistry of large molecules*. New York: Interscience Publishers.
28. Coultate, T. P. (2002). *Food: The chemistry of its components*. Cambridge: Royal Society of Chemistry.
29. Voronova, M. I., Surov, O. V., & Zakharov, A. G. (2010). *Interactions of aqueous-organic mixtures with cellulose*. New York: Nivinka/ Nova Science Publishers.
30. Blashfield, J. F. (2001). *Phosphorus*. Austin, TX: Raintree Steck-Vaughn.
31. Sommers, M. A. (2008). *Phosphorus*. New York, NY: Rosen Central.
32. Garnett, H. (2006). *Phosphorus*. Saskatoon, Sask: Thistledown Press.
33. Canada. (1984). *Phosphorus*. Ottawa, Ont: Technical Services Branch, Environmental Protection Programs Directorate, Environmental Protection Service.
34. Beatty, R. (2004). *Phosphorus*. Selangor: Federal Publications.
35. Martynov, V., & American Petroleum Institute. (1973). *Phosphorus*. Washington: American Petroleum Institute.
36. United States. (1970). *Phosphorus*. Washington, D.C..
37. Fixen, P. E., & Potash & Phosphate Institute. (1996). *Phosphorus*. Norcross, Ga: Potash & Phosphate Institute.
38. van, W. J. R. (1966). *Phosphorus and its compounds: Chemistry*.
39. Dumas, A., & Dumas, A. (2017). *The man in the iron mask*.
40. Tocci, S. (2005). *Iron*. New York: Children's Press.
41. Munro, R. (2015). *Iron*.
42. National Research Council (U.S.). (1979). *Iron*. Baltimore: University Park Press.
43. Morris, N. (2006). *Iron*. North Mankato, MN: Smart Apple Media.
44. Murray, P. (2002). *Iron*. North Mankato, MN: Smart Apple Media.
45. Griffin, R. J. (1962). *Iron*. Sydney: Issued under the authority of J.B. Simpson, Minister for Mines [by V.C.N. Blight, Govt. Printer.
46. Cook, J. D. (1980). *Iron*. New York [u.a.: Churchill Livingstone.
47. Franck, I. M., & Brownstone, D. M. (2003). *Iron*. Danbury, Conn: Grolier.

48. Retno Peni Sancayaningsih. (1997). *Sporulatton Of Some Amf Isolates On Several Media And Host Plants*. Yogyakarta] : Universitas Gadjah Mada.
49. Janna Beckerman, & Quello, Kacie Louise. (2009). *Screening for fungicide resistance in orchard fruit: Biochemical and molecular analyses*. Purdue University.

CHAPTER

7

FUNGAL GENETICS

This chapter includes the applied and basic genetics of fungi, comprising the structures that stay to make the fungi as significant exemplary creatures for genetical investigation. This section comprises new molecular methods in different fields like the growth of fungi as "manufacturing works" for remote gene products and the examination of determinants of fungal pathogenicity. It likewise includes the possessions of fungal viruses (hypoviruses) in overwhelming virulence of pathogens virulence and the characters of extra-chromosomal genes in senescence linked to aging.

PLACE OF FUNGI IN GENETICAL RESEARCH

For over 60 years organisms have been real devices for traditional genetical research since they have a blend of components unmatched by different eukaryotes:

- They are anything but difficult to develop in research facility conditions and they finish the life cycle in a brief span.
- Most growths are haploid so they are anything but difficult to transform and to choose for mutants.
- They have a sexual stage for examination of the isolation and recombination of qualities, and every one of the results of meiosis can be recovered in the haploid sexual spores.
- They create agamic spores so that hereditarily uniform populaces can be built up and kept up.

Notwithstanding these focuses, organisms are famously reasonable for biochemical investigations as a result of their basic supplement prerequisites, and on the grounds that "traditional hereditary qualities" has given astounding physical maps of the chromosomal qualities. [1] Concentrates on one growth specifically – Neurospora crassa – prompted the established idea of "one quality, one chemical", for which Beadle and Tatum got the Nobel Prize in 1945. Nonetheless, it is more exact to state that "one quality can encode one chemical" – the circumstance is entangled in light of the fact that quality joining jumps out at evacuate noncoding introns in the pre-flag-bearer RNA. [2]

At the season of composing, the genomes of about 200 living beings have been sequenced – for the most part microscopic organisms and archaea, additionally the genomes of "mouse and man." The "top notch draft" genome arrangements of ten parasitic fungi have been distributed, including Saccharomyces cerevisiae, Neurospora crassa, Emericella nidulans, Schizosaccharomyces pombe, the rice impact pathogen Magnaporthe grisea, and a wood-spoiling growth, Phanerochaete chrysosporium. The initial four of these are Ascomycota with very much mapped chromosomes, giving a premise to joining established and sub-atomic hereditary qualities. [3]

NEUROSPORA AND MENDALIAN HEREDITARY QUALITIES

Neurospora crassa is a standout amongst the most seriously contemplated parasitic fungi for genetical research. It is one of four (conceivably five) Neurospora species that are heterothallic, requiring haploid strains of two distinctive mating sorts, named and a, for sexual propagation. [4] Basically, it starts when a female open hypha named a trichogyne is treated by a "male" spore, the spermatium, of inverse mating sort. The cells that will in the long run turn into the asci are isolated by septa.

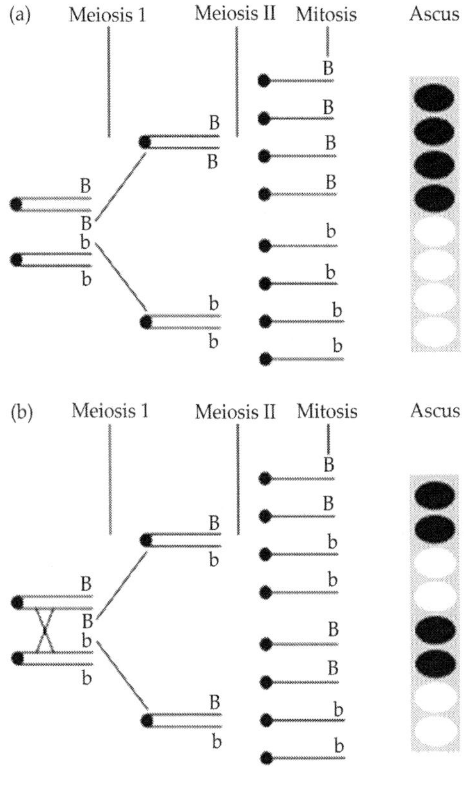

Figure- 7.1: Segregation of spore color genes through the first division of meiosis in Ascomycota.

These ascus mother cells contain two haploid cores, one of each mating sort. The cores circuit to shape a diploid, and the ascus stretches. Meiosis inside every ascus brings about the creation of four haploid cores, and each of these experiences one round of mitosis to deliver eight cores, which are bundled into eight ascospores, directly masterminded inside every ascus. [5] The example of quality isolation in an ascus can be trailed by making crosses between strains that contrast in biochemical components or spore coat shading. For instance, the example of meiosis in a growth heterozygous at a locus that decides spore coat shading. The allele B codes for dim spores and the allele b codes for pale spores. (Note that every chromosome comprises of two chromatids, connected to a centromere, yet just a single arm of every chromosome is appeared). [6]

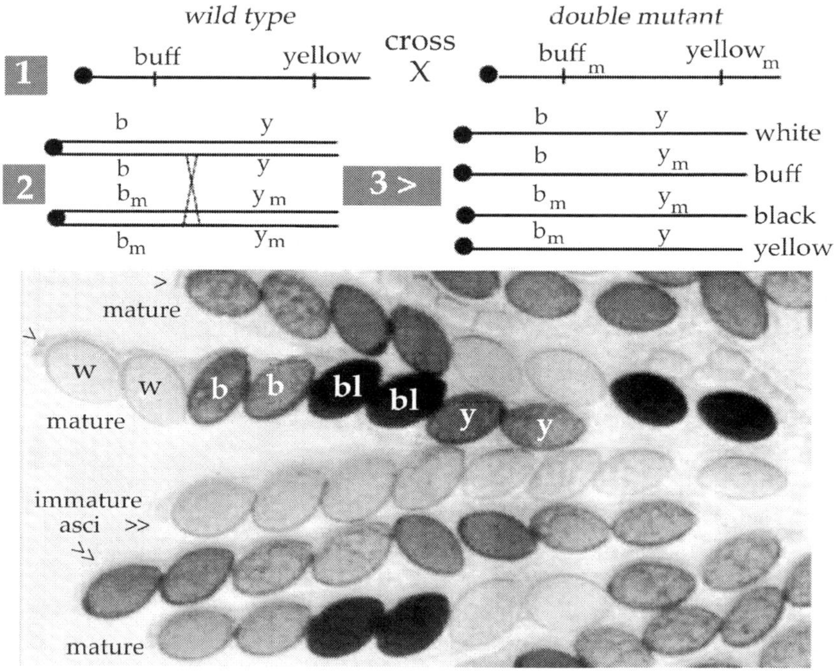

Figure- 7.2: Sordaria crushed perithecium expressing numerous asci comprising eight ascospores.

Amid the principal meiotic division the chromosomes particular. The chromatids at that point isolate in the second meiotic division, and this is trailed by mitosis, prompting an ascus containing a direct plan of four dark ascospores and four pale ascospores. The ensuing example of spore coat shading is distinctive in the last ascus.

Typically, there would be a few hybrid occasions (chiasmata, solitary chiasma) on any one arm of a chromosome, however this would best be distinguished by intersection strains that vary at three distinct loci – for instance, loci X, Y, and Z – on one arm of the chromosome. Comprehensively, the possibility of a hybrid occasion happening between any two quality loci on a chromosome relies upon the physical separation between these loci. Likewise, the possibility of traverse between a quality locus and a centromere relies upon the separation between these. [7]

So it is conceivable to build physical maps of the relative places of various quality loci on any one arm of a (chromosome mapping) by making rehashed crosses including diverse quality loci. (This is not precisely genuine in light of the fact that the recurrence of traverse has a tendency to be bring down close to the centromere and higher close to the closures of the chromosomes – the telomeres – yet it allows the request of qualities to be resolved.). To give a "genuine" case of the examples of quality isolation and recombination in an ascus, The aftereffects of a cross between two parental strains of Sordaria brevicollis, a parasite firmly identified with Neurospora crassa. One strain has the wild-sort alleles for buff-hued (b) and yellow-shaded (y) ascospores on one arm of the chromosome. The other strain has transformations at both of these quality loci, demonstrated as bm and ym, where the subscript m signifies a change. [8]

In this specific illustration, the result of such a cross would ordinarily create either buff spores (with the alleles b and ym) or yellow spores (with the alleles bm and y). Be that as it may, the b and y loci are adequately far separated that there is around 20% likelihood of a hybrid happening between these loci (see name 2) so that after meiosis and a resulting round of mitosis, the eight develop ascospores will show four unique hues (see mark 3). [9] Spores with the b and y alleles are white, those with b and ym are buff-shaded, those with bm and ym are dark, and those with bm and y are yellow hued. Three of the asci are marked "develop" – the spores are named w (white), b (buff), bl (dark), and y (yellow). Each of these asci demonstrates an alternate example of spore isolation, contingent upon which chromatids were included in chiasma arrangement.

Structure and Organization of the Fungal Genome

The genome of a living being incorporates all the hereditary data, not only the qualities encoded by the core. In organisms the genome frequently incorporates four separate parts: the chromosomal qualities, the mitochondrial qualities, plasmids (and versatile hereditary components), and contagious infection qualities, which are really occupant hereditary components. Each of these can contribute fundamentally to the phenotype of fungi. [10]

Chromosomes and Chromosomal Genes

To recap quickly, most organisms are haploid, however the Oomycota are diploid, a couple of growths can interchange amongst haploid and diploid substantial stages, and a few yeasts (e.g. Candida albicans) are for all time diploid. A few growths and parasite like living beings have polyploid arrangement (e.g. Allomyces spp. what's more, Phytophthora spp., including P. infestans). This can be appeared by recoloring the hyphae with a fluorochrome, for example, DAPI, which intercalates in the A-T-rich locales of DNA, and afterward measuring the fluorescence of individual cores under a magnifying instrument. [11] Fluorescence increments in a stepwise way with each expansion in the quantity of chromosome sets.

The chromosomes of almost all parasitic fungi are little and very dense. They are hard to number by regular microscopy of recolored cells in light of the fact that the atomic layer holds on amid the vast majority of the mitotic cycle. [12] However tallies have now been acquired for a few growths by a mix of cytology, linkage investigation (which empowers qualities to be alloted to specific chromosomes), and heartbeat field electrophoresis of removed chromosomes. The haploid chromosome tally of most growths and parasite like living beings appears to lie in the vicinity of 6 and 16, however can be as low as 3 or as high as 40.

Table- 7.1: Reported numbers of chromosomes in various fungus

Fungi	**Chromosome count**
Oomycota	
Phytophthora spp.	9-10
Achlya spp.	3, 6, 8
Saprolegnia spp.	8-12
Pythium	10 or 20
Chytridiomycota	
Allomyces arbuscula	16
A. javanicus	14 (variable in hybrids and polyploids)
Ascomycota	
Schizosaccharomyces pambe	3

Neurospora crassa	7
Saccharomyces cerevisiae	16
Emericella nidulans	8
Coccidioides posadasii	4
Trichophyton rubrum	4
Magnaporthe grisea	7
Basidiomycota	
Filobasidiella neoformans	11
Schizophyllum commune	11
Coprinus cinereus	13
Puccinia Kraussianna	30-40

The atomic genome size of parasitic fungi is little in correlation with numerous different eukaryotes. For instance, the genome size of Saccharomyces cerevisiae is a little more than 12 Mb (12 megabase sets, or 12 million base sets), and that of Schizophyllum collective (Basidiomycota) is around 37 Mb. Some detailed genome sizes of other growths. The qualities for S. cerevisiae and Schizophyllum collective are just around three and eight times bigger than the genome of Escherichia coli (4 Mb) and significantly littler than the genome of the natural product fly Drosophila (165 Mb) or people (around 3000 Mb). [13] Some portion of the explanation behind the little genome size of parasitic fungi is that they have little multicopy (repeated) DNA. It speaks to just around 2–3% of the genome in Emericella nidulans and around 7% in S. cooperative.

The repeated DNA codes chiefly for cell segments that are required in extensive sums – ribosomal RNA, exchange RNA and chromosomal proteins. Be that as it may, an unusually expansive measure of the genome of the fleece mold pathogen Bremia lactucae (Oomycota) is monotonous (65% of the aggregate genome of around 50 Mb), the explanation behind this is obscure. Parasitic fungi decipher a generous measure of the atomic DNA into delivery person RNA – an expected 33% in S. collective and 50–60% in S. cerevisiae. Contrasted and different eukaryotes, in this way, they have moderately small noncoding (repetitive) DNA. Growths take after different eukaryotes in that their protein-encoding qualities contain noncoding DNA arrangements named introns. The introns are interpreted into mRNA yet are extracted before the mRNA is converted into proteins.

Be that as it may, the introns of growths are short (frequently around 50–200 base sets) contrasted and those of higher eukaryotes (regularly 10,000 base sets or more), and S. cerevisiae is strange on the grounds that it has not very many introns. [14]

MITOCHONDRIAL GENES: ORDINARY CAPACITY AND ASSOCIATION IN MATURING

Mitochondria contain a little round particle of DNA. The measure of the mitochondrial genome fluctuates, from as meager as 6.6 kb (kilobase sets) in people to more than 1 Mb in plants. Contagious mitochondrial genomes are frequently in the scope of 19–121 kb, for instance 70 kb in S. cerevisiae, and 50 kb in Schizophyllum collective. Any varieties are expected for the most part to the measure of noncoding material, since all mitochondrial DNAs code for similar things: a few segments of the electrontransport chain (counting cytochrome c and ATPase subunits), some basic RNAs of the mitochondrial ribosomes, and a scope of mitochondrial exchange RNAs. Both the atomic and the mitochondrial qualities are expected to create finish, practical mitochondria.

Table- 7.2: Approximate sizes of fungi and fungus like organisms

Fungus and fungus like organisms	Genome size (Mb)
Aspergillus fumigatus	30
A. niger	30
Candida albicans	16
Filobasidiella	21
Emericella	28
Neurospora crassa	40
Phanerochaete chrysosporium	40
Saccharomyces cerevisiae	12
Schizosaccharomyces pombe	14

The mitochondrial DNA of parasitic fungi has gotten uncommon consideration in connection to maturing, in light of the fact that in a few filamentous growths (Podospora, Neurospora, Aspergillus) a solitary transformation in a solitary mitochondrion can prompt senescence of the entire settlement, when the mutant quality causes the progressive uprooting of wild-sort mitochondrial DNA. [15] The strains of Podospora that show the senescence phenotype can be kept up inconclusively as over and over subcultured youthful provinces, yet they quit developing, end up noticeably senescent and kick the bucket after they have been developed consistently for around 25 days. It has for some time been realized that a non-atomic "infective component" is included, in light of the fact that nonsenescent strains (which never experience senescence) secure the capacity to senesce when their hyphae anastomose with senescence-inclined strains.

Additionally, mitochondria were ensnared on the grounds that the onset of senescence could be put off inconclusively by developing strains within the sight of sublethal measurements of inhibitors of mitochondrial DNA blend or mitochondrial protein combination, however senescence happened when the inhibitors were evacuated. Later examinations demonstrated that DNA from strains that were experiencing senescence could be changed into protoplasts of solid strains, and the protoplast descendant's senesced quickly. The reason for this is by all accounts a 95 kb plasmid which regularly exists as an essential piece of the mitochondrial DNA of sound strains or of adolescent (pre-senescent) societies of senescent strains. In any case, as the senescent strains age this DNA is extracted from the mitochondrial genome, turns into a shut round particle, and self-repeats, causing senescence. [16]

Exactly how it does this is still in question, yet the plasmid indicates DNA homology with an intron in one of the mitochondrion qualities – the quality that codes for a subunit of cytochrome-c-oxidase, a catalyst fundamental for typical capacity of the respiratory electron transport chain. [17] Senescent strains of Podospora need cytochrome-c-oxidase movement, maybe on the grounds that the plasmid embeds in the mitochondrial DNA, prompting interruption of quality capacity or causing mitochondrial quality modifications. Brokenness of cytochrome-coxidase or different parts of electron transport would be deadly for Podospora in light of the fact that this organism appears to be not able develop anaerobically by maturing sugars.

Transposable Elements and Plasmids

Plasmids as a rule are shut round atoms of DNA with the capacity to duplicate self-sufficiently in a cell. Nonetheless, they can likewise be straight DNA particles if the finishes are "topped" (like chromosomes) to keep their debasement by nucleases. Plasmids or plasmid-like DNAs have been found in a few parasitic fungi. The most outstanding illustration is the "two-micron" plasmid of S. cerevisiae, supposed as a result of its 2 μm length as found in electron micrographs. [18] This plasmid is a shut roundabout particle of 6.3 kb, and it is strange on the grounds that it is found in the core, where it can be available in up to 100 duplicates. It has no known capacity, yet in the past it was utilized to build "vectors" for quality cloning in yeast. Most different plasmids of growths are found in the mitochondria. The best described are the straight DNA plasmids of Neurospora crassa and N. intermedia.

They demonstrate a level of base succession homology to the mitochondrial genome, recommending that they are faulty, extracted sections of the mitochondrial qualities. Be that as it may, some other mitochondrial plasmids of Neurospora are shut round particles with almost no homology to the mitochondrial genome. They have a variable "unit" length of around 3–5 kb (in various cases) and the units can join make a beeline for tail to frame bigger rehashes. None of these contagious plasmids has any known capacity, so they dislike bacterial plasmids that code for anti-infection resistance, pathogenicity, or the capacity to corrupt pesticides, and so forth. [19]

Transposons (transposable components) are short areas of DNA that stay in the chromosome however encode proteins for their own replication. They deliver RNA duplicates of themselves, and they encode the compound, turn around transcriptase, which integrates new duplicates of DNA from this RNA format, like the activity of retroviruses, for example, HIV. The new duplicates of DNA would then be able to embed at different focuses in the same or different chromosomes, prompting adjustments in quality expression. Transposons appear to be uncommon in filamentousparasitic fungi, yet there are a few sorts in S. cerevisiae. The best concentrated of these are the chromosomal Ty components, introduce in around 30 duplicates in yeast cells. Oliver (1987) depicted the known and conceivable parts of Ty components. Notwithstanding a part in changing quality expression, they could effectsly affect chromosomal revisions when the "delta successions" on the closures of these components join with each other. These transposable components appear to have no capacity, with the exception of self-propagation. The mating-sort qualities of S. cerevisiae are transposable casettes, causing mating-sort exchanging. In any case, mating-sort exchanging can't happen in Neurospora crassa, in light of the fact that individual haploid strains comprise of just a single mating sort.

VIRAL GENES AND VIRUSES

Fungus were first found in the 1960s, related with "La France" illness of the developed mushroom Agaricus bisporus. (The name was instituted by British mushroom cultivators, mirroring the understanding cordiale that has since a long time ago described Anglo-French relations!) In this infection the natural product bodies are contorted and the organic product body yield is poor. Electron micrographs of both the hyphae and the natural product bodies demonstrated the nearness of numerous isometric infection like particles (VLPs), thought to be the reason for the issue. VLPs were then found in other parasitic fungi, and by the 1980s they were known in more than 150 species, including agents of all the major contagious gatherings (Buck 1986). With a couple of remarkable special cases, nonetheless, the nearness of VLPs is not related with any conspicuous issue, so most parasitic infections appear to be symptomless. Concentrates on a scope of parasitic fungi have demonstrated that contagious infections (or VLPs) have comparable fundamental components. [21]

They are isometric particles, 25–50 nm measurement, with a genome of twofold stranded RNA (dsRNA), a capsid made out of one noteworthy polypeptide, and they code for a dsRNA-subordinate RNA polymerase for replication of the viral genome.

The genome measure is variable. Indeed, even inside a solitary parasite it ranges from around 3.5 to 10 kb. Now and again this variety is because of inside cancellations of a full-length atom, yet in different cases the genome is partitioned between various capsid particles. [23]

In most growths the VLPs are discovered rarely in hyphal tips, however they can happen as crystalline clusters in the cytoplasm of more seasoned hyphal districts, frequently nearly connected with sheets of endoplasmic reticulum that encase the totals.

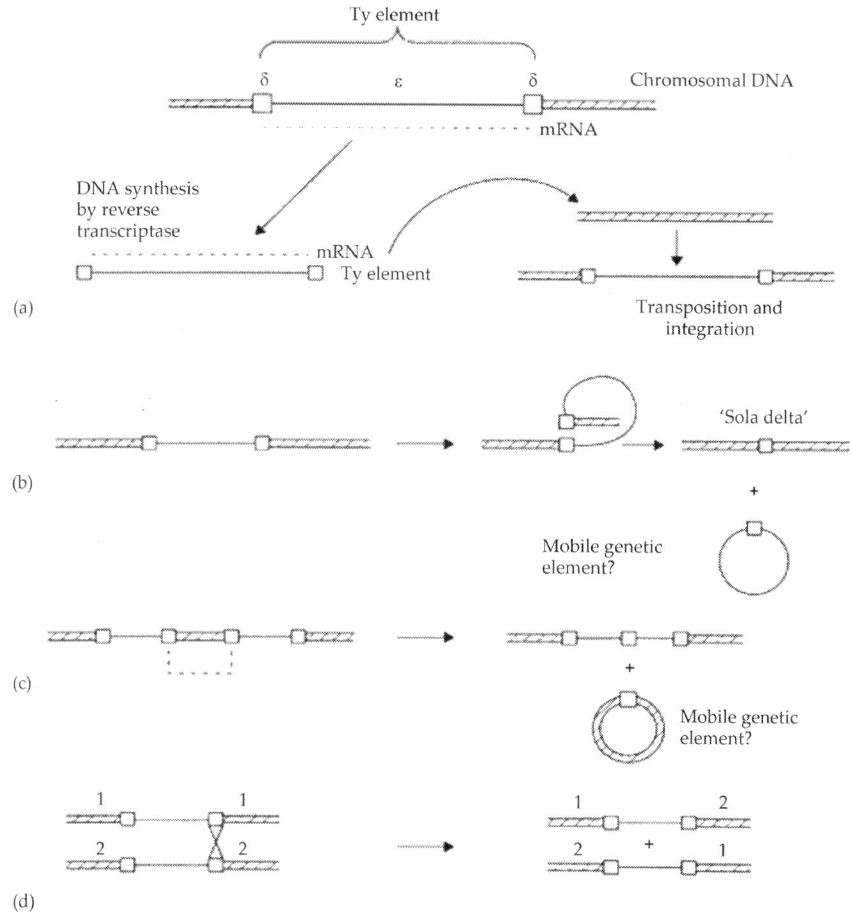

Figure- 7.3: Ty elements of yeast.

The regular methods for transmission of VLPs is through the cytoplasm amid hyphal anastomosis (hyphal combinations) or by section into the agamic spores. VLPs can likewise enter the sexual spores of some Basidiomycota and in Saccharomyces, yet this is by all accounts uncommon in the sexual spores of mycelial Ascomycota.

VLPs are inhabitant hereditary components of parasitic fungi since they have no normal instrument for intersection species hindrances. For quite a long while this made an issue in deciding their capacities, on the grounds that the relationship amongst VLPs and phenotypic characters was just correlative. In any case, two noteworthy advancements have changed this and opened the field to basic examination. To start with was the disclosure that infection like dsRNA can be available in growths notwithstanding when VLPs are missing. In these cases it can be accepted that the infection has lost the capacity and the need, to deliver a capsid. Second, protoplasting strategies and change frameworks have now been created for a few parasitic fungi, so that dsRNA can be extricated, decontaminated and brought into protoplasts, or reciprocal DNA (cDNA) can be gotten from dsRNA in vitro and after that changed into protoplasts. These methodologies, examined later in this section, have demonstrated that the viral dsRNA of Saccharomyces and a few different yeasts can make the cells create executioner poisons which follow up on different strains of similar species. [22]

In the chestnut curse growth Cryphonectria parasitica, additionally talked about later, the dsRNA causes a stamped lessening in pathogenic harmfulness, making hypovirulent strains that can conceivably be utilized to control the genuine chestnut scourge illness. In acknowledgment of the one of a kind properties of viral dsRNA and its part in diminishing pathogenic harmfulness, another name has been endorsed for this gathering of infections – the hypovirus gathering (Hypoviridae).

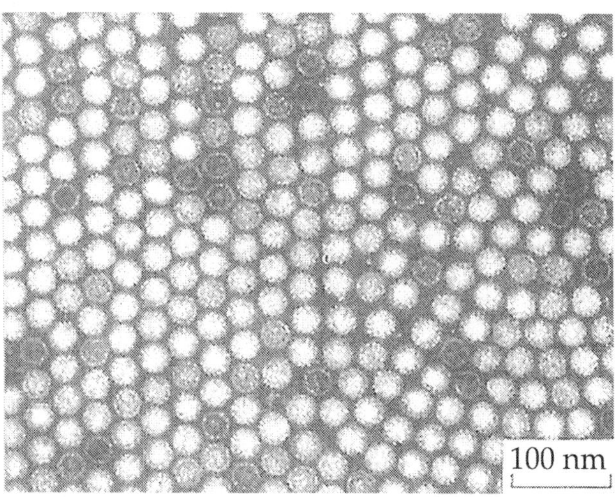

Figure- 7.4: Isomeric virus like particles extracted from hyphae of Colletotrichum sp.

GENETIC VARIATION IN FUNGI

Non-sexual Variety: the Noteworthiness of Haploidy

Change is the premise of all variety, yct transformations are communicated and recombined in various routes relying upon the science of a life form. A standout amongst the most critical elements of parasitic fungi is that they have a haploid genome, while all other real gatherings of eukaryotes are diploid. [24]

Haploid life forms ordinarily uncover every one of their qualities to choice weight. Any change will either cause lost wellness, or an expansion in wellness (e.g. anti-infection or fungicide resistance). This can be helpful for the time being however the inconvenience is that haploid life forms can't aggregate changes that are not of quick specific esteem. Diploid living beings have precisely the inverse components. Transformations frequently are passive to the wild sort, so they are not quickly communicated, rather they collect and can be recombined in different courses amid sexual intersection.

Be that as it may, mycelial organisms ordinarily have a few haploid cores in a typical cytoplasm, thus passive changes can be protected from determination weight, being supplemented by the wild-sort cores. Mycelial growths can likewise open their qualities to determination weight occasionally – at whatever point they create uninucleate spores or when hyphal branches create from just a single "originator" core. At the end of the day, mycelial parasitic fungi have a large portion of the upsides of both haploidy and diploidy. This is not valid for the Oomycota, which are diploid. [25] The circumstance is distinctive again for yeasts in light of the fact that these develop as uninucleate cells. A few yeasts (e.g. Candida spp.) are forever diploid, and even Saccharomyces develops as a diploid yeast in nature, attributable to mating-sort exchanging.

Non-Sexual Variety: Heterokaryosis

Heterokaryosis is characterized as the nearness of at least two hereditarily distinctive cores in a typical cytoplasm (hetero = diverse, karyos = bit, or core). Parasitic fungithat show this are named heterokaryons, rather than homokaryons which have just a single atomic sort. [26] Heterokaryons can emerge in two ways. To begin with, when a transformation happens in any of the cores of a hypha and the changed core multiplies alongside the wild-sort cores. This must happen all the time, yet a steady, utilitarian heterokaryon will grow just if the hereditarily distinctive cores multiply in the apical cells so that all the recently shaped hyphae contain both sorts. A moment path in which heterokaryons emerge is by tip-to-tip combination (anastomosis) of the hyphae of two strains. Once more, the cores would need to multiply in the apical cells to frame a stable heterokaryon. [27]

Table- 7.3: Impacts of growth medium composition on the ratio of nuclear forms in a heterokaryon

Composition of medium		**% of nuclei in the heterokaryon**		**Relative growth rates of homokaryons A and B**
Minimal nutrients	Apple pulp	Type A	Type B	A : B
0	100	8.6	91.4	0.47 : 1
20	80	7.8	92.2	0.53 : 1
40	60	11.1	88.9	0.54 : 1
60	40	12.7	87.3	0.67 : 1
80	20	13.5	86.5	1 : 1
100	0	51.8	48.2	1.56 : 1

Most exploratory examinations on heterokaryosis have included the matching of strains with characterized changes, for example, amino corrosive auxotrophs. The heterokaryon will then carry on as a prototroph, equipped for developing on negligible medium. The most intriguing element in these cases is that the proportion of atomic genotypes can fluctuate inside wide cutoff points and is affected by ecological conditions. [28] Along these lines, from a certain point of view, a solitary heterokaryotic strain can change the recurrence dissemination of its atomic sorts in light of choice weight. A heterokaryon of the apple-decay parasite, Penicillium cyclopium was built by permitting two diverse homokaryons to intertwine. At that point the heterokaryon was developed on agar containing diverse extents of apple mash or negligible supplement medium. The proportion of atomic sorts (which we will call An and B) in the heterokaryon was surveyed by testing irregular examples of uninucleate (homokaryotic) spores from the state. At the point when the heterokaryon was developed on apple-mash medium the extent of B-sort cores was high, however as the measure of apple mash was brought down so the extent of A-sort cores expanded, and drastically so when the heterokaryon was developed on insignificant medium. [29] Different analyses of this sort have demonstrated that the atomic proportion in a heterokaryon can fluctuate by up to 1000: 1 in either heading. On the off chance that this occurs at all usually in nature it would contribute fundamentally to persistent variety and determination of the best adjusted atomic proportion.

HOW DO HETEROKARYONS SEPARATE?

Heterokaryons can separate in two ways – either amid the creation of uninucleate spores, or when branches emerge that contain just a single atomic sort. Numerous normal organisms create uninucleate spores, regularly by rehashed mitotic division of a "mother" core in a phialide – Aspergillus, Penicillium, Trichoderma, Gliocladium, and so forth. Multinucleate spores can likewise be delivered from phialides, if a solitary core enters the creating spore and after that partitions to create a few cores. For instance, this is seen in numerous Fusarium species. Be that as it may, some other parasitic fungi(e.g. Neurospora) deliver conidia specifically from multinucleate hyphal tips or buds, and these spores will be either homokaryotic or heterokaryotic, contingent upon whether the phones that created them were homokaryotic or heterokaryotic. [30]

Heterokaryons additionally separate if a branch emerges that, by shot, contains just a single atomic genotype. This branch can deliver additionally branches and inevitably offer ascent to a homokaryotic segment of the state. [31] On the off chance that the homokaryon is supported more than the heterokaryon in the overall condition then it will grow to possess logically a greater amount of the settlement edge, if not supported it will be smothered. In a case of this, where a parasitic province was at first obscurely pigmented (with day by day zones of white elevated hyphae). Two light-hued areas soon created and due to their speedier development they logically extended. This kind of sectoring is frequently observed in contagious provinces – either as contrasts in pigmentation or sporulation. In any case, areas can likewise vary in pathogenicity or biochemical properties, and these frequently go undetected.

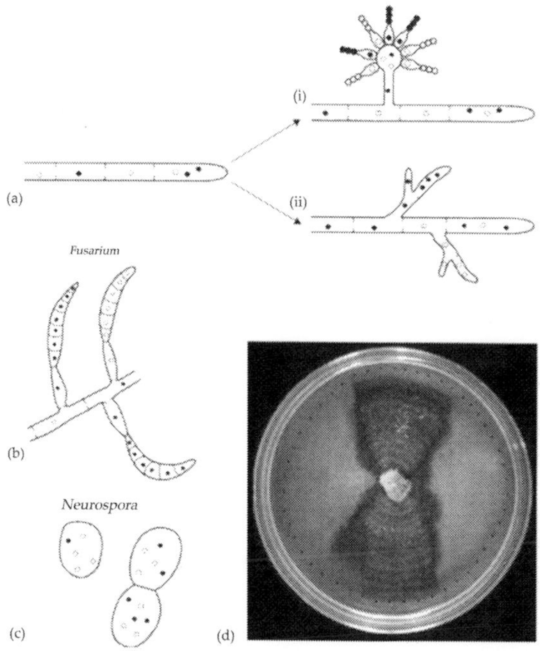

Figure- 7.5: Homokaryosis and heterokaryosis.

Essentialness of Heterokaryosis

Heterokaryosis is a conceivably effective wonder, however alert is required on the grounds that the degree of heterokaryosis in indigenous habitats is generally obscure. Numerous research facility thinks about have included matched auxotrophic mutants, and these can be viewed as "constrained heterokaryons" – there is a solid determination weight to keep up the heterokaryotic condition. [32] Likewise, there are noteworthy obstructions to the production of heterokaryons in nature, on the grounds that numerous organisms have atomic encoded "heterokaryon inconsistency" (het) quality loci. For instance, Neurospora crassa has no less than 10 such loci, with two alleles at every locus, so there are conceivably 210 (1024) diverse "vegetative" similarity gatherings (VCGs). [33] Pairings of strains of various VCGs prompt hyphal combination at the purposes of contact, trailed by various degrees of cytoplasmic incongruence, depending in any event incompletely on the quantity of het loci that any two strains have in like manner. Normally, the cytoplasm of the intertwined cells passes on bringing about a reasonable outline zone between contradicting states.

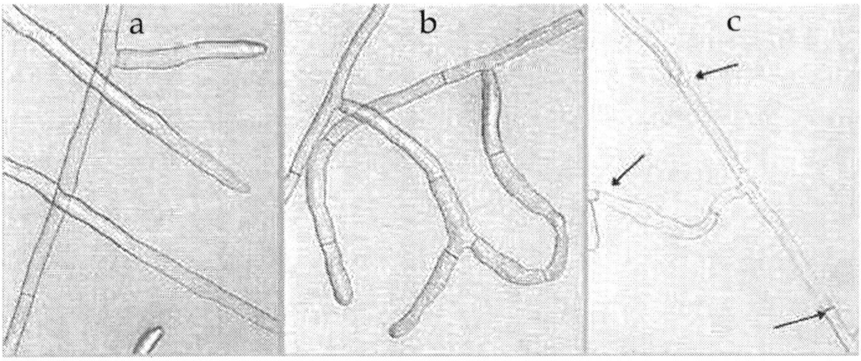

Figure- 7.6: Anastomosis reactions of Rhizoctonia.

Non Sexual Variety: Parasexuality

Numerous normal mitosporicparasitic fugi, for example, Aspergillus, Penicillium, Fusarium, and Trichoderma, appear to be to a great extent to have relinquished sexual propagation, on the grounds that their sexual stages are not found in research center conditions and are discovered just occasionally, if by any means, in nature. For instance, the regular growths, Aspergillus fumigatus and *A. niger* have never been found to deliver a sexual stage. This leads us to solicit whether they have elective components from hereditary recombination, and the appropriate response is by all accounts that they do have this potential through a system named parasexuality (or the parasexual cycle). [34]

Parasexuality was found by (Pontecorvo 1956), amid ponders on heterokaryosis in Emericella (Aspergillus) nidulans. He developed a heterokaryon from two parental strains that had markers at two quality loci (we will call the strains Ab and aB) and was investigating the homokaryotic spores created by the heterokaryon. Obviously, the greater part of the spores had cores of the "parental" sorts, either Ab or aB, however a noteworthy number were observed to be recombinants (AB or stomach muscle) and their frequencies were too high to be clarified by change. Clearly, the qualities had recombined in the heterokaryon, in spite of the fact that this can't happen by heterokaryosis alone in light of the fact that the cores stay as particular substances paying little mind to how they are blended in the cytoplasm. [43] Facilitate examination drove Pontecorvo to propose a parasexual cycle, including three phases:

Diploidization: Occasionally, two haploid cores breaker to frame a diploid core. The instrument is to a great extent obscure, and this is by all accounts a moderately uncommon occasion, however once a diploid core has been framed it can be exceptionally steady and separation to shape encourage diploid cores, alongside the ordinary haploid cores. Hence the heterokaryon comprises of a blend of the two unique haploid atomic sorts and in addition diploid combination cores.

Mitotic chiasma development: Chiasma arrangement is basic in meiosis, where two homologous chromosomes break and rejoin, prompting chromosomes that are crossovers of the parental sorts. It can likewise happen amid mitosis however at a much lower recurrence in light of the fact that the chromosomes don't match in a normal plan. By the by, the outcome will be a similar when it occurs – the recombination of qualities. [35]

Haploidization: Occasionally, nondisjunction of chromosomes happens amid division of a diploid core, with the end goal that one of the little girl cores has 2n + 1 chromosomes and alternate has 2n – 1 chromosomes. [36] Such cores with fragmented products of the haploid number are named aneuploidy (rather than euploid cores, with n or finish products of n). [44] They have a tendency to be temperamental and to lose advance chromosomes amid consequent divisions. So the 2n + 1 core would return to 2n, though the 2n – 1 core would continuously return to n. Predictable with this, in E. nidulans (n = 8) cores have been found with 17 (2n + 1), 16 (2n), 15 (2n – 1), 12, 11, 10, and 9 chromosomes. [37]

It must be stressed that each of these occasions is moderately uncommon, and they don't constitute a customary cycle like the sexual cycle. In any case, the result would be comparable. Once a diploid core has framed by combination of two haploid cores from various guardians, the parental qualities can possibly recombine. Also, the chromosomes that are lost from an aneuploid core amid its inversion to euploid could be a blend of those in the parental strains.

Noteworthiness of Parasexuality

Parasexuality has turned into a significant instrument for modern mycologists to create strains with sought blends of properties. However its essentialness in nature is to a great extent obscure and will rely upon the recurrence of heterokaryosis, dictated by cytoplasmic inconsistency boundaries. Accepting that heterokaryosis occurs, we can inquire as to why a few (clearly fruitful) growths have surrendered a proficient sexual system of hereditary recombination for a more arbitrary and apparently less effective process. The appropriate response may be that the parasexual occasions can happen whenever amid typical, substantial development and without any preconditions like those for sexual generation. Albeit each phase of the parasexual procedure is generally uncommon, there are a large number of cores in a solitary province, so the odds of the parasexual cycle happening inside the state all in all might be very high. [38]

Sexual Variety

Sex is the real system for delivering hereditary recombinants, through traverse (chiasma arrangement) and free combination of homologous chromosomes amid meiosis. A parasite, for example, Emericella (Aspergillus) nidulans, with eight chromosomes, could produce 28 (i.e. 256) diverse chromosome blends by autonomous grouping alone. This would rely upon a proficient outcrossing system.

Numerous growths are heterothallic (outcrossing), requiring the combination of cells of two distinctive mating sorts. However, some are homothallic (e.g. most Pythium spp., around 10% of Ascomycota, and a couple of Basidiomycota) and some display auxiliary homothallism – the sexual spores are binucleate with one core of each mating sort (e.g. Neurospora tetrasperma and Agaricus bisporus). Another variety is found in Saccharomyces cerevisiae and the indirectly related parting yeast, Schizosaccharomyces pombe. Both of these experience mating-sort exchanging. [39] These minor departure from the typical components of outcrossing have likely developed in light of the fact that the sexual spores of organisms work as torpid spores to survive unfavorable conditions. At any rate for the time being, survival is more critical than sex!

Atomic ways to Deal with Populace Structure

Parasitic species are dynamic substances. Their populaces section by geological separation or the advancement of substantial inconsistency boundaries, at that point the pieces veer by hereditary float or because of nearby choice weight. [40] This is especially valid for the clonal mitosporic organisms, since strains of various vegetative similarity gatherings (VCGs) are segregated for all time from each other. Numerous sexual species additionally have VCGs yet the mating-sort qualities abrogate the physical inconsistency qualities so that strains of various VCGs can mate.

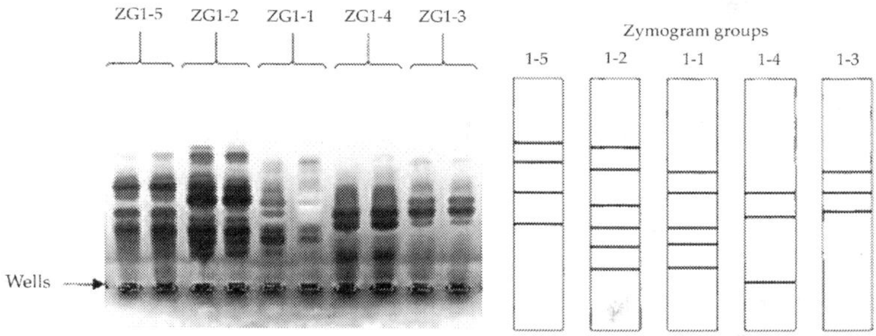

Figure- 7.7: Five diverse groups of zymogen of Rhizoctonia strains caused bare patches of wheat in Austrailia.

When all is said in done, organisms have excessively couple of morphological components for distinguishing proof of populace subunits, so biochemical and sub-atomic instruments must be utilized for this. [41] One approach is to think about the electrophoretic banding examples of proteins on gels, utilizing either add up to protein separates or distinctive structures (isozymes) of specific catalysts, for example, pectic chemicals, envisioned on the gels by shading responses with the compound substrate. These zymograms reflect irregular transformations in the DNA encoding the protein, in spite of the fact that not at the compound dynamic site which is much rationed. Around 30% of the amino corrosive changes coming about because of transformations will influence the net charge on the protein and therefore adjust its electrophoretic versatility. Since these progressions are irregular, they have a tendency to amass after some time and along these lines mirror the historical backdrop of a populace. As one functional case, MacNish et al. (1993) utilized a blend of pectic zymogram and VCG writing to recognize distinctive subgroups of the dirt borne growth Rhizoctonia solani that causes exposed fix (hindering) sickness of wheat in Australia. All contagious confines from inside every ailment fix were of indistinguishable VCG + zymogram gathering, however unique fixes regularly spoke to various VCG + zymogram gatherings. [42]

The patches can blend, yet never cover in light of the fact that the subgroups appear to be not able attack one another"s region. The patches (and in this way the parasitic populaces) are likewise powerful – they can grow, contract or even vanish in various trimming seasons. At the point when the sickness fix vanishes the organism can never again be found in the soil.

Contagious populace structure can likewise be examined by the utilization of confinement compounds (endonucleases) on DNA removed from cells. Any one limitation protein (e.g. BamH1) will cut the DNA at particular "target" focuses. For instance, BamH1 cuts DNA at locales where the nucleotide succession GGATCC happens (with CCTAGG on the integral strand of DNA), giving parts of various lengths that band on gels as per their size. The banding designs are named confinement section length polymorphisms (RFLPs). [45] They resemble fingerprints, reflecting aggregations of point transformations that change single nucleotides inside the succession perceived by the limitation chemical, or chromosomal revisions that change the relative places of these arrangements.

Diverse catalysts give distinctive RFLP designs since they cut the DNA at various destinations. They likewise give distinctive quantities of parts, since some perceive four-base successions which are more typical than, say, six-base groupings. Additionally, these chemicals can be utilized on mitochondrial DNA, which is more exceptionally rationed than the aggregate DNA, so extraordinary levels of affectability can be chosen to break down both minor and real changes inside a parasitic populace. Kohn in 1995 depicted a decent case of this approach for contrasting the intercontinental and intracontinental clonal subgroups of the plant pathogen Sclerotinia sclerotiorum in both wild and rural plant groups. [46]

The polymerase chain response (PCR) can be utilized to create symptomatic tests for particular growths. The least difficult and most basic approach includes the irregular intensification of DNA from a rough DNA separate by including an arbitrary groundwork made out of, for instance, 10 nucleotides. This tempers to a corresponding nucleotide arrangement on the separated DNA. At that point a DNA polymerase reaches out along the DNA, perusing the arrangement of bases along from the preliminary. This DNA is then intensified amid 25–40 progressive rounds of PCR. [47] The method is known as RAPD, articulated "rapids" (irregular opened up polymorphic DNA). Then again, the DNA that codes for specific proteins can be focused with preliminaries in view of information of the incomplete amino corrosive arrangement of the protein. Regardless, chose sections of the opened up DNA (in single-stranded shape) can be reasonably labeled and utilized as indicative tests that will tie to proportionate single-stranded DNA successions in concentrates of an example. [48]

Tests of this sort are accessible industrially for identifying a few individual plant pathogens (Fox 1994). For examination of populace structure, both RFLP what's more, RAPD have been utilized to recognize the diverse pathogenic strains of Ophiostoma spp. that cause Dutch elm malady, following the starting point and advance of the current Dutch elm malady pandemics (Pipe et al. 1995). As a further case, PCR-based techniques have been created to identify, and measure, the levels of a few critical mycotoxins in nourishment items (Edwards et al. 2002). Atomic instruments are being produced quickly in clinic look into research centers, for the sub-atomic writing of organisms that cause human ailments. One of the primary drivers for this is to have the capacity to recognize singular strains or subgroups inside a contagious animal varieties, and subsequently to help epidemiological investigations. [49]

The qualities encoding ribosomal RNA are generally utilized for distinguishing parasitic fungi and for developing phylogenetic trees. Ribosomal RNA (rRNA) is required in such huge adds up to deliver the cell ribosomes that there are various duplicates of the rRNA qualities, frequently organized couple however isolated from each other by untranscribed spacers. Each single rRNA quality has coding data for the three sorts of rRNA found in eukaryotic ribosomes (18S, 28S), additionally contains other significant data, particularly in the inside interpreted spacers (ITS) and remotely deciphered spacer (ETS). The rRNA quality at first creates a prerRNA, which at that point experiences handling, including the extraction of the spacer areas, to deliver the three "develop" rRNAs. [50] The 18S rRNA has changed adequately finished transformative time to be utilized as a sort of "sub-atomic clock." But the ITS areas are significantly more factor, and can frequently be utilized to recognize distinctive species. This is done through PCR, utilizing groundworks that quandary to the 5′ end of every DNA strand and advance towards the 3′ end. The nucleotide successions in the ITS locales of the built DNA would then be able to be contrasted and the known groupings of various parasitic fungi. [51]

Connected Sub-Atomic Hereditary Qualities of Organisms

In this area we think of some as cases of atomic methodologies for understanding parasitic conduct or for immediate, functional applications. The cases must be illustrative however a large number of them speak to earth shattering work and cover some naturally intriguing issues.

Generation of Heterologous Proteins in *Saccharomyces Cerevisiae*

The accessibility of effective vectors has empowered *S. cerevisiae* to be utilized as an "industrial facility" for the results of numerous outside (heterologous) qualities. A standout amongst the most critical items to have risen up out of this is an antibody against hepatitis B. [52] It comprises of one of the viral coat proteins, Hepatitis B surface Antigen (HBsAg), and was created by changing yeast cells with the viral quality encoding this protein. Hepatitis B antibody was the main hereditarily designed immunization to be endorsed for use in people (it contains no qualities). Teacher Sir Ken Murray was knighted for his work in building up this antibody. Numerous different proteins have been delivered tentatively from yeast, including cellulases, amylases, interferon, epidermal development figure, and β-endomorphin. [53]

Be that as it may, there have additionally been issues in utilizing Saccharomyces for heterologous protein generation. Specifically, yeast has a moderately poor capacity to expel introns from outside qualities (its own particular introns are few and little), so it is most effective when changed with reciprocal DNA (cDNA) gotten in vitro from the dispatcher RNA of a protein (the introns are grafted out amid the preparing of mRNA). Yeast likewise neglects to perceive the promotor arcas of the qualities of other parasitic fungi, and it doesn't generally loyally glycosylate remote proteins. This can be critical in light of the fact that few bioactive proteins are glycoproteins that rely upon the sugar chains for their movement.

These troubles have served to exhibit that S. cerevisiae is hereditarily very not the same as the mycelial growths. For instance, E. nidulans can perceive the promoter groupings of the qualities of other parasitic fungi and furthermore can extract their introns. [54] It might be conceivable to utilize E. nidulans or the splitting yeast Schizosaccharomyces as another option to Saccharomyces for heterologous protein creation. In any case, as a rule, it is presently suspected that the best approach is to utilize cell lines identified with the normal maker creature – mammalian cell lines for mammalian quality items. [55]

Distinguishing proof of qualities for Plant Pathogenicity and Separation

We found how separation particular mRNAs were recognized in Schizophyllum cooperative by looking at the mRNA profiles of societies developed in conditions where fruitbodies were or were not created. The particular mRNAs would then be able to be utilized as formats to create cDNA in vitro. This cDNA, created from marked nucleotides, turns into a test for authoritative to corresponding arrangements of removed chromosomal DNA. [56] Along these lines a particular quality can be recognized regardless of the possibility that nothing is thought about the fundamental hereditary qualities of a parasite. By utilizing such procedures, Wessels and his collaborators could portray an exceptional gathering of proteins, the hydrophobins, which have significant parts in contagious science and separation. These procedures of "turn around hereditary qualities" have likewise made it conceivable to perform focused on quality disturbance. For the parasitic quality of intrigue, a cDNA is created in vitro and disturbed to make it nonfunctional. At that point it is changed into the parasite, to supplant the first quality by homologous recombination. The accompanying case demonstrates the energy of this procedure.

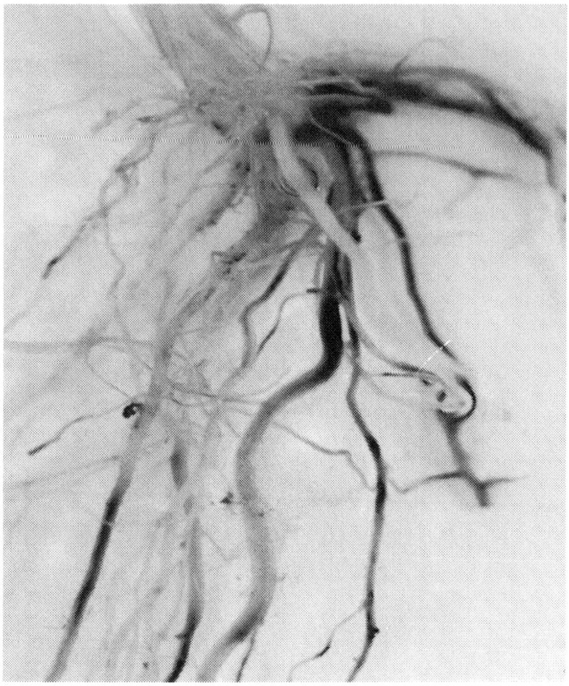

Figure- 7.8: Wheat root infected by fungus, Gaeumannomyces graminis.

Parts of Pre-Shaped Inhibitors in Plant Imperviousness to Pathogens

The take-all parasite of grains, *Gaeumannomyces graminis*, develops on the surface of oat roots by dull "runner hyphae" at that point enters the root cortex and enters the vascular framework, causing interruption of the sugar-directing phloem cells, and blockage of the water-leading xylem vessels. In the event that enough roots are executed along these lines amid the developing season then the plants kick the bucket rashly, with genuine loss of grain yield. *G. graminis* has two primary pathogenic structures – assortment tritici (GGT) which assaults wheat roots yet not oat roots, and assortment avenae (GGA) which assaults both wheat and oats. This distinction was clarified in the 1960s when oat attaches were found to contain pre-framed inhibitors. [57]

The strongest of these inhibitors is avenacin A, a saponin (cleanser like compound) which consolidates with sterols in the parasitic layer, making particle penetrable pores. In research center culture both GGT and GGA develop promptly in watery concentrates of wheat roots, yet GGT is completely smothered by fluid concentrates of oat roots while GGA is unaffected by them. [58] The rcason was proposed to be that GGA detoxifies avenacin, cutting the terminal sugars from the particle, by creating a glycosidase compound named avenacinase. This protein would in this way be a key pathogenicity determinant, enabling GGA to stretch out its host range to oats.

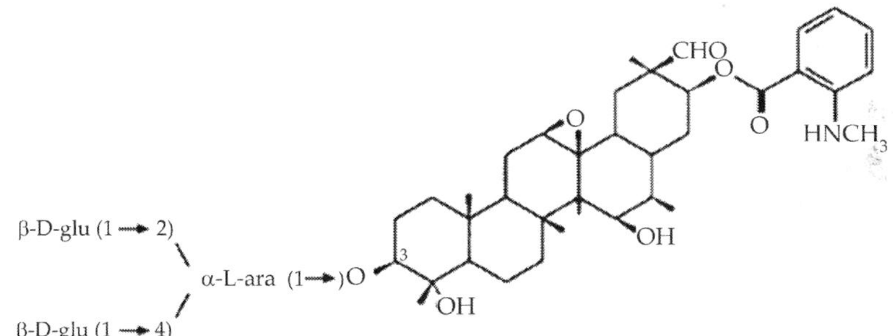

Avenacin A-1

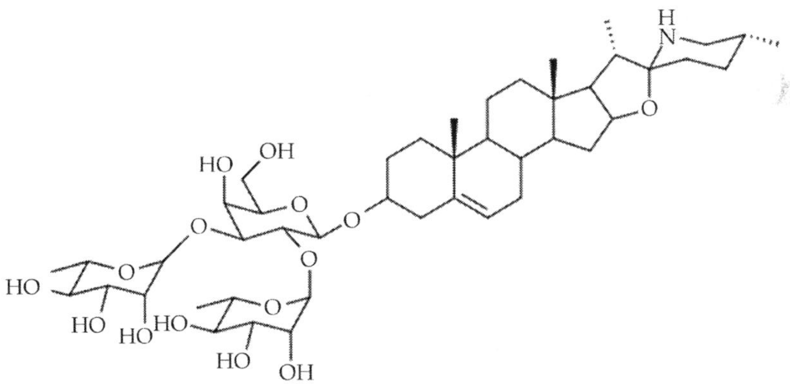

Tomatine

Figure- 7.9: Two saponins structures that are resistant compounds in plants.

This was appeared to be the situation by focused disturbance of the avenacinase quality. GGA was changed with a checked, upset cDNA of the quality, which embedded at the quality site by homologous recombination and made the parasite lose its pathogenicity to oats, while holding its ordinary pathogenicity to wheat. Yet, when the checked cDNA had embedded somewhere else in the genome (nonhomologous recombination) the growth was still completely pathogenic to oats. Stretching out from this work, Osbourne et al. inspected other host–pathogen frameworks where saponins have been involved as plant resistance variables. Specifically, tomatoes are known to contain the saponin α-tomatine, and a pathogen of tomatoes, Scptoria lycopersici, is known to detoxify this by separating a solitary sugar from the atom. [59] The chemical mindful (tomatinase) was observed to be fundamentally the same as avenacinase: it was perceived against avenacinase counter acting agent, and cDNA of avenacinase hybridized with DNA segments of S. lycopersici, apparently by perceiving the quality for tomatinase. This may be clarified by the way that all the saponin-detoxifying compounds are β-glycosidases, with maybe some normal structure, reflected in the DNA successions. Be that as it may, avenacinase (from GGA) had next to no capacity to detoxify tomatine, and tomatinase (from S. lycopersici) had immaterial capacity to detoxify avenacin. So it appears that these pathogenic parasitic fungi have developed saponindetoxifying compounds with very particular movement against the saponins of their hosts. [60]

YEAST EXECUTIONER FRAMEWORK

With this theme we come back to the part of viral dsRNA. Singular types of no less than eight genera of yeasts (Saccharomyces, Candida, Kluyveromyces, and so forth.) have been found to contain executioner strains. These strains discharge a protein that executes different strains of similar species however not of irrelevant species. The poison ties to a receptor on the mass of defenseless cells, at that point goes to the layer where it causes spillage of H^+, and the cells bite the dust attributable to loss of transmembrane potential and disturbance of amino corrosive take-up, K+ adjust, and so on. The poisons are steady just at low pH and are thought to give a huge preferred standpoint to executioner strains over nonkiller strains of similar species in acidic situations.

In Kluyveromyces lactis the poison is encoded by a direct DNA plasmid, however in all different cases it is encoded by dsRNA. Both the executioner and nonkiller strains can contain infection like particles (VLPs) so this component alone does not connect with executioner movement. Notwithstanding, these particles are observed to be of two sorts: the dsRNA in the "L sort" encodes the infection coat protein, while the dsRNA in the "M sort" codes for the poison. The M dsRNA relies upon the L dsRNA for the coat protein, however the L dsRNA can happen alone with no impact on the cells.

The atomic science of the executioner framework in S. cerevisiae has been examined seriously and has indicated why the poison makers are not influenced by their own poison. They deliver the poison as an extensive antecedent protein which experiences changes amid section through the secretory framework to create a protoxin. This protoxin is at last separated at the cell layer, to discharge the dynamic poison however leave some portion of the particle in the film. This remaining part appears to associate with a poison receptor, making the cell impervious to dynamic poison in the outer condition. Comparable dsRNA executioner frameworks are found in the yeast period of Ustilago maydis (Basidiomycota) which causes filth infection of maize. An executioner framework may likewise happen in the take-all organism Gaeumannomyces graminis, on the grounds that some dsRNA-containing strains can particularly hinder the development of different resist low pH. However the instrument for this situation stays obscure. [61]

Hypovirulence of Plant Pathogens: the Control of Chestnut Scourge

Chestnut scourge, caused by the growth Cryphonectria parasitica (Ascomycota), has wrecked a large number of trees in North America, following the primary report of its event in the New York Zoological Garden in 1904. Its dynamic spread over the eastern portion of the USA diminished the local American chestnut (Castanea dentata) practically to the status of an understorey bush, though it was before a heavenly "high backwoods" tree. In any case, late examinations raise the prospect that chestnut curse may be controlled by hereditary control of hypovirus-related dsRNA, talked about underneath.

Cryphonectria contaminates through injuries in the bark of chestnut trees, and after that spreads in the cambium, logically supporting the stem and executing the plant over the disease point. This sickness was likewise a huge issue in European sweet chestnut trees (Castanea sativa) in the 1940s, yet in Italy in the mid 1950s some vigorously pervaded sweet chestnut estates started to recoup suddenly. The sores in these trees quit spreading round the trunks, and strains of *C. parasitica* detached from them indicated unusual components. They developed gradually and sporadically on agar, were white instead of the typical orange shading, delivered altogether less conidia, and indicated just low harmfulness when twisted immunized into trees. This low harmfulness (hypovirulence) could be transmitted to different strains amid hyphal anastomosis on agar, so it was begat transmissible hypovirulence. None of the hypovirulent strains contained "ordinary" infection like particles, however all the hypovirulent strains contained dsRNA while destructive strains had no dsRNA. Electron micrographs demonstrate that this dsRNA is contained in adjusted or clubshaped, film bound vesicles in the cytoplasm and that, not at all like numerous VLPs, these can happen in noteworthy sums in the hyphal apices of *C. parasitica*.

Figure- 7.10: Sunken canker on a tree of Quercus frainetto.

The unconstrained malady decrease in Italy drove French laborers to build up an exceedingly effective natural control program. Hypovirulent strains of *C. parasitica* were refined in the research facility and vaccinated at the growing edges of ulcers in the field. Inside a brief span the ulcers quit developing, and just hypovirulent strains could be recouped where once the harmful strain had been. This transmissible change of phenotype was constantly joined by the transmission of dsRNA.

Following the accomplishment of this program in Europe, endeavors were made to bring hypovirulent strains of Cryphonectria into the USA. In any case, they gave just halfway and confined illness control, on the grounds that the pathogen populace in the USA comprises of various vegetative similarity gatherings (VCGs) that breaking point the characteristic exchange of dsRNA, inferable from cytoplasmic passing when the strains anastomose. In one early examination around 0.5 hectare of characteristic chestnut woods in the USA was found to contain no less than 35 VCGs. All the more as of late the VCGs have been appeared to be in a persistent condition of flux: samplings of indistinguishable trees more than quite a while demonstrated that a portion of the transcendent VCGs declined, while new ones emerged (Anagnostakis 1992). The European populace of C. parasitica is substantially more uniform as far as VCGs, and this likely records for the achievement of the biocontrol program in Europe. Advance in understanding the part of dsRNA in Cryphonectria was altogether postponed by the absence of a reasonable change framework for this organism.

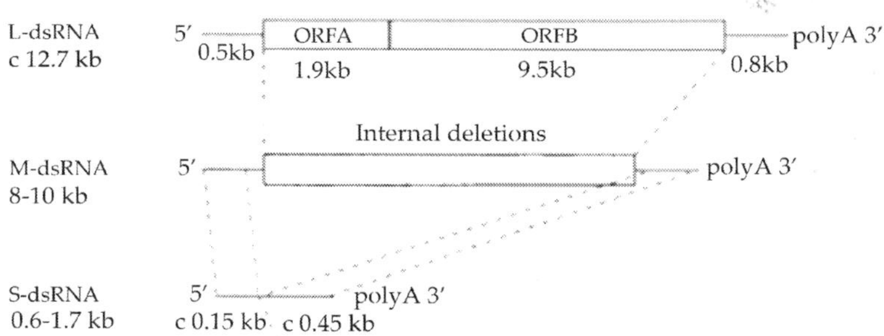

Figure- 7.11: Double-stranded RNA of hypovirulent strains of C. parasitica.

Be that as it may, when a framework was in the end created it prompted fast advance (Nuss 1992). The dsRNA of C. parasitica was observed to be profoundly factor, with lengths falling into three wide size extents – S (little), M (medium), and huge (L, around 12.7 kb). As appeared in, every one of these sorts have a similar terminal districts – a poly-A tail at the 3′ end and a 28-nucleotide preserved grouping at the 5′ end. They vary predominantly in the level of inner erasure. The L-shape is by all accounts the full-length particle, and the littler structures are flawed (apparently functionless) subordinates which gather in the hyphae. This high level of inconstancy of RNA genomes is not uncommon in light of the fact that there is no viable editing framework for RNA, to guarantee that it is duplicated steadfastly.

With the advancement of a change framework, Nuss and his associates could create a cDNA from the full-length dsRNA and change it into harmful strains. It made the harmful strains progress toward becoming hypovirulent and to display every one of the components normal of hypovirulent strains – moderate development, pale shading, and decreased sporulation (the hypovirulence-related qualities). In this manner, dsRNA was demonstrated unequivocally to be the reason for hypovirulence. The cDNA could likewise be utilized for sub-atomic investigation of the dsRNA, which was appeared to comprise of two open perusing outlines (ORFs) – ORF An of 622 codons (nucleotide triplets) and ORF B of 3165 codons. At the point when a cDNA duplicate of ORF A was brought into Cryphonectria, it caused lost pigmentation, a decrease of sporulation and a lessening of laccase action. Be that as it may, the changed strain stayed destructive. By differentiate, ORF B codes for a polyprotein that was found to contain themes normal for RNA-subordinate RNA polymerases and RNA helicases. In this way, the essential capacity of ORF B could be the replication and upkeep of the dsRNA, causing the loss of pathogenic destructiveness, while ORF An encodes a large portion of the hypovirulence-related qualities (diminished sporulation, and so on.) that are conceivably disadvantageous in a biocontrol strain, lessening its natural wellness. It might be conceivable, along these lines, to control the cDNA in vitro with the goal that it has just the most alluring qualities for biocontrol.

It is remarkable that the cDNA, when changed into Cryphonectria (making the parasite be hypovirulent), turned out to be steadily coordinated in the chromosomal genome so it was duplicated alongside the other chromosomal qualities. This cDNA was kept up for the duration of the life cycle, notwithstanding entering the sexual spores, while dsRNA rarely enters the sexual spores of Cryphonectria or other Ascomycota. This would imply that biocontrol strains could be delivered with perpetual, stable hypovirulence, subject to editing like whatever remains of the chromosomal qualities.

Current developments

The inquiry emerging from the steady reconciliation of "hypovirulencc" cDNA is: can there still be cytoplasmically transmitted hypovirulence? The appropriate response is by all accounts "yes" in light of the fact that the chromosomally coordinated cDNA is interpreted into dsRNA instead of into single-stranded mRNA (as in whatever remains of the genome) and this dsRNA amasses in the cytoplasm, where it can be transmitted between strains amid hyphal anastomosis. This still leaves unanswered the subject of how dsRNA smothers destructiveness. Preparatory proof proposes that it does as such by downregulating a portion of the typical chromosomal qualities, including the quality for creation of laccase, a compound required in lignin breakdown and which is probably going to be huge for an organism that colonizes wood.

At the point when the hypovirulence cDNA (gotten from dsRNA of C. parasitica) was changed into other cankerforming Cryphonectria species, and into a less firmly related Endothia species, it neglected to change over these to hypovirulence. Be that as it may, when a similar cDNA was utilized as a layout to deliver RNA and this was transfected into the other parasitic fungi, it offered ascend to full-length dsRNA in the cytoplasm and caused both a stamped lessening of harmfulness and an adjustment of the development rate and pigmentation (Chen et al. 1994). The achievement accomplished with RNA yet not cDNA is by all accounts clarified by the way that C. parasitica produces RNA from the cDNA yet then grafts this RNA to erase a 73-base grouping before the RNA can go about as a layout for dsRNA generation.

In outline, this work on the hypovirus arrangement of Cryphonectria has raised many issues of essential enthusiasm for parasitic hereditary qualities and additionally holding the possibility of growing new ways to deal with plant ailment control. Nonetheless, late investigations in Switzerland (Hoegger et al. 2003) recommend the requirement for a wary approach, in light of the fact that the main method for transmitting the hypovirus is by anastomosis, and both the atomic DNA and the mitochondrial DNA of the benefactor (biocontrol) strain are brought into the earth amid this procedure. The mitochondrial DNA of the benefactor strain was observed to be transmitted in almost one-portion of the treated infections. The atomic DNA likewise persevered in the treated infections however it didn't spread past them. The real issue that remaining parts to be settled is the long haul security of presenting hereditarily altered strains of pathogenic parasitic fungi into the earth, particularly if the atomic DNA is exchanged and expands the hereditary differences of the pathogen.

Back to Genome

The term genomics was authored in 1986 to depict the mapping, sequencing and examination of genomes, a definitive objective being to comprehend the structure, work, association, connections, and development of qualities. While many sequenced genomes of microscopic organisms and archaea have been distributed, there are moderately few distributed genome groupings ofparasitic fungi, albeit a few draft arrangements are accessible. All the genome successions produced by legislative or open financing bodies are made accessible on the web so analysts can think about various genomes and scan for likenesses and contrasts.

The fundamental strategies of genome sequencing are generally basic, and large portions of the methodology are computerized. Basically, the DNA to be sequenced (named the layout) is denatured by warmth or soluble base to create single-stranded DNA, and DNA polymerase is utilized to blend DNA in light of the nucleotide succession in the format. DNA polymerases require a district of twofold stranded DNA to start amalgamation. This is given by including a short single-stranded DNA particle – a preliminary, with a DNA grouping corresponding to the layout DNA. The preliminary ties to the format to shape a short area of twofold stranded DNA, from which whatever is left of the layout DNA is blended.

For sequencing ventures, generally long layout DNA arrangements are set up by cutting the DNA haphazardly with limitation proteins and embeddings them into plasmids or different vectors, for example, cosmids (cloning vectors that take after plasmids yet are bundled into λ phage capsids which can convey embeds up to 40 kb). The numerous lengths of sequenced DNA from various locales of the genome are then scanned for covering areas. Along these lines the DNA can be collected into contigs, speaking to ceaseless scope of the nucleotide grouping of entire chromosomes or locales of chromosomes. The genome grouping is then hunt down trademark elements, for example, promoters, translation start destinations, and "stop" codons, to reason the places of protein-encoding qualities. Practically speaking, a significant part of the era of genome groupings depends intensely on PC programs.

Most genome sequencing ventures include the entire genome shotgun approach, in which short successions are produced haphazardly instead of methodicallly and are then gathered into contigs. Normally, all aspects of the arrangement is secured no less than five times and regularly ten times (by covering of the sequenced areas). The outcome is a "top notch draft" of the entire genome, yet regularly around 2% of the genome can't be mapped precisely – where there are visit rehash groupings and DNA areas of high G + C (Guanine + Cytosine) content.

WHAT DO WE PICK UP FROM ENTIRE GENOME SUCCESSIONS?

Many advantages accumulate from looking at the genomes of various growths, or from contrasting contagious genomes and those of different living beings. To give only a couple of cases:

The nucleotide groupings of qualities can be utilized to foresee the protein successions, so robotized inquiries, for example, BLAST (fundamental neighborhood arrangement look apparatus) can recognize homologous qualities in various creatures (and any developmental changes in those qualities over various timeframes).

Nucleotide successions that don't show up in momentum databases may speak to qualities with unfamiliar capacities – the premise of "quality mining" for potential new proteins of business intrigue.

Many parts of human, creature and plant infection are as yet uncertain, so the explanation of qualities controlling these procedures could give new bearings to handling these issues.

Most importantly, as more quality capacities are found they add to the whole of learning, and since the vast majority of this data is unreservedly accessible the revelation of a recently described quality in one living being can "fill the crevices" in the genomes of different creatures. The sequencing of the principal eukaryotic genome (Saccharomyces cerevisiae) discharged in 1996, demonstrated that about portion of the open perusing outlines (ORFs) had no unmistakable homologs in distributed databases. This example has been rehashed time and in the genomes sequenced since that time. Genomics is "Huge Science." The sequencing of S. cerevisiae included 90 look into research facilities, and the paper portraying the genome arrangement of Neurospora crassa had 77 creators. Given the size of these undertakings, and the a large number of contagious genomes that could conceivably be sequenced, it is critical to organize and facilitate sequencing endeavors. As one case, the contagious research group of the USA has, since the year 2000, attempted expansive discussion and distributed a progression of "White Papers" on the Fungal Genome Initiative. Of 15 competitor parasitic fungi at first proposed, seven are as of now being sequenced. In 2003 the second White Paper incorporated a rundown of 44 extra parasitic fungi with accentuation on bunches of related species to advance similar genome examination:

In a field that is moving so quickly, and where subsidizing choices have still to be made, the FGI site (address above) is the most solid wellspring of data. In any case, Table 9.4 gives brief points of interest of the first 15 entries to outline the justification behind such sequencing endeavors. It included agents of all the major contagious phyla (Chytridiomycota, Zygomycota, Ascomycota, and Basidiomycota) and creatures in three classifications – those of medicinal importance, those of business essentialness, and those that would add to comprehension of advancement and parasitic differing qualities.

Huge Discoveries from the *Neurospora Crassa* Genome Succession

The astounding draft succession of the *N. crassa* genome was finished in 2003 and speaks to a point of reference – the finish of over 60 years of research on a standout amongst the most hereditarily very much portrayed parasitic fungi. The succession still needs additionally itemized work, to check potential errors and to join the current contigs, however as of now it has uncovered new data, including the recognizable proof of qualities possibly connected with light flagging and auxiliary digestion.

The primary components of the sequenced N. crassa genome are appeared in Table 9.5 (from Galagan et al. 2003). Among the more remarkable focuses is the anticipated nearness of more than 10,000 protein-encoding qualities, the greater part of which code for proteins of more than 100 amino acids. In any case, 41% of the Neurospora proteins need noteworthy matches to any of the known proteins out in the open databases, and 57% of Neurospora proteins need huge matches to qualities in either Saccharomyces cerevisiae or Schizosaccharomyces pombe. Another intriguing element of Neurospora is that it has the most stretched out scope of genome barrier components known for any eukayotic living being. One of these is a procedure obviously one of a kind to parasitic fungi, named repeatinduced point change (RIP). Tear was found in Neurospora quite a long while prior, as a procedure that viably avoids genome development.

The duplication of qualities is broadly perceived to be in charge of transformative advancement, in light of the fact that the quality duplicate is allowed to change and can in the long run expect another capacity while the first quality holds its capacity. However, this procedure is hindered in Neurospora amid the haploid dikaryotic period of the sexual cycle. The RIP procedure distinguishes and changes both duplicates of a copied quality by making various transformations from G-C A-T sets, and regularly prompts DNA methylation which causes quality hushing in Neurospora (as DNA methylation likewise does in warm blooded animals). The impacts of this can stretch out to adjoining qualities past the copied groupings. Predictable with this part of RIP, N. crassa has a bizarrely low extent of qualities in mutigene families, in connection to its genome size, and it has no very comparative quality sets. A few different lines of proof help the view that genome advancement in N. crassa has been generally captured since the obtaining of RIP eventually in its developmental history.

Table- 7.4: Fungi firstly suggested to practice the base of a synchronized genome sequencing effort

Organisms by category	Significance	Estimated genome (Mb)
Medicine		
Cryprococcus	Basidiomycota: It cause fatal meningitis in humans.	24 Mb on 11 chromosomes
Trichophyton rubrum	Ascomycota: Most common fungal infection of human skin.	12 Mb on 4 chromosomes
Rhizopus oryzae	Zygomycota: It cause infection of humans.	36 Mb

Commerce		
Magnaporthe grisea	Rice blast disease: Model fungal pathogen of plants.	40 Mb on7 chromosome
Aspergillus flavus	Ascomycota: Cause human aspergillosis and a source of aflatoxin.	40 Mb on 8 chromosome
Fusarium gra-minearum	Mitosporic fungus: Produce mycotoxins and cause head blight on wheat	40 Mbon 9 chromosome
Fungal diversity		
Ustilago maydis	Basidiomycota: Model of plant-pathogenic interac-tions.	20 Mb
Paxillus involutus	Basidiomycota: Easily manipulated in laboratory conditions and make symbi-otic associations with many trees.	40 Mb

It is proposed that RIP goes about as a guard against "narrow minded DNA," consequently ensuring the genome. Predictable with this is the way that no in place versatile components were recognized in the genome grouping, and 46% of dreary nucleotides can be distinguished as relics of portable components. Arrangement examinations with other Neurospora spp. would address the more extensive essentialness of these discoveries. Despite the fact that *N. crassa* has a great cluster of genome resistance components, it is not one of a kind in having protection frameworks. For instance, the numerous anastomosis gatherings and vegetative similarity loci in Rhizoctonia solani and Cryphonectria parasitica most likely serve comparative capacities in obstructing the spread of possibly harming portable infections or other hereditary components.

Communicated Succession Labels and Microarray Innovation

Communicated succession labels (ESTs) give a contrasting option to entire genome sequencing and are especially helpful for discovering qualities that are communicated in particular conditions (as such, the qualities that are "exchanged on" at various circumstances or in various conditions). This innovation can likewise be utilized as a snappy and modest way to deal with finding new qualities and for building genome maps.

ESTs are little lengths of DNA (around 200–500 nucleotides) delivered by sequencing one or both closures of a communicated quality. The procedure includes inspecting the pool of detachment RNA (the result of quality expression) and utilizing this mRNA as a format to integrate a correlative DNA (cDNA), utilizing the chemical switch transcriptase. cDNA is a great deal more steady than mRNA, and it has the extra favorable position of containing just the coding locales, since introns have been evacuated by grafting amid the characteristic handling of delegate RNA. Once a solitary strand of cDNA has been framed from the RNA layout, this RNA is processed by a ribonuclease. At that point a corresponding strand of DNA is integrated, to create twofold stranded DNA.

Substantial quantities of ESTs can be produced effectively and can consequently be collected into contigs or contrasted and the numerous ESTs of different living beings now accessible on electronic databases, giving a methods for distinguishing diverse sorts of quality. As an outline of the energy of such systems, Martina et al. (2003) utilized ESTs to think about the sorts of qualities communicated by mycelial societies of two mycorrhizal organisms (Laccaria bicolor and Pisolithus microcarpus) and furthermore to look at the qualities communicated in cooperative tissues of Pisolithus mycorrhizaz. By correlation with EST databases, a few of the ESTs could be allocated to useful gatherings, for example, qualities required in digestion, cell guard, or cell structure. In any case, the larger part of ESTs demonstrated no homology to referred to qualities (as is additionally valid for entire genome arrangements).

A comparative EST examination was accounted for by Freimoser et al. (2003), looking at two subspecies of the basic creepy crawly pathogenic growth Metarhizium anisopliae – one subspecies (M. anisopliae) has a wide host go and the other (M. anisopliae acridum) is a particular grasshopper pathogen. Both strains were developed under conditions that augment the emission of creepy crawly fingernail skin debasing chemicals, and a shockingly high extent of ESTs could be doled out to practical classes. Both subspecies had ESTs for all intents and purposes all the pathogenicity-related qualities cloned to date from M. anisopliae.

DNA MICRO-ARRAY INNOVATION

Basically, a DNA microarray is a little strong help, for example, a magnifying instrument slide or a nylon film onto which DNA is spotted (or printed by mechanical systems) to create hundreds or thousands of minor spots speaking to various sorts of DNA, organized in a particular request. Flag-bearer RNAs from a specimen to be investigated are then used to create cDNAs, which are fluorescently marked so that any authoritative of the cDNA to the immobilized spots on the microarray (through correlative base-matching) can be checked consequently by laser innovation. There are a few sorts of microarray, intended for various purposes, yet the fundamental ones for the most part measure one of the accompanying:

- Changes in the level of quality expression, checked naturally by blending tests of a test DNA (labeled with red fluorescence) and control DNA (labeled with green fluorescence) so that the laser separates between the levels of articulation of the test and control DNA.
- Changes in genomic additions and misfortunes, again distinguished by various levels of red or green fluorescence relying upon the quantity of duplicates of a specific quality.
- Mutations in DNA, regularly including just a single or a couple of nucleotides.

The majority of these applications are utilized as a part of prescription, including sickness conclusion, tranquilize advancement and following illness improvement, yet they are similarly pertinent to numerous parts of essential science and they are ending up noticeably perpetually broadly utilized in light of the fact that a similar fundamental procedures should be possible inexpensively by manual spotting of DNA onto strong backings. They can uncover examples of differential or facilitated articulation of qualities in any organic framework. [62]

SUMMARY

- This section covers applied and basic genetics of fungi, comprising the structures that remain to make fungi as most significant model organisms for purpose of genetic research for more than 60 years. It happens because fungi can grow easily in labortaries, having short life cycle and have both types of reproduction, sexual and asexual. Sexual reproduction is significant to study stages of meiosis. *Neurospora crassa* had used to study classical genetics.

- Organization and structure of the fungal genome has been given which includes detailed study of chromosomal genes and chromosomes, mitochondrial genes, and transposable elements and plasmids. Genetic alterations in fungi comprises significance of haploidy, heterokaryosis, parasexuality and sexual variations.
- Here we also discussed applied molecular genetics of fungi, comprising of production of heterologous protein. Methods of gene identification for plant differentiation and pathogenicity, parts of inhibitors in resistance of plants to pathogen, hipovirulence of plant pathogens, microarray technology and expressed sequences tags also discussed in this unit.

REFERENCES

1. Nester, E. W., & Verma, D. P. S. (1993). *Advances in Molecular Genetics of Plant-Microbe Interactions, Vol. 2: Proceedings of the 6th International Symposium on Molecular Plant-Microbe Interactions, Seattle, Washington, U.S.A., July 1992*. Dordrecht: Springer Netherlands.
2. Visser, J.A.G.M., de, & Hoekstra, R.F. (n.d.). *Synergistic epistasis between loci affecting fitness: evidence in plants and fungi*. (Genetical Research 71 (1998) 1.)
3. Grognet, Pierre, Lalucque, Hervé, Malagnac, Fabienne, & Silar, Philippe. (2014). *Genes That Bias Mendelian Segregation*. Public Library of Science.
4. Gomollon-Bel, Fernando, Marca, Eduardo, Hurtado-Guerrero, Ramon, Delso, J. Ignacio, Tejero, Tomás, & Merino, Pedro. (2015). *Design and synthesis of fungical transglycosidase inhibitors*.
5. In Peters, J. A. (1959). *Classic papers in genetics*. Englewood Cliffs, N.J: Prentice-Hall.

6. Ahluwalia, Kavita B. (2009). *Genetics*. New Age International.
7. Klug, W. S. (2012). *Concepts of genetics*. San Francisco: Pearson Education.
8. Levine, L. (1971). *Papers on genetics: A book of readings*. Saint Louis: C.V. Mosby Co.
9. In Gabriel, M. L., & In Fogel, S. (1955). *Great experiments in biology*. Englewood Cliffs, N.J: Prentice-Hall.
10. Scott, D. (2012). *Solutions manual for Introduction to genetic analysis, tenth edition*. New York: W H. Freeman & Co.
11. In Ayala, F. J., & In Avise, J. C. (2014). *Essential readings in evolutionary biology*.
12. Dunlap, J. C., & Wu, C. (2002). *Homology effects*. San Diego: Academic Press.
13. Mitchell, M. B., Mitchell, H. K., & Tissieres, A. (n.d.). *Mendelian and Non-Mendelian Factors Affecting the Cytochrome System in Neurospora Crassa**.
14. Mishra, N. C., & Tatum, E. L. (n.d.). *Non-Mendelian Inheritance of DNA-Induced Inositol Independence in Neurospora*.
15. Erwitt, E., & Sheed, W. (1978). *Recent developments: Photographs*. New York: Simon & Schuster.
16. Thomsen, C. W., & Blanchet, R. (2005). *Hollywood: Recent developments*. Stuttgart: Ed. Menges.
17. Society of Automotive Engineers. (1986). *Aerodynamics: Recent developments*. Warrendale, PA: Society of Automotive Engineers.
18. Gaonkar, A. G. (1995). *Food processing: Recent developments*. New York: Elsevier.
19. Sherman, I. W., & Sherman, I. W. (2012). *The Malaria Genome Projects: Promise, Progress, and Prospects*. London, UK: Imperial College Press.
20. 7th Siena meeting: From genome to proteome: back to the future. (January 01, 2007).
21. Roth, P. (2014). *The end of back pain: Access your hidden core to heal your body*.

22. Calder, N. (2003). *Magic universe: The Oxford guide to modern science*. Oxford: Oxford University Press.
23. Luk`i˜a˜nenko, S., & Bliss, L. (2014). *The genome: A novel.*
24. Liu, Z. (2007). *Aquaculture genome technologies*. Ames, Iowa: Blackwell Pub.
25. Harris, E. E. (2015). *Ancestors in our genome: The new science of human evolution.*
26. Holland, C. A., & Farkas, D. H. (2008). *DNA from A to Z & back again: A primer on genetics and molecular medicine for ... everyone.* Washington, DC: AACC Press.
27. Corey, M. A. (1994). *Back to Darwin: The scientific case for Deistic evolution*. Lanham, Md: University Press of America.
28. Rémi, J. (2007). *The circadian surface of Neurospora crassa: From physiology to molecular mechanisms.*
29. Pacific Northwest National Laboratory (U.S.), United States. & United States. (2015). *Draft Genome Sequence of Neurospora crassa Strain FGSC 73*. Washington, D.C: United States. Dept. of Energy. Office of Science.
30. .Davis, R. H. (2000). *Neurospora: Contributions of a model organism*. New York: Oxford University Press.
31. Li, L. (2007). *G protein coupled receptors in Neurospora crassa.*
32. Wang, L. C. (1979). *Nitrogen regulation, purification and characterization of uricase in Neurospora crassa.*
33. Lee, K. J. (1987). *The heat shock responses of Neurospora crassa.*
34. Baker, C. L. (2010). *Post-translational regulation of the Neurospora crassa circadian system.* .
35. Beth, D. A. (1998). *Chittin biosynthesis in Neurospora crassa: Structure and function of chitin synthase genes*. Rehovot: publisher not identified.
36. Dettmann, A., Seiler, S., Pöggeler, S., & Jahn, R. (2012). *Regulation and communication between the NRD kinase COT1, the MAK2 MAP kinase and the Striatin complex in Neurospora crassa*. Göttingen: Niedersächsische Staats- und Universitätsbibliothek Göttingen.
37. Bueche, J. A. (2010). *Post-translational modifications of phosducin-*

like protein BDM-1 and their implications in G-protein signaling and virulence of fungal plant pathogen Cryphonectria parasitica.

38. Basso, D. L. (2002). *Viruses and transposons of the chestnut blight fungus, cryphonectria parasitica: Tools to examine virulence, fungal, diversity, and evolution.*
39. Gao, S. (1994). *Characterization of an endopolygalacturonase produced by Cryphonectria parasitica.*
40. Anagnostakis, S. L. (1985). *Diversity and population dynamics in natural populations of Endothia (Cryphonectria) parasitica (Murr.) Anderson (Barr).*
41. Rivera-Velázquez, G. (2011). *Circadian rhythms in the chestnut blight fungus Cryphonectria parasitica.*
42. University of Guelph. Dept. of Environmental Biology., Mooij, D. L., & University of Guelph. School of Graduate Studies. (1997). *A Prescriptive Approach to Using Hypovirulence as a Biological Control of Chestnut Blight Caused by Cryphonectria Parasitica. (Thesis).*
43. Dunn, M. M. (1992). *Characterization of the chestnut blight pathogen, Cryphonectria parasitica, in southern Ontario.* Ottawa: National Library of Canada = Bibliothèque nationale du Canada.
44. Zhang, B. (2012). *Transgenic American chestnut (Castanea dentata) expressing oxalate oxidase developed through Agrobacterium-mediated co-transformation shows reduced Cryphonectria parasitica necrosis.*
45. Chang, N. (1994). *Marking chromosomes of Cryphonectria parasitica with a hygromycin B resistance gene.*
46. Li, P. C. H., Sedighi, A., & Wang, L. (2016). *Microarray Technology: Methods and Applications.* .
47. In Li, P. C. H., In Sedighi, A., & In Wang, L. (2016). *Microarray technology: Methods and protocols.*
48. Falciani, F. (2007). *Microarray Technology Through Applications.* Hoboken: Taylor & Francis.
49. Russell, S., Meadows, L. A., & Russell, R. R. (2009). *Microarray technology in practice.* Amsterdam: Elsevier.

50. Müller, U. R. (2005). *Microarray technology and its applications: Including 16 color plates*. (Microarray Technology and Its Applications.) Berlin: Springer.
51. Li, S. (2009). *DNA Microarray Technology and Data Analysis in Cancer Research*. Singapore: World Scientific.
52. Mocellin, S. (2007). *Microarray technology and cancer gene profiling*. Austin, Tex: Landes Bioscience/Eurekah.com.
53. Hamdan, M., & Righetti, P. G. (2005). *Proteomics today: Protein assessment and biomarkers using mass spectrometry, 2D electrophoresis, and microarray technology*. Hoboken, N.J: John Wiley & Sons.
54. Matson, R. S. (2013). *Applying genomic and proteomic microarray technology in drug discovery*. Boca Raton: Taylor & Francis.
55. Lindroos, K. (2002). *Accessing genetic variation by microarray technology*. Uppsala: Acta Univ. Upsaliensis.
56. Hundewale, N. (2006). *CAD tools for DNA micro-array design, manufacture and application*.
57. Spuhler, P. S. (2012). *High-throughput detection of DNA orientation and conformation for characterization of protein-DNA interactions*.
58. Tang, C. (2005). *Automatic phenotype structure mining underlying gene expression profiles*. .
59. Matson, R. S. (2009). *Microarray methods and protocols*. Boca Raton: CRC Press.
60. Zhang, K. (2008). *Inference of nonparametric hypothesis testing on high dimensional longitudinal data and its application in DNA copy number variation and micro array data analysis*. Manhattan, Kan: Kansas State University.
61. Lezar, S. (2013). *Assessment and development of microarray-based DNA fingerprinting in Eucalyptus grandis and related species*.
62. Wang, Yijin. (2010). *Gold nanoparticles based DNA chips utilizing CMOS technologies*.

CHAPTER

8

FUNGI AS PATHOGEN OF PLANTS

Fungus is pre-renowned as pathogen of plant. Coarsely 70% from all of the main diseases of crop are produced through the fungus-alike Oomycota, or fungi. One of the utmost disreputable instances is produced through Phytophthora infestans (Oomycota), potato late blight which distressed crops of potato in Ireland in the 1840s bringing to extensive scarcity. It is expected that greater than half a million people expired of malnourishment in that age, and alike quantity migrated to the North America and Europe. Although of numerous efforts to regulate it through fungicides and plant breeding, it is also currently possibly severe disease anywhere potatoes are grown. It's most current illustration is the Great Bengal Starvation of 1943, which is predictable to have charge the lives of several 2 million people owing to catastrophe of the rice crops produced through a pathogen causing leaf-spot, Helminthosporium oryzae (presently recognized as Cochliobolus miyabeanus, and Bipolaris oryzae). Fungi too reason for numerous severe ailments of amenity trees and landscape. Dutch elm sickness, produced through Ophiostoma ulmi and the related species O. novo-ulmi, removed continually through

Britain, North America, and continental Europe terminated the last century, destroying the populations of elm. Likewise Cryphonectria parasitica affected the chestnut blight, has distressed the chestnut tree of Native American. [1]

This ailment cleared crosswise over a significant part of the eastern piece of the USA, after it was first recorded in the New York Zoological Garden in 1904. The radiant American chestnut timberlands have now for all intents and purposes vanished and are spoken to just by a bush like understory layer. Cinnamomi root decay, caused by Phytophthora cinnamomi, is another case of a presented pathogen, at present debilitating huge tracts of characteristic eucalypt woods in Australia. [2] What's more, at the season of composing, sudden oak demise caused by Phytophthora ramorum is assaulting the common oak forests in the beach front mist belt of northern California and southern Oregon. This "new sickness," first recorded in 1996, has now spread too many parts of Europe and is causing genuine caution. It can possibly annihilate a significant number of Britain's old forests. These are only a couple of cases of numerous that could be refered to. Taken together, the pandemics in trim plants, manors, and regular forests have caused endless harm. In this part we concentrate on the exercises of plant-pathogenic growths and their few adjustments for causing illness.

THE SIGNIFICANT SORTS OF PLANT-PATHOGENIC ORGANISMS: SETTING THE SCENE

It is useful in the first place a few definitions:

- A parasite can be characterized as a life form that increases all or part of its healthful prerequisites from the living tissues of another creature – the host. As it were, parasitism portrays a nourishing relationship.
- A pathogen can be characterized as a life form that causes malady. It is quite often a parasite, yet the refinement between a parasite and a pathogen is vital in light of the fact that parasitic fungi don't generally cause evident or genuine malady. [3]

There are a large number of plant-pathogenic growths, so we have to amass them somehow that mirrors their key science. The framework we will utilize depends on two vital components: (i) the sort of dietary connection between the pathogen and the host, and (ii) regardless of whether the pathogen has an expansive host extend or is have specific.

As far as the nourishing relationship, we recognize necrotrophic pathogens and biotrophic pathogens. Nectrotrophs murder the host tissues, more often than not by straightforwardly attacking them, or by delivering poisons or degradative catalysts, and after that feast upon the tissues that they slaughter (Greek: nekros = dead, trophos = a feeder). [4] By differentiate, biotrophic pathogens feast upon living host tissues (Gr. profiles = life), regularly by delivering extraordinary supplement engrossing structures that take advantage of the host's tissues. A third gathering – hemibiotrophic pathogens – at first develop as biotrophs yet then attack and murder the tissues. A portion of the Phytophthora species, for example, P. infestans carry on along these lines. [5]

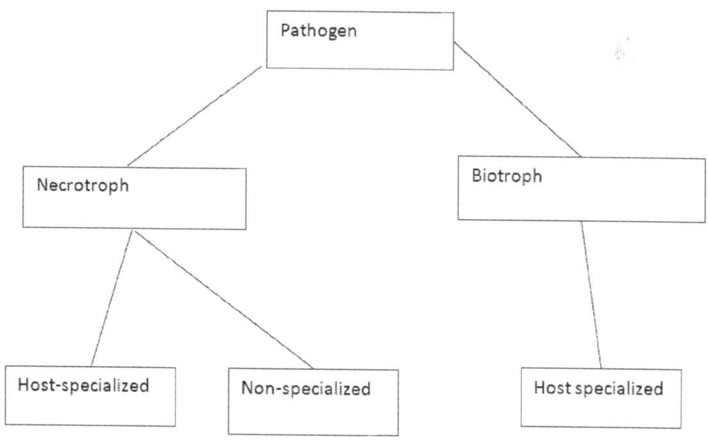

Figure- 8.1: Pathogens diversity.

The auxiliary qualification is between pathogens of expansive host go and those that are have specific, with a tight host extend. Pathogens of wide host extend typically assault youthful or senescing tissues, or plants whose resistance is traded off by ecological components. By differentiate, have particular pathogens are adjusted to defeat the particular barrier systems of plants, for example, the fungitoxic intensifies that many plants deliver to avoid disease. [6] The biotrophic pathogens are fundamentally have specific. In this way, we can classify all pathogens in the accompanying basic plan, which will shape the reason for whatever is left of this section. We will talk about agent cases of the significant sorts of plant malady, to develop a photo of the differing qualities of plant-pathogenic parasitic fungi and their exercises.

Necrotrophic Pathogens of Juvenile or Traded Off Hosts

Countless assault the tissues of youthful plants, causing seed decays or seedling illnesses. Other growths normally assault the more established, senescing tissues of plants, but then others assault plants that are traded off by dry spell or different anxiety variables. [7] Despite the fact that these growths don't have an extremely advanced method of parasitism, they can cause genuine monetary harm. A few cases are talked about beneath.

Seed decays and Seedling Pathogens

A trademark scope of fungal growths because seed decays or seedling infections, once in a while canceled damping ailments since they are frequently connected with wet soil conditions. For instance, numerous Pythium spp. (Oomycota) assault the root tips of seedlings, and strains of Rhizoctonia solani frequently assault the youthful shoot base, simply above or beneath the dirt level. Other parasitic fungiwith comparable conduct incorporate Fusarium spp. which cause seed-decays and seedling maladies of oats. [8]

Every one of these parasitic fungi react quickly to the nearness of seed or seedling exudates, and attack the youthful tissues that have minimal intrinsic host resistance. To help check this issue, seeds are typically covered with broadspectrum fungicides, for example, maneb, thiram, or dinocap which meddle with a few essential metabolic procedures of organisms. Be that as it may, there is an expanding pattern to supplant fungicidal seed coatings with organic control specialists. A few of these are currently accessible monetarily. They incorporate antibioticproducing microscopic organisms, for example, Bacillus subtilis, utilized generally in the USA for cotton and a few other field harvests, and Pseudomonas chlororaphis, utilized broadly in Scandinavia to secure against seedling infections of grains. [9] A few growths are showcased for biocontrol of seedling ailments, including Trichoderma spp. what's more,

Clonostachys rosea (under its previous name Gliocladium catenulatum). Every one of these items function admirably in proper conditions, yet seedling pathogens can taint quickly, before a natural control operator can end up plainly settled. For instance, soil-borne sporangia of the normal seedling pathogen Pythium ultimum can grow inside 1–2 hours in light of unpredictable metabolites discharged from sprouting seeds (most likely ethanol yet perhaps at the same time acetaldehyde), and the seed tissues can be intensely colonized inside 6–12 hours of planting in Pythiuminfested soil. Biocontrol specialists are probably not going to act sufficiently quickly to control such pathogens. [10]

Athelia Rolfsii

The sclerotium-framing parasite Athelia (Sclerotium) rolfsii (Basidiomycota) is a standout amongst the most crushing seedling pathogens in the hotter parts of the world where precipitation is occasional. The sclerotia of A. rolfsii make due in soil amid the dry season and are activated to develop after the principal downpours. At that point the parasite develops on the re-wetted trim buildups from the past season and uses these deposits as a nourishment base to assault the recently sown harvest, regularly spoiling the basal stem tissues inside 3–4 weeks. [11] Extensive ranges of grains, and different products, can be wrecked by this growth.

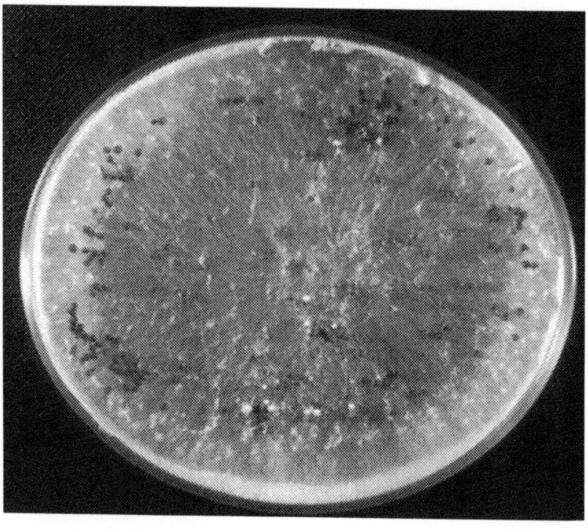

Figure- 8.2: Athelia rolfsii on agar plate.

The way to comprehension *A. rolfsii* lies in the conduct of its sclerotia, as appeared by the analysis in. On the off chance that sclerotia are gathered from research center societies they display hyphal germination, whereby just a couple of hyphae rise and these can taint plants at just a short separation from the sclerotia (0.5 cm). In any case, sclerotia that have been air-dried show eruptive germination when remoistened – all the sclerotial saves are utilized to create a mass of hyphae, which are sorted out into mycelial lines and can contaminate a plant at up to 3.5 cm separate [12]. The sclerotia of this growth are much more successful within the sight of newly deteriorating natural issue (e.g. trim deposits from the past season, which begin to deteriorate after the primary downpours). Unstable mixes, particularly methanol, are discharged from these naturally deteriorating deposits, because of the activities of pectic catalysts that held on in the buildups. These unpredictable mixes trigger the sclerotia to develop eruptively and furthermore fill in as supplements, orientating development towards the wellspring of methanol. In these conditions, the sclerotia can contaminate seedlings even 6 cm away.

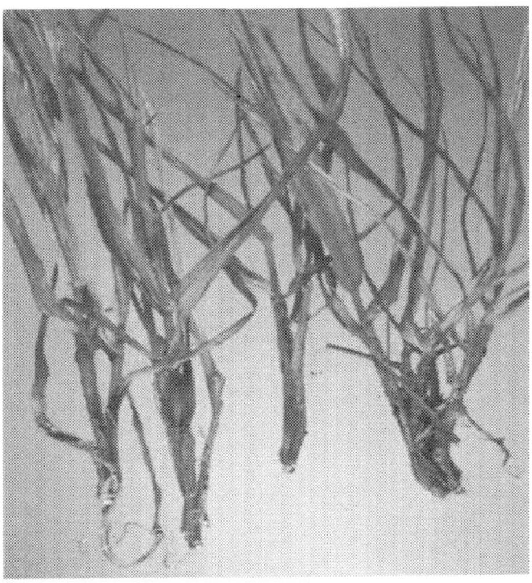

Figure- 8.3: A. rolsfii rotted young wheat plants below soil level.

These analyses appear to parallel intently the conduct of the organism in editing frameworks, where A. [13] rolfsii typically colonizes the yield buildups and utilizations these as a sustenance base to help contamination of the basal stem tissues of youthful plants. It has a to a great degree forceful method of assault, creating pectic chemicals to corrupt the plant tissues, helped by the generation of overflowing measures of oxalic corrosive. [14]

Rhizoctonia Solani

Rhizoctonia solani (sexual stage Thanatephorus cucumeris, Basidiomycota) is another basic seedling pathogen that assaults the basal stem tissues of plants. It is a systematically complex growth – a species-total comprising of strains that are separated by atomic fingerprinting, by have go, and by their capacity to anastomose (combine) with each other when provinces are restricted on agar plates. Some of these strains caused hindering infections of oats, for example, whole malady of wheat on the substantial, dark earth soils of Northern Transvaal, South Africa. [15] In this illness the plants develop ordinarily at in the first place, however expansive patches of hindered plants begin to show up around 3–4 weeks in the wake of sowing, and from this time onwards the hindered plants make minimal further development, while whatever is left of the yield develops typically. The foundations of the hindered plants bear globule like masses of hyphae at around 7– 10 cm beneath the dirt surface, where the free developed soil meets the fundamental mud soil that has not been furrowed. From these hyphal dots the growth enters and murders the roots, prompting the hindering disorder. Comparable hindering infections of oats are found in the compacted sandy soils of parts of Australia, where the roots have dark colored, skewer pointed closures, caused by decaying of the root tips. [16]

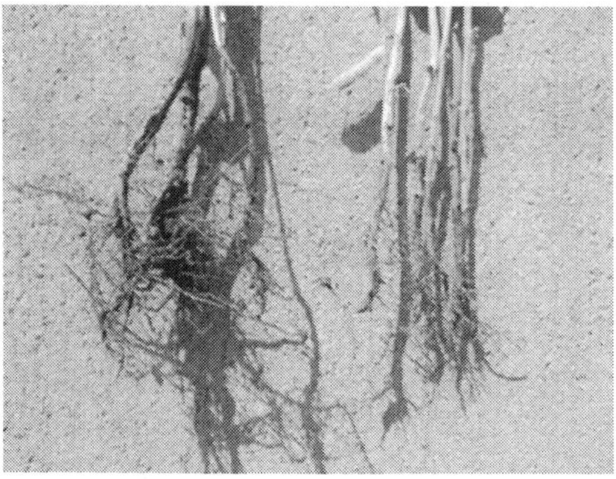

Figure- 8.4: Damage to soybean roots exhibiting damage from Rhizoctonia solani.

Ailments of this sort have a typical cause: the compacted soil layer blocks root development, which is probably going to expand supplement exudation from the root tips, advancing intrusion by Rhizoctonia. Also, Rhizoctonia develops as a broad mycelial organize in the lower, undisturbed soil layer. The ailment patches speak to the regions where this contagious system is available underneath the dirt surface, and these patches are spread dynamically along the furrow lines, expanding in degree from year to year. [17]The best method for controlling these illnesses is by profound furrowing, despite the fact that this is from time to time monetary on substantial dirt soils. In any case, on the sandy soils of Australia the utilization of a tine that tears the dirt around 10–12 cm beneath the level at which the seed is sown can enable the roots to infiltrate profoundly and significantly enhance edit yields.[18]

Decrease and Replant Sickness

A few perpetual natural product crops, including strawberry, apple, cherry, and avocado, demonstrate a dynamic decrease in yield as the plants age, until the point when generation is not any more monetary. In the event that the declining plants are evacuated and supplanted by others the new plants regularly develop inadequately incredible. These decay and replant illnesses are firmly related, showing a typical cause. As we found, the youthful "feeder roots" of plants frequently have a short life expectancy before they senesce and are supplanted by others. In the decrease or replant sicknesses this harmony between the rates of root generation and root rot is adjusted, with the goal that root tip rot surpasses the rate of root tip creation, and in the end the product bites the dust. [19]Types of Pythium (e.g. P. sylvaticum) are emphatically involved in the decrease of plantation yields, for example, apples. Likewise, Phytophthora cinnamomi (Oomycota) is one of the primary reasons for decrease in avocado plantations and it additionally causes a dieback illness of local eucalypt vegetation in Australia, raising genuine worries for the eventual fate of this one of a kind gathering of plant species. One of the real lines of confirmation that involves Pythium and Phytophthora spp. as the reason for decay and replant ailments is that the acylalanine fungicides act particularly against Oomycota, and when these fungicides are connected to handle trims tentatively (yet at uneconomic levels) they can cause awesome changes in plant development. Comparable investigations propose that Pythium spp. cause up to 10% yield diminishment in wheat edits in the Pacific Northwest of the USA, despite the fact that the products are obviously sound. These cases outline two essential focuses. [20]

- First, the root tips and youthful tissues stay powerless to disease for the duration of the life of a plant. Along these lines, even develop trees can decrease if the rate of pathogen instigated root rot surpasses the rate of root-tip generation.[21]
- Second, a significant number of these pathogens demonstrate some level of host-adjustment. For instance, Pythium graminicola and P. arrhenomanes are distinctively, yet not only, related with graminaceous has (the grass family), and are involved in the dynamic decrease of sugar-stick crops. Trial thinks about contrasting "host" with "nonhost" (nongrass) plants unequivocally propose that the host adjustment of these two parasitic fungi is connected to their proficient zoospore encystment on have plants. Also, Phytophthora sojae is a host-adjusted pathogen of soybean harvests, and its zoospores demonstrate solid appreciation for the flavonoids discharged from soybean roots.[22]

Infections of Senescence: the Stalk-Decay Pathogens

A few growths that harm seedlings can likewise develop on senescing plant tissues, abusing the declining host resistance. Vital cases incorporate the stalk-spoil pathogens of maize, some of which (e.g. Macrophomina phaseolina) assault many yields though others are very host-particular towards maize (Diplodia maydis, Gibberella zeae, Colletotrichum graminicola). Regardless, their comparable conduct concerning senescence enables them to be considered as a gathering. Stalk-spoil pathogens are to a great degree basic where maize and sorghum are developed, yet they cause genuine sickness in just a few destinations or developing seasons, though in different locales or seasons they cause almost no harm. [23]The reason is that they react to the plant's physiology, and especially to any anxiety conditions that incline the plants to disease. These anxiety variables are numerous and changed. They incorporate impermanent dry season stretch, poor light conditions (cloudy climate), mineral supplement inadequacy, or bug (stem-borer) harm. [24]

Indeed, even an incredibly abnormal state of grain-setting can prompt stalk-decaying by making an overwhelming interest on the plant's supplement holds. On the off chance that any of these conditions happens at a basic stage in the harvest's development, amid grain-filling, at that point the entire product can be contaminated by stalk-decay pathogens, bringing about unexpected passing and crumple, with overwhelming loss of yield.[25]

A binding together speculation to clarify this kind of sickness was produced right on time in the 1900s and was refined by Dodd (1980). The filling of the grain in maize and sorghum puts an overwhelming interest on plant sugars, and just around 80% of this sugar request can be provided by "current" photosynthesis in the clears out. The other 20% is provided from sugar stockpiling saves, mainly in the base of the stem. As the plant approaches development its save sugars are dynamically drained, and any anxiety conditions that incidentally decrease the rate of photosynthesis at this stage make more sugars be expelled from the stem tissues, prompting untimely senescence. [26] At that point, the pitifully parasitic stalk-spoil organisms that were at that point show yet were kept within proper limits can attack and demolish the stalk base. This is anything but difficult to show tentatively, on the grounds that stalk spoil never happens if the creating grain is expelled to wipe out the sugar push, yet this is not really a functional arrangement! Rather, producers must utilize their experience to foresee the probability of stress conditions on their ranches and change the manure and plant dividing in like manner. Every one of the growths that cause infections of this sort are particular, characteristic trespassers of senescing plant tissues. [27]

Pathogens of Fruits: the Roles of Pectic Enzymes

Aging products of the soil sugar-rich or starch-rich plant tissues are regularly decayed by growths.One basic case is Botryotinia fuckeliana (better known by its abiogenetic stage, Botrytis cinerea) which causes dark shape of delicate organic products, for example, strawberries, raspberries, and grapes. This parasite regularly starts disease by developing on the senescing blossom remains (it is likewise an issue in the business cut bloom industry) and afterward attacks the living tissues of the maturing natural product. [28]By differentiate, a few other natural product decay growths taint through minor injuries in the organic product surface, caused by creepy crawlies or by harm amid reaping and pressing. Cases of these parasitic fungi incorporate the "green molds" Penicillium italicum and P. digitatum which normally cause the decaying of oranges, P. expansum which ordinarily spoils apples away, Sclerotinia fructigena which decays apples, peaches, and some different natural products, regularly through injuries caused by winged animals and wasps before the organic product is collected, and types of Rhizopus which spoil many sorts of organic product, including pears. [29]

Figure- 8.5: Apple rot done by fungal (Penicillium expansum) attach.

The fascinating component of these organisms is that they cause diverse sorts of decay, going from quick spreading watery decays to firm, dry decays. In all cases they can't decay the natural product until the point that it goes through the stage named climacteric, when the organic product starts to age. This is the phase at which the corrosive substance decreases and sugar levels begin to rise.[30]

The parts of Pectic Catalysts in Pathogenesis

The regular component of all natural product spoil sicknesses is that the tissues are decayed by pectic compounds that debase the center lamella (the establishing layer) between adjoining plant cell dividers. [31]The pectin of the center lamella comprises of chains of α-connected galacturonic corrosive deposits, some of which are methylated, and blended polymers of galacturonic corrosive, mannose, and lesser measures of different sugars. The pectic chemicals are of three noteworthy sorts.

- Pectin methyl esterase (PME) is regularly created by the plant itself and it demethylates pectin. PME appears to assume little part in the breakdown of pectin, however the methanol discharged by this chemical can go about as a germination trigger and potential carbon hotspot for Athelia rolfsii (see prior). [32]
- Pectic lyase (PL) is a chain-part protein with both endo-and exo-acting structures. It cuts the bonds between the sugar deposits distinctively, taking out water amid the procedure.
- Polygalacturonase (PG) is another chain-part compound with both endo-and exo-frames. Not at all like PL, this protein is a hydrolase – it utilizes water to add H^+ to one sugar buildup and OH^- to the nearby deposit amid cleavage of the bond.[33]

Parasitic fungi's are actuated to orchestrate pectic proteins within the sight of pectic mixes, and the chemicals most likely act in show. This was demonstrated when the qualities encoding the proportionate proteins of the bacterium Erwinia carotovora (the reason for potato dark leg malady) were designed independently into E. coli. The beneficiary cells required at least one exo-PL, two endo-PLs, and one endo-PG with a specific end goal to spoil the potato tissues. Pectic compounds of parasitic fungiexist as a scope of isomers which isolate as indicated by size and net charge amid gel electrophoresis. The subsequent pectic zymograms can be utilized to recognize populace subgroups of organisms, however the isomers may not vary fundamentally in work. [34]

At first it was felt that the key part of pectic compounds in pathogenesis was to isolate cells, causing disturbance of the fine protoplasmic spans (plasmodesmata) that connection cells through their dividers. Later examinations propose different parts. Pectic proteins can be straightforwardly poisonous to plant protoplasts in vitro, and a portion of the fractional breakdown results of pectin – the oligomers made out of 4 or 5 galacturonic corrosive buildups that are created by endo-PG – are particularly lethal, causing fast cell demise. [35]

Plant cells additionally contain a PG-repressing protein, maybe as a general resistance component. As the plant tissues age, the pectic mixes are progressively cross-connected with calcium, which renders them more impervious to corruption, maybe by limiting access by the endo-acting chemicals. For instance, the hypocotyls (basal stem tissues) of bean seedlings (Phaseolus vulgaris) demonstrate a checked increment in imperviousness to contamination by Rhizoctonia solani around 2 weeks after seed germination, corresponding with calcification of the center lamella.[36] The pathogens that decay more established plant tissues (Athelia rolfsii, and Sclerotinia sclerotiorum which spoils the seed heads of harvests, for example, sunflower) defeat this by delivering a lot of oxalic corrosive, which they likewise create in research center culture. Oxalic corrosive can join with calcium, expelling it from relationship with the pectic mixes and rendering these vulnerable to enzymatic assault.[37]

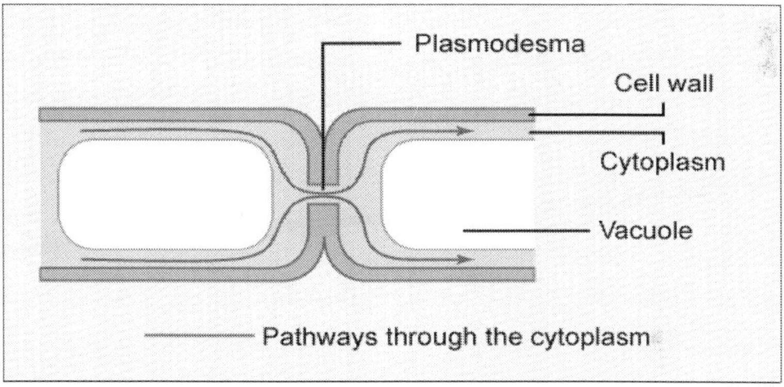

Figure- 8.6: Junction among plant cells.

Host-Specific Necrotrophic Pathogens

As opposed to the pathogens of youthful or traded off hosts, numerous necrotrophic pathogens are have particular. They have advanced systems for attacking the typical sound tissues of plants by defeating the particular barrier components of their hosts. These protection instruments can incorporate physical hindrances to infiltration or fungitoxic chemicals, and they can either be preformed or shaped because of contamination. In the accompanying areas we examine the trademark components of these host-particular pathogens.[38]

The Part of Inoculum: An Examination of Botrytis fabae and *B. Cinerea*

Botrytis cinerea and *B. fabae* are firmly related organisms, however *B. cinerea* attacks the traded off tissues of many plants, while *B. fabae* is a host-particular pathogen of expansive bean (Vicia faba). Both growths deliver ovoid conidia, however those of *B. fabae* are bigger (20–25 μm long) than those of *B. cinerea* (10–15 μm), speaking to an about sixfold distinction in spore volume. As we found, bigger spores have a higher productivity of impaction onto leaves, particularly onto wide leaves, for example, those of wide bean plants. Vast spores likewise have more endogenous nourishment stores to help disease. [39]

Figure- 8.7: Leafs affected by Botrytis spp.

The spores of both B. fabae and B. cinerea develop when put in water drops on the surface of expansive bean leaves, however B. cinerea frequently neglects to enter the leaf fingernail skin while B. fabae infiltrates and causes a quick host reaction – the passing of the entered cells and their prompt neighbors, giving an expansive necrotic spot. This response to contamination is named the touchy reaction, and it is related with a collection of fungitoxic metabolites in the dead cells, regularly adequate to anticipate advance disease. [40]

The distinction in beginning entrance by the Botrytis species is to a great extent clarified by their supplement saves. *B. cinerea* will infiltrate and cause a necrotic spot if sugars are added to the inoculum drop, or now and then regardless of the possibility that the typical leaf surface microorganisms are executed by anti-microbials so that B. cinerea can utilize the leaf-surface exudates. Then again, matured spores of B. fabae, taken from old parts of agar provinces, can neglect to infiltrate in light of the fact that their endogenous stores have been drained, yet they enter if provided with exogenous sugars. [41] Consequently, *B. cinerea* has advanced a procedure as a general intruder of senescent tissues: it produces numerous more spores from a given measure of asset yet these spores rely upon supplements discharged from traded off host tissues. B. fabae produces less spores from a similar measure of asset, yet each of them has a noteworthy shot of starting disease of the sound host. Such contrasts in the infectivity of pathogens can be evaluated, to foresee the probability of contamination in various conditions. B. cinerea and B. fabae when water drops containing diverse quantities of spores were set on wide bean takes off. [42]

Disease was scored as the nearness of nonappearance of a necrotic spot underneath each water drop. The outcomes are plotted as the logarithm of spore measurement (spore number) against the rate reaction (disease, communicated as a probit esteem, where the probit of 5 = half). This sort of plot changes over a skewed sigmoid bend into a straight line, and the evaluated measurements of spores that gives half likelihood of disease (the ED50 esteem) can be perused from the diagrams. [43] A similar sort of plot is utilized generally in plant and creature examinations to get LD50 esteems for deadly poisonous quality of chemicals, and so forth. The hostspecialized parasite B. fabae was found to have an ED50 of 4 in water drops, relating to a 16% shot of any single spore having the capacity to start contamination. By differentiate, B. cinerea had an ED50 of around 500, relating to a 0.14% shot of a solitary spore starting disease. Analyses of this sort give data significant to handle conditions. For instance, a few distinctive Colletotrichum spp. create leaf spots on elastic leaves in Malaysia. By vaccinating spores in beads it was conceivable to anticipate: (i) the species well on the way to taint from low spore numbers and (ii) the age at which the leaves are most infectible, before the advancement of a thick leaf fingernail skin keeps any further disease. [44]

The Parts of Tissue-Debasing Proteins

Numerous necrotrophic organisms create tissue-corrupting proteins as an essential instrument of pathogenesis, similar to the pectic catalysts talked about before. In any case, the grass family, Gramineae, is by all accounts bizarre on the grounds that grasses have distinctive divider grid parts contrasted and numerous different plants. Rather than pectin in the center lamella they have blended polymers of β-1,4-xylan and α-1,3-arabinose, with lesser measures of non-cellulosic β-1,4-and β-1,3-connected glucans. Cooper et al. (1988) explored the proteins delivered by three stem-base pathogens of grains – Rhizoctonia cerealis which causes sharp eyespot illness, Fusarium culmorum which causes fusarium foot spoil, and Tapesia yallundae which causes "genuine" eyespot sickness.[45] At the point when these parasitic fungiwere developed on oat cell dividers in lab culture, each of the three growths created arabinase, xylanase, and β-1,3-glucanase, however just little measures of pectic compounds. By differentiate, when pathogens of nongrass plants (Fusarium oxysporum, Verticillium albo-atrum) were developed on dividers of their hosts they created a lot of pectic catalysts. [46]

In this way, the grain pathogens appear to be adjusted to corrupt the commonplace divider parts of their hosts. However, the astonishing element in this investigation was that the oat pathogens created arabinase and glucanase constitutively, without have divider material, and the generation of these catalysts was not curbed by sugars[47] These are uncommon components in light of the fact that polymerdegrading catalysts when all is said in done are both inducible and repressible. The possible clarification is that these hostspecialized organisms become just on oats, so they have adjusted to create the important divider debasing chemicals constitutively.

Plant Defense against Host-Specialized Necrotrophic Pathogens

Plants have a few potential lines of barrier against attacking pathogens. These protections incorporate both physical obstructions and fungitoxic mixes in the plant tissues. [48]The primary physical hindrance to disease of many plants is the fingernail skin on the leaf or stem surface, and in such manner the catalyst cutinase has been believed to be imperative for contagious contamination. There is conditional and correlative confirmation for this. In any case, when focused quality interruption was utilized to square cutinase generation by three leaf pathogens, there was no decrease in pathogenicity of two of the growths and just a fractional diminishment in a third parasite. A conceivable clarification is that a portion of the pathogenicity qualities may just be communicated amid disease and not in research center culture. On the other hand, there may be different types of these qualities, some of which are communicated in vitro and others just in the host condition. This case outlines the trouble of unwinding the frequently complex host resistance frameworks of plants. [49]

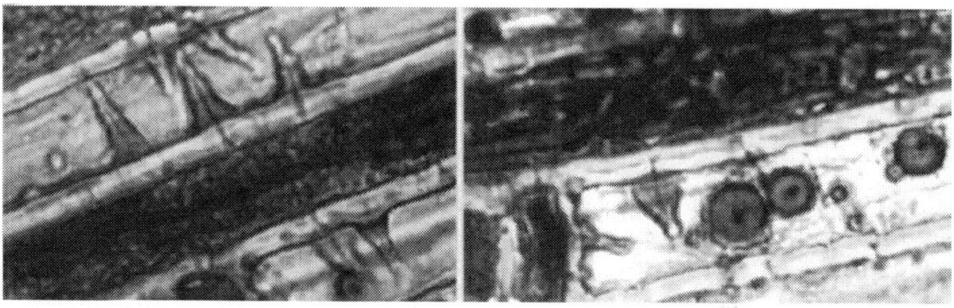

Figure- 8.8: Lignified papillae formed in reaction to attack of wheat root cells through the fungus, Gaeumannomyces graminis.

Notwithstanding preformed physical obstructions, plants can react to contamination by building up a papilla – a confined thickening of the cell divider at the purpose of endeavored intrusion. In outrageous cases the papilla keeps on creating as the hypha develops, with the goal that it encases the infiltrating hypha. In electron micrographs, papillae are viewed as enormous limited aggregations of divider material in which layers and other worsen cytoplasmic segments have been caught. [50] Cytochemical recolors for the most part uncover the nearness of callose in papillae, however in the Gramineae the papillae regularly indicate recoloring responses for lignin and suberin. The essentialness of papillae in plant resistance may be to defer attacking hyphae while different guards are actuated. The lignin-like antecedents in papillae of the Gramineae might be especially vital, in light of the fact that lignin can frame edifices with the divider polymers, keeping their enzymatic debasement. Ride and Pearce (1979) showed this for grain leaves immunized with Botrytis cinerea (not a pathogen of oats) and a few other growths that incited the creation of lignified papillae amid endeavored entrance. [51]The ordinary leaf cell dividers were processed effortlessly by business divider debasing compounds, however the papillae and the encompassing "haloes" of lignified divider material were not processed. Along these lines, lignification and the improvement of papillae could have three resistance-related capacities:

- to expand the divider thickness that a growth must enter while relying upon its endogenous vitality holds, to render the divider impervious to assimilation by the parasitic catalysts, and to give a locally lethal condition caused by the phenolic forerunners of lignin and suberin.[52]

Oxidative Burst

One of the soonest occasions in the endeavored entrance of leaves and other over the ground parts of plants is an oxidative blasted, comparable to the oxidative barged in phagocytes. It has been contemplated for the most part in connection to the excessively touchy reaction – the fast demise of a gathering of cells in light of restricted intrusion – yet it is by all accounts a general response of plant cells to injury caused by endeavored parasitic attack or compound elements. [53]Sheep et al. (1994) have demonstrated that it includes a greatly quick (2–3 moment) creation of H2O2 at or close to the plant cell surface, by the response of oxygen with NADPH to deliver superoxide:

$O_2 + NADPH \rightarrow O_2^- + NADP +$

At that point O_2^- is transmuted to H_2O_2 by a plasma layer oxidase. H_2O_2 causes cross-connecting of the plant cell divider proteins, which could fortify the divider against assault by divider corrupting chemicals.

Chemical Defenses

Plants create an extensive variety of auxiliary metabolites for guard against attacking life forms. Some of these are preformed mixes, on the whole named phytoanticipins. [54] They are available either as dynamic mixes or as antecedents that are quickly changed over to dynamic mixes because of contamination. A moment classification of guard mixes are named phytoalexins. Once more, they are optional metabolites yet they are combined once more in light of disease. Nonetheless, there is not generally an unmistakable contrast between these two gatherings of mixes. The phytoanticipins and phytoalexins are broadly accepted to be noteworthy parts of plant safeguard frameworks, at any rate as in they can keep most potential trespassers from causing dynamic infection of plants. Be that as it may, have particular pathogens can here and there defeat the barriers of their particular hosts, and we think of some as agent cases of this underneath. [55]

Phytoanticipins

A portion of the best confirmation for a part of phytoanticipins has originated from investigations of the take-all organism of oats Gaeumannomyces graminis, and related examinations on the tomato pathogen Septoria lycopersici. These host-particular organisms can overcome the phytoanticipins of their host plants by creating catalysts (avenacinase and α-tomatinase) that detoxify the phytoanticipins, though growths that are not particularly adjusted to these host plants are executed by these mixes.[56] Phytoanticipins might be included in a few different plant–fungus communications. An exemplary illustration is the resistance of red-and yellow-cleaned onions to Colletotrichum circinans, the reason for "onion smear" infection, though white-cleaned onions are defenseless. The colors themselves are not vital but rather they are related with large amounts of the phenolic phytoanticipins, catechol and protocatechuic corrosive. [57]

Phytoalexins (warding-off compounds)

Phytoalexins are unmistakable from phytoanticipins in light of the fact that they are low-atomic weight mixes created because of disease. More than 350 phytoalexins have been accounted for over a scope of harvest plants, particularly dicotyledonous plants, for example, beans, potatoes, peas, and cotton. The majority of them are flavonoids, phenolics, or terpenoids, orchestrated by three noteworthy auxiliary metabolic pathways: the acetic acid derivation malonate course, the acetic acid derivation mevalonate course, and the shikimic corrosive course, like the optional metabolic pathways of growths. [58]

A significant collection of confirmation recommends that phytoalexins go about as resistance mixes – in any event in securing plants against general assault by organisms. These mixes gather quickly amid the easily affected reaction, and at development inhibitory levels in or around the dead or biting the dust tissues. In any case, they can likewise be detoxified by pathogens that are particular to singular plant species, so there is a dynamic association between the speed of host reaction and the capacity of a parasite to detoxify the protection mixes. A comparable restricted collection of phytoalexins can regularly be actuated by simulated injuring or utilization of a dangerous synthetic to the plant surface. It is constantly gone before by an oxidative burst, and H_2O_2 is believed to be one of the underlying signs that enacts the barrier qualities required in phytoalexin generation. [59]

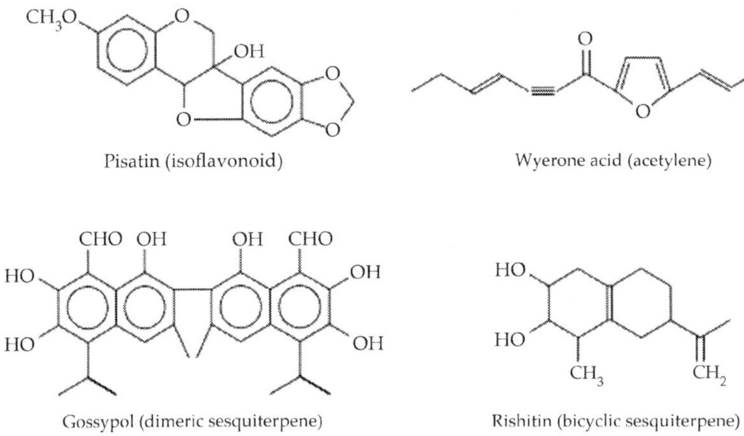

Figure- 8.9: Four phytoalexins: Rishitin, Gossypol, Wyerone, and Pisatin.

The succession of occasions prompting phytoalexin collection has been examined by the revelation that low-atomic weight mixes comprising of a couple of β-connected sugar deposits can go about as effective elicitors of the excessively touchy reaction. These oligosaccharides are discharged from the extracellular polysaccharides of parasitic hyphae, presumably by the activities of β-glucanase compounds that dwell in plant cell dividers.[60] At the point when connected to plant cells or tissues in vitro, the elicitors cause an early oxidative burst and the enlistment of plant catalysts of the phenylpropanoid pathway, including phenylalanine smelling salts lyase which is included in phytoalexin generation, and union by the plant of chitinase and β-1,3-glucanase which can corrupt contagious dividers.

Plainly, this course of cell reactions makes a threatening domain for a potential pathogen. So how do have adjusted necrotrophic pathogens overcome or evade these safeguards? Concentrates on the Botrytis–broad bean connection have demonstrated that both *B. fabae* and *B. cinerea* can cause an excessively touchy reaction in fitting conditions and the phytoalexin wyerone corrosive gathers in the dead cells. [61] Be that as it may, *B. fabae* is more tolerant of wyerone corrosive than is *B. cinerca* in Vitro, and B. fabae additionally causes less wyerone corrosive to gather in the injury, maybe by detoxifying it. In this manner, *B. cinerea* never spreads encourage from the necrotic spot, though *B. fabae* is checked briefly yet then can spread through the encompassing tissues, slaughtering them logically. The subsequent extensive dark colored sores offer ascent to the name for this ailment, spot. The instruments of detoxification of phytoalexins have been accounted for some other growths. The pathogen of pea plants, Nectria haematococca, detoxifies pisatin by demethylation. Different strains of this pathogen that contaminate bean plants detoxify one of the bean phytoalexins kievitone by hydration. [62]

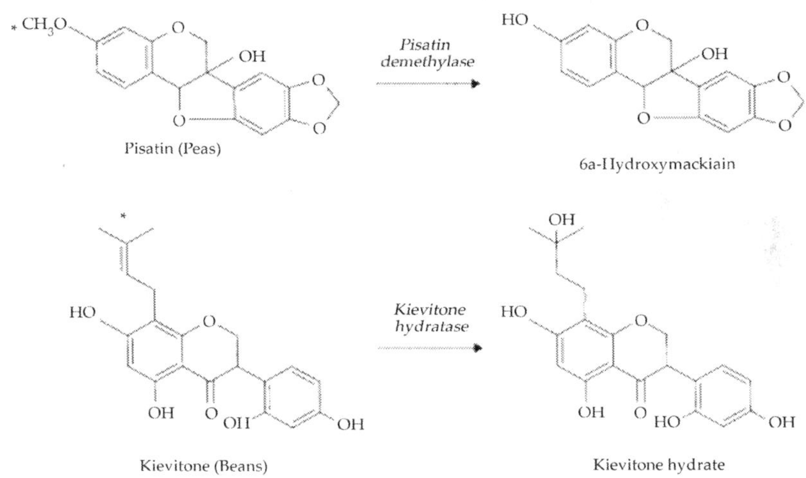

Figure- 8.10: Examples phytoalexins detoxification.

In both cases the capacity of the pathogen to taint the plant is related with the capacity to detoxify the phytoalexin. In addition, when the pisatin demethylase quality from *N. haematococca* was changed into Cochliobolus heterostrophus, a maize pathogen, this organism turned into a pathogen of pea takes off. In this way, there is solid proof that host-adjusted pathogens have advanced components for defeating the inhibitory impacts of their hosts' phytoalexins, and even the procurement of a solitary phytoalexin-detoxifying quality can change the host scope of an organism.[63] There is, nonetheless, a turn to this story. In the Nectria–bean communication, all strains of N. haematococca were found to deliver the hydrating catalyst, kievitone hydratase, however a few strains did not discharge it from the hyphae. These strains were nonpathogenic, rather than strains that emitted the protein. In the Nectria–pea collaboration the connection amongst pathogenicity and creation of the detoxifying compound was bolstered by many lines of proof, yet focused on disturbance of the quality for pisatin demethylase prompted strains that were as yet pathogenic. Exploring all the work on detoxification of phytoanticipins and phytoalexins, VanEtten et al. (1995) proposed that pathogens may have developed a few components for beating the host resistances, so that some of these systems, albeit still communicated, might be repetitive. [64]

SYSTEMIC ACQUIRED RESISTANCE AND INDUCED SYSTEMIC RESISTANCE

Prior in this part, we noticed that plants can react to assault from necrotrophic pathogens by mounting an easily affected reaction, in which the contaminated cells and their prompt neighbors bite the dust quickly and aggregate fungitoxic mixes. Despite the fact that this reaction is limited, it prompts the generation of flagging particles that are prepared all through the plant, so that the plant is, as a result, inoculated. Any consequent contamination will prompt expanded resistance or lessened articulation of infection. [65]

This marvel is named systemic gained resistance (SAR). It includes the enactment of an arrangement of qualities that blend pathogenesis-related proteins, including β-glucanases and chitinase which can assault the cell dividers of attackingparasitic fungi. The headache medicine like compound salicylic corrosive. It has been distinguished that the flag atom required in SAR, and this is bolstered by the finding that salicylic corrosive can be connected exogenously and will incite the blend of pathogenesisrelated proteins, prompting upgraded plant resistance. Indeed, salicylic corrosive is advertised monetarily for this reason. The amazing point is that SAR presents a summed up imperviousness to an extensive variety of pathogens, including parasitic fungi, microbes, and infections. [66]

A moment sort of instigated resistance creates in light of colonization of plant roots by particular nonpathogenic strains of rhizosphere microscopic organisms, for example, Pseudomonas fluorescens. Be that as it may, this guard framework is interceded by an alternate arrangement of flagging mixes, including jasmonic corrosive and ethylene. This sort of actuated resistance is named rhizobacteria-interceded incited systemic resistance (ISR). L [67]ike SAR, initiated systemic resistance acts against a few distinct sorts of pathogen, including strains of Fusarium oxysporum, the fleece mold pathogen Peronospora parasitica (Oomycota), and the bacterial pathogens Xanthomonas campestris and Pseudomonas syringae. Along these lines, there are two unmistakable flagging pathways, one intervened by salicylic corrosive and one by jasmonic corrosive/ethylene, and both give imperviousness to pathogenic assault. Late examinations recommend that both of these pathways can act all the while, on the grounds that there is no critical crosstalk between them. In this manner, the blend of SAR and ISR could present an upgraded level of plant resistance.[68]

VASCULAR WILT DISEASES

Vascular wither pathogens cause probably the most destroying plant sicknesses. Their trademark method of contamination is to enter the water-leading xylem vessels, either by methods for vectors, for example, bark scarabs on account of Dutch elm illness (Ophiostoma ulmi and O. novo-ulmi) or by means of wounds, or by attack of the youthful root tips. At that point these organisms multiply in the xylem as spores or yeast-like maturing cells and are conveyed upwards in the water stream.[69] The spores wind up plainly caught on the punctured end dividers of the xylem vessels, where they sprout, become through the pores, and deliver facilitate spores that are conveyed continuously upwards. Plants react to intrusion by the vascular wither organisms in a few ways:

- Balloon-like swellings (called tyloses) can swell into the xylem from parenchyma cells nearby the xylem vessels, this is by all accounts an underlying reaction before other protection systems become possibly the most important factor.[70]
- Pectic gels are expelled into the xylem, through circumscribed pits where the vessel divider is thin on the grounds that it

comprises just of the essential divider, not overlaid by the auxiliary cellulosic divider.

- Phenolic mixes are discharged into the vessels from phenol store cells. The phenolic mixes are then oxidized and polymerize, balancing out the gels and to make a fungitoxic situation.
- Phytoalexins gather in the vessels, apparently as a component of a general anxiety reaction.[71]

On the off chance that the host reacts sufficiently quick to attack of the xylem vessels, the contamination will be contained and the plant will endure no noteworthy harm. In any case, if the host reaction is moderate the parasite can logically colonize the vessels, causing quick shrinking and plant passing. The fascinating element of vascular shrink parasitic fungi is that they stay restricted to the nonliving xylem vessels until the point when the plant is vigorously infected, and at exactly that point do they attack the living parenchyma cells, creating thick-walled resting structures which inevitably come back to the dirt when the tissues deteriorate.[72]

Figure- 8.11: Vascular wilt disease of smokebush.

Vascular wilt in Fungi

The three fundamental parasitic fungi that cause vascular shrink ailments are Fusarium oxysporum, Verticillium albo-atrum, and V. dahliae. Of these, *F. oxysporum* is the most critical on the grounds that it comprises of more than 80 strains that show particular pathogenicity to specific yield species (Armstrong and Armstrong 1981). These strains are named "exceptional structures" (formae speciales), shortened to ff. ssp (plural) or f. sp (particular). For instance, the vascular wither pathogen of bananas is named F. oxysporum f. sp. cubense, and the vascular shrink pathogen of tomato is F. oxysporum f. sp. lycopersici. Every one of these strains seem to be indistinguishable, and all make due by delivering chlamydospores (spores gotten from individual hyphal compartments by amassing supplements and building up a thick, safe cell divider). These spores can make due in soil for a long time, yet most likely likewise develop and keep up their populace by developing in a saprotrophic mode in the rhizosphere. Verticillium albo-atrum and V. dahliae likewise cause genuine shrivel illnesses, for example, verticillium shrink of jumps and numerous different plants. In any case, they appear to be less host-particular than F. oxysporum, in light of the fact that strains disengaged from one host will regularly contaminate a few other random has in fake immunizations.[73]

Pathogenicity Determinants

The run of the mill indications of vascular shrink illnesses more often than exclude the dynamic yellowing and demise of leaves, joined by early crumple of the petioles (leaf stalks) so that the leaves hang down – a component named epinasty. These indications unequivocally recommend a part of poisons in the infection disorder. Fusaric corrosive was one of the primary potential poisons to be recognized. It is known to be created by F. oxysporum in research facility culture. It has additionally been identified in ailing plants, and when connected to plant protoplasts in vitro it builds their porousness to particles. Nonetheless, there is still uncertainty about the part of fusaric corrosive in vivo in light of the fact that its focuses in xylem liquid are frequently low. An additionally persuading part has been built up for a poison named cerato-ulmin, delivered by Ophiostoma ulmi (the reason for Dutch elm sickness). At the point when the bases of cut elm shoots are inundated in arrangements of cerato-ulmin they build up the ordinary leaf indications of Dutch elm illness. Cerato-ulmin is a hydrophobin, and hydrophobin particles suddenly gather into water-repellent layers at an air–water interface. Consequently they could have a critical part in confining water stream in xylem vessels. [74]

Role of Environment Factors

Natural variables assume a noteworthy part in vascular wither ailments. Early confirmation of this was found in the significant banana manors of Central America, where soils could be sorted as either "long-life" or "short-life." In long-life soils, banana ranches could stay gainful for a long time, with little proof of ailment, though short-life soils ended up plainly ineffective inside a couple of years of building up the product. Concoction examinations uncovered a general connection between's the diverse sorts of dirt mineral and the life of banana ranches. Long-life soils have a tendency to have a higher extent of montmorillonite dirts, instead of the kaolinite muds of short-life soils. This has been connected to a higher populace of hostile microbes, on the grounds that montmorillonite dirts hold more supplements and make a greater pH for microorganisms. [75]

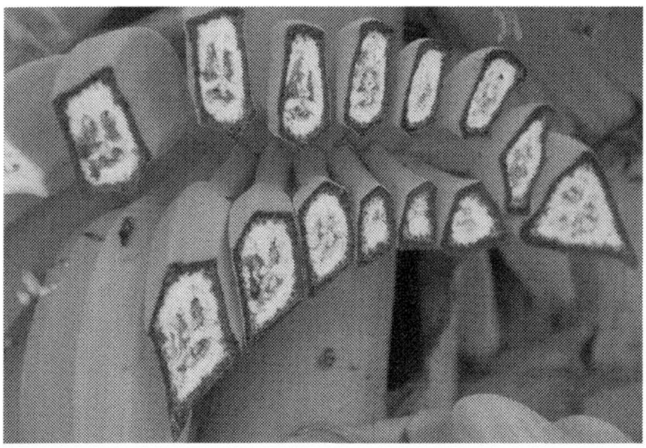

Figure- 8.12: Panama disease of banana.

At one phase the dynamic decrease of banana estates in Central America undermined the whole eventual fate of banana creation in that district, which depended on a solitary cultivar, "Gros Michel," that is profoundly powerless to Panama malady. The risk was turned away by the possibility revelation of a more wiltresistant cultivar of the "Cavendish" sort. Attributable to its incentive as an ailment safe cultivar, "Cavendish" bananas were just planted on the more great destinations in Central America, and they have remained diseasefree for over 50 years. In any case, the circumstance is distinctive in the subtropics (South Africa, Taiwan, Canary Islands, and Northeast Queensland) where the Cavendish cultivars have been presented. A number of these plants have now kicked the bucket from Panama ailment, corresponding with the rise of another pathotype (race 4) of F. oxysporum f. sp. cubense. The conditions in these nations are negligible for banana generation on the grounds that the winter temperatures drop to 12°C or lower. Bananas quit developing at this temperature (confirm by their inability to make any further leaf development), yet *F. oxysporum f. sp. cubense* can at present develop at 12°C in vitro. This could have encouraged the development of the new pathotype.

Fusarium Wilt-Suppressive Soils

The expressions "long-life" and "short-life" for Fusarium wither suppressive soils have now been supplanted by the terms malady suppressive and sickness favorable soils. Numerous cases of infection suppressive soils have now been accounted for vascular wither ailments and different pathogens over the world (Schneider 1982). In all cases there is clear proof that microorganisms add to suppressiveness, on the grounds that suppressive soils can be made helpful by sanitization, and on the other hand the presentation of a moderately little measure of suppressive soil (10% or less) to purified soil or even to ordinary handle soil can make a dirt suppressive.

The absolute most fascinating cases of ailment concealment are found in destinations, for example, the Chateaurenard area of France, where vegetables have been developed for quite a long time with next to zero sickness. In these destinations there is solid confirmation that nonpathogenic strains of Fusarium oxysporum contend with the pathogenic strains for natural substrates. Likewise, in these suppressive soils the chlamydospores develop ineffectively, the germtubes develop inadequately, the pathogen populace in the dirt decays all the more quickly, and a considerably higher inoculum level is expected to cause sickness. These dirts are observed to be suppressive to an extensive variety of formae speciales of F. oxysporum, yet are not suppressive to other, irrelevant pathogens, for example, Verticillium dahliae. Endeavors are currently being made to abuse nonpathogenic strains of F. oxysporum as business organic control operators of the pathogenic fusarium wither growths. This is not without its threats, since despite everything we know minimal about the variables that administer the host ranges – or the pathogenicity – of vascular wither fusaria.

The Enigma of Vascular Wilt Fusaria

The vascular wither fusaria are mysterious on the grounds that, from one viewpoint, they cause the absolute most destroying plant illnesses and demonstrate a high level of hostspecificity, yet then again they become just in the nonliving xylem tissues (after passage through the juvenile root tips or through injuries) and just attack the living plant tissues when the plant starts to senesce. Essentially, these growths carry on as frail parasitic fungi which, for the majority of their life in the plant, rely upon the low levels of natural supplements that hole into the xylem vessels. In a few regards this conduct is like the method of development of the nonpathogenic contagious endophytes which develop inadequately in the divider spaces and intercellular liquids of plants without making evident mischief the host (talked about underneath). Endophytism is described by a fundamental similarity between the colonizing organism and the host cells.

Reliable with this, a portion of the communications between vascular wither fusaria and their host plants – for instance, F. oxysporum f. sp. pisi on peas – are administered by a quality for-quality relationship run of the mill of biotrophic plant pathogens, talked about later in this part.

Smut Fungi

The muck growths are individuals from the Basidiomycota and pick up their name from the way that they create a great many dark, dingy spores, known as teliospores. More than 1000 types of filth growths are known to parasitize plants, yet the types of most monetary significance are those that assault grain crops, including free muck of wheat (Ustilago nuda), maize muck (Ustilago maydis), and stinking filth of wheat (Tilletia caries), so-named on the grounds that the spores have a fishy odor because of the nearness of trimethylamine. The trademark highlight of muck organisms is that they grow gradually inside the plant tissues, regularly creating just little measures of hyphal development, however later in the developing season they deliver huge quantities of diploid teliospores, which create set up of the oat grain or other conceptive structures. For instance, in free muck of grains the entire blooming spike is supplanted by a mass of spores, which are uncovered when a thin sheath which contains these spores separates at development. Stinking filth is comparable, yet the grains have all the earmarks of being typical in light of the fact that they are encompassed by the seed coat (pericarp) and the spores are just discharged when the grain is sifted. Plants wind up plainly tainted by free filth while growing spores enter the creating ovary, so that the subsequent seeds as of now contain the mycelium.

Figure- 8.13: Barley loose smut.

At that point, when the seeds develop the growth continuously colonizes the stem tissues and in the long run enters the blooming spike where it replaces the grain with a spore mass. Contamination by stinking filth is to some degree distinctive: the spores can get by for up to 15 years, and they grow in the meantime as the grain, enter the base of the shoot (coleoptile), and after that continuously colonize the tissues, driving at long last to the generation of a spore mass. In different mucks the procedure of improvement is diverse once more. For instance, in Uromyces anemones (a filth parasite that develops on anemone and buttercup) the spore masses create underneath the surface of the stem or leaf before they develop and emit through the plant epidermis. At the point when the teliospores sprout, they experience meiosis to create a haploid stage named the promycelium. This produces uninucleate sporidia, proportional to basidiospores. At that point plasmogamy (cell combination) of sporidia of good mating sorts offers ascend to a dikaryon, and this will in the end deliver the mass of teliospores. A fascinating component of filth organisms is that the dikaryotic stage is by all accounts obligately parasitic on plants, however the monokaryotic, uninucleate sporidia increase as yeast-like sprouting cells. These are nonpathogenic and can be refined on standard research facility media.

FUNGAL ENDOPHYTES AND THEIR TOXINS

Lately the leaves and underlying foundations of many sorts of plant have been appeared to harbor organisms that develop as indication less endophytes. These organisms don't enter the plant cells, rather they develop as moderately inadequate hyphae in the spaces between the cells or inside the dividers of the plant cells, using low levels of supplements in the intercellular liquids and causing no undeniable harm. From various perspectives, this conduct is like the route in which the vascular shrink fusaria colonize the low levels of supplements in the xylem liquids, talked about prior.

The noteworthiness of parasitic endophytes lies in the way that they create an extensive variety of optional metabolites of potential business significance. Furthermore, a portion of the endophytes present abiotic stretch resistance on their plant has, or go about as bug antifeedants. In the discourse that tails we will concentrate on one specific gathering of contagious endophytes – those that develop in some field grasses and have over and over been appeared to diminish creepy crawly harm (Bony et al. 2001). The grass-related endophytes become gradually in research facility culture and don't deliver sporing stages.

Indeed, some of them have most likely lost this capacity out and out in light of the fact that they generally develop inside a plant and are exchanged from era to era by developing into the seed coat. This makes it simple to look at the conduct of endophyte-contaminated and sans endophyte populaces of grasses. The few disconnects that have been initiated to sporulate in culture deliver a straightforward conidial organize named Neotyphodium (beforehand called Acremonium) and looking like the conidial phase of the ergot growth Claviceps purpurea. These "clavicipitaceous endophytes" (so-named in light of the fact that they are identified with the ergot parasite) are found in a few essential field grasses, for example, Lolium (ryegrass), Festuca (fescue), and Dactylis glomerata (cocksfoot) in the USA, Europe, and New Zealand. They develop inside the leaf tissues yet their development is by all accounts firmly controlled on the grounds that it is synchronized with development of the leaves, and frequently stops when the leaves develop (Clay 1989). Endophyte-pervaded lasting ryegrass has been appeared to contain a few mycotoxins, including three that are especially imperative:

- The soaked aminopyrrolizidine alkaloids, for example, lolitrem B, has been appeared to be in charge of the harming "ryegrass amazes" sickness of munching sheep and dairy cattle in New Zealand. These alkaloids can achieve levels of up to 2% in the grass biomass.
- Ergovaline, which causes sicknesses, for example, "fescue foot" and "fescue toxicosis" in touching creatures.
- Peramine, a tripeptide that is repellant and dangerous to creepy crawlies, however not to well evolved creatures.

The generation of these mixes is presumably supported by the moderate, substrate-constrained development of endophytes inside plant tissues, since auxiliary metabolites naturally aggregate in conditions that limit typical development. The numerous impacts of these mycotoxins raise situations. From one perspective, endophyte-contaminated grass assortments can posture genuine dangers to nibbling domesticated animals. Then again, there are many reports that endophytes increment the anxiety resistance of grasses, and present an upper hand over sans endophyte grasses by stopping creepy crawly harm. The endophyte Neotyphodium coenophialum is found in tall fescue field and creates ergovaline in the plant tissues. Ergovaline causes a scope of indications in brushing creatures, including hyperthermia, weight reduction, lessened rates of pregnancy, diminished drain generation, and (in stallions) birth deformities and premature birth.

The endophyte N. lolii in lasting ryegrass produces lolitrem B as its significant poison. This neurotoxin is in charge of a condition called ryegrass amazes, where creatures encounter tremors and loss of coordination. The potential degree of these issues in the USA was uncovered by a study of stallion pastures (USDA 2000). Broadly, 61.6% of inspected pastures tried positive for endophytes, and an aggregate 28.5% tried positive for poison. Without mediation, the disease rates would increment with time in light of the fact that contaminated grasses can outcompete non-tainted grasses.

Since the clavicipitaceous endophytes never leave the host plant, field and turf directors have choices. For munching creatures, a field can be restored with affirmed without endophyte seed, though administrators of decorative or recreational turf can choose to sow endophyte-tainted seed for enhanced grass life and creepy crawly resistance. This has generated a specialty advertise in the USA, where "endophyte-improved grasses" are accessible through garden focuses and seed stores for use in comfort fields, greens, sports turf, and so on.

PHYTOPHTHORA RAMORUM: SUDDEN OAK DEATH

Another types of Phytophthora was found in Germany and the Netherlands from 1993 onwards, and was depicted formally in 2001 as P. ramorum. It makes harm the branches of Rhododendron and, less regularly, Viburnum brambles, now and again prompting demise of these bushes. This malady is currently spreading on decorative rhododendrons and viburnums in tree nurseries crosswise over Europe. In the mean time, in the waterfront haze belt of northern California and southern Oregon, a sudden passing of oak trees was happening, and this ailment was additionally observed to be caused by P. ramorum. It assaults a few sorts of oak, however primarily the "live oaks" and "tan oaks" which are a piece of the common vegetation of the seaside bush/tree group. The organism assaults the plants in two ways – either by creating injuries on the leaves and terminal shoots of an extensive variety of bushes, prompting shoot dieback, or by delivering ulcers close to the bases of the trees, where the cambium (which lies just underneath the bark and delivers the yearly rings of new wood) is logically crushed.

The side effects of this incorporate breaking of the bark and drainage of dull, gooey sap from the bark close to the base of the tree – an indication named gummosis. In cutting edge phases of the sickness the trees bite the dust all of a sudden and drastically, consequently the name, sudden oak demise. In any case, despite the fact that the demise of the trees happens all of a sudden it is gone before by a moderate, dynamic develop of disease underneath the bark. This can regularly go unnoticed until the point when the trees abruptly create propelled side effects.

Figure- 8.14: Sudden oak death.

There is a nearby parallel between P. infestans and P. ramorum, on the grounds that both have been brought into parts of the world where they didn't happen earlier and both have caused across the board harm. Current proof proposes that the European populace of P. ramorum comprises just of the A1 mating sort, while the North American populace is of the A2 mating sort. This firmly demonstrates isolate wellsprings of birthplace of the two populaces, yet the characteristic land starting point of the organism stays obscure. Here there is a parallel with the spread of another forceful Phytophthora species, P. cinnamomi, which assaults many plants over the world, however the land starting point of this growth is obscure. It is making genuine harm the eucalypt vegetation of Australia. At whatever point a growth attacks a zone of regular vegetation it is likely that the parasite has been presented from somewhere else, on the grounds that plants in normal groups co-develop with the inhabitant pathogens.

BIOTROPHIC PLANT PATHOGENS

Biotrophic plant pathogens are portrayed by the way that they have an expanded nutritious association with living host cells, as opposed to necrotrophic pathogens which murder the plant tissues. The accomplishment of this sort of parasitism relies upon two fundamental elements:

- The capacity to abstain from evoking host cell demise.
- A methods for securing a persistent supplement supply from the living host tissues. In this manner, biotrophs are from numerous points of view the embodiment of effective parasitic fungi– they bolster from the host cells without murdering them.

There are many sorts of biotrophic growth and organism like life forms, since this method of parasitism has advanced autonomously on a few events. The real gatherings of biotrophs incorporate the rust organisms (Basidiomycota), the fine buildup growths (Ascomycota), the fleece molds (Oomycota), and the intracellular plasmodiophorids (protists of dubious ordered liking, however unmistakably particular fromparasitic fungi). Notwithstanding these, the endophytic growths develop basically as biotrophs. An exemplary illustration is the growth Fulvia (Cladosporium) fulvum which causes blue form ailment of tomato plants. Not at all like the pervasive Cladosporium spp. that develop as epiphytes on leaves, F. fulvum enters the host stomata from a growing spore and structures a broad intercellular system in the leaf, before rising up out of the stomata as conidiophores which discharge promote spores. It has no undeniable methods for sustaining from the living host cells aside from by rummaging leachates, yet it may upgrade the rate of spillage and furthermore utilize a portion of the cell divider polymers. A strain of F. fulvum can just encourage from the host on the off chance that it abstains from prompting a touchy reaction, and this is represented by "avirulence" qualities of the pathogen and comparing resistance qualities in the host. This quality for-quality association with the host plant (Honee et al. 1994) is usually found in biotrophic plant pathogens, and is talked about later. In the accompanying areas we consider a few cases of biotrophic plant pathogens.

Haustorial Biotrophs

Numerous biotrophic parasitic fungi start contamination from spores that arrive on the host surface and after that grow and enter through the stomata. The underlying occasions include the exact introduction of germ-tube development, utilizing signals, for example, plant surface geology and supplement slopes. The hyphae at that point develop between the host cells, and append to singular cells by delivering a haustorial mother cell. From this, a tight infiltration hypha becomes through the host cell divider however it doesn't enter the film. Rather, the host cell film invaginates to suit the attacking organism, which forms into a haustorium that is constantly isolated from the host cytoplasm by an extrahaustorial layer. The develop haustorium is encompassed by a liquid grid, and there is a tight "seal" in the neck area to keep the liquid from getting away.

One of the outcomes of this in trial work is that the entire haustorial complex (encompassing layer, liquid network, and the haustorium itself) can be confined by processing the host cell divider with proteins. From thinks about on these segregated haustorial buildings of rust and fine buildup growths it has been demonstrated that the extrahaustorial film permits the free section of sugars and amino acids. The divider and layer of the plant cell can likewise be required to permit the section of supplements through fitting film transporters. In this way, the vital interface for supplement take-up by the growth is probably going to be the haustorial film.

In a model proposed by Voegele and Mendgen (2003), sugars and amino acids are accepted to be taken into the haustorium in relationship with H^+ particles by layer symporters, utilizing the proton thought process constrain created by ATPase. The sugars are then utilized for biosynthetic responses inside the haustorium. Fructose (inferred by hydrolysis of sucrose to glucose and fructose in the extrahaustorial grid) is changed over to mannitol, which is translocated into the parasitic hyphae and at last into the creating spores. The net outcome is that supplements are ceaselessly pulled back from the living plant cells, to help rehashed cycles of sporulation on the host surface. A significant part of the monetary significance of rust and fine mold organisms comes from this constant withdrawal of supplements from the plant to help sporulation of the growth.

Rust Fungi

Rust parasitic fungi pick up their name from the unmistakable rustcolored spores, named uredospores, which are created in plenitude for dispersal amid the late spring months. More than 4000 rust species have been depicted on harvest or wild plants. Many rust parasitic fungi require two separate hosts to finish their life cycle. For instance, Puccinia graminis (dark stem rust of wheat) requires both an oat and another host (barberry). Be that as it may, other rust parasitic fungi, for example, mint rust (Puccinia menthae), finish their life cycle on a solitary host. Beneath, we will consider the life cycle of P. graminis on the grounds that it is a standout amongst the most monetarily critical species, particularly in North America. It likewise shows the greater part of the commonplace sporing phases of rust organisms.

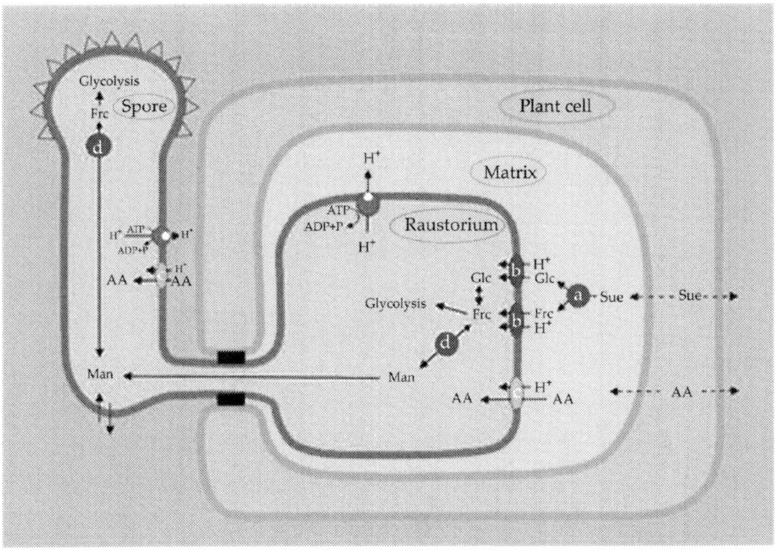

Figure- 8.15: Sugar and amino acid redistribution and uptake in rust fungi.

Black Stem Rust of Wheat (Puccinia graminis)

We will start this cycle at the dispersal stage, when rust pustules blasted through the epidermis of oat or grass leaves in early spring and discharge substantial quantities of the trademark rust-hued uredospores, from structures called uredinia. Each uredospore is single-celled however contains two cores, one of each mating similarity sort, so the spores are dikaryotic.The uredospores are scattered by wind and can experience a few cycles of contamination over the span of a season, bringing about real pestilences. Towards the finish of the season the creation of uredospores stops, and rather the pustules (now named telia) deliver two celled teliospores. These likewise are dikaryotic, with two cores in every cell, except the cores breaker to shape diploid cells, and the growth overwinters in this frame.

Figure- 8.15: Puccinia graminis effect on Barberry.

Ahead of schedule in the spring every phone of the teliospore grows to create a short mycelium. The diploid core moves into this and experiences meiosis, prompting the generation of four monokaryotic basidiospores (spores with just a single haploid core). These spores will just grow further on the off chance that they arrive on a barberry leaf.

At that point they enter the leaf and, a couple of days after the fact, deliver carafe formed spermogonia. These have responsive hyphae and furthermore discharge numerous little spermatia. Each spermogonium produces spermatia of a solitary mating sort and they are oozed from the neck of the spermogonium in a sugary liquid. Flies and different bugs are pulled in to this, and when they visit a spermogonium of an alternate mating sort they exchange the spermatia. The spermatia of one mating sort at that point meld with open hyphae of the inverse mating sort and exchange the core through a pore in the divider.

This core at that point separates and relocates through the monokaryotic hyphae that developed from the basidiospores, delivering a dikaryotic mycelium. At this point, the mycelium has become through the thickness of the leaf and creates a further sporing structure that emits through the lower epidermis – an aecium containing aeciospores. The dikaryotic aeciospores can just taint wheat, and they offer ascent to diseases on which uredospores will grow, in this manner finishing the cycle. A potential answer for breaking the cycle of dark stem rust is destroy barberry hedges so that the parasite can't taint its substitute host. A noteworthy barberry annihilation program was endeavored in North America in the most recent century, however was unsuccessful in light of the fact that the plague spreads continuously northwards from the early-sown products in Mexico to progressively later-sown harvests in the all the more northerly districts of the USA.

POWDERY MILDEW FUNGI

The fine buildup growths (Ascomycota) cause huge scourges in many seasons on the off chance that they are not treated with fungicides, and can be especially genuine in dry, hot summers. A few product and wild plants are tainted by these growths, which by and large are hostspecific. Regular cases incorporate fine buildup of roses (Sphaerotheca pannosa), of gooseberries (S. mors-uvae), and of hawthorn shrubberies (Podosphaera clandestina). The vast majority will have seen at least one of these parasitic fungi since they deliver white, fine malady pustules on the surfaces of many leaves and organic products.

Figure- 8.16: Powdery mildews on plants.

In any case, the most monetarily harming species is Blumeria graminis (already called Erysiphe graminis), which causes fine buildup of oats. Like the rust parasitic fungi, *B. graminis* experiences numerous cycles ofninfection in a solitary season and can make genuine harm a scope of oat crops. A solitary "stategically" situated sore at the base of a grain leaf can adequately keep that leaf from being a net exporter of supplements. This can influence root development and additionally shoot development, since roots are moderately poor contenders for plant acclimatizes, thus the impacts of this organism are exacerbated in dry soil conditions. However, strangely, the spores of fine buildup growths can't taint in wet leaf conditions, just in drier conditions.

Downy Mildews (Oomycota)

The wool buildup pathogens are haustorial biotrophs with comparative conduct to the rust and fine mold growths despite the fact that they require more sticky conditions for contamination and sporulation. Cases incorporate Bremia lactucae on lettuce, Plasmopara viticola on grape vine, and Pseudoperonospora humuli on jumps. In any case, there is a degree of conduct in the Oomycota, on the grounds that some Phytophthora spp. which are firmly identified with the wool molds start their parasitic stage as haustorial biotrophs however then murder the host tissues and spread inside them as necrotrophs (e.g. *P. infestans* on potato and *P. sojae* on soybean).

Figure- 8.17: Downy mildews on leaves.

Pathogen Recognition: the Gene-for-Gene Hypothesis

In many editing frameworks, vast zones of land are sown with a solitary yield assortment (cultivar) that has been reared for imperviousness to every known race of the pathogen. Choice weight at that point works on the pathogen to defeat this resistance by change. Since a huge number of spores are discharged by the significant pathogens every year – even in a solitary field – there is a solid probability of the resistance separating. This is particularly valid for crops that have been reproduced for single real quality resistance (R quality resistance) rather than "field resistance" that depends on the consolidated exercises of a few "minor" qualities. The famous plant pathologist H.H. Flor, working with flax rust (Melampsora lini) in the 1950s, proposed a basic plan to clarify the connection between significant quality resistance and the event of sickness – the quality for-quality theory. In light of broad research of the hereditary qualities of both the host plant and the pathogen, he demonstrated that for each quality that presents resistance (R, a hereditarily overwhelming characteristic) in flax plants, there is an integral quality that gives avirulence (AVR, again a hereditarily predominant attribute) in the pathogen. Along these lines, the result of a host–pathogen communication can be outlined in the accompanying table:

Table- 8.1: Genotype of pathogen

		Host Genotype	
Pathogen Genotype		RR or Rr	Rr
	AVR avr or AVR AVR	Disease resistant	Susceptible to disease
	avr avr	Susceptible to disease	Susceptible to disease

Where RR = homozygous safe, Rr = heterozygous safe, rr = homozygous helpless, AVR = homozygous avirulent, avr = heterozygous avirulent, avr = homozygous harmful. From this table we can see that resistance happens just in blends including the predominant R allele and the overwhelming AVR allele. The least complex clarification would be that the protein result of the R quality collaborates with the protein result of the AVR quality, prompting the touchy reaction – a quick cell passing that would keep a biotroph from creating. Basically, the R quality result of the host would resemble a receptor, collaborating with the AVR quality item (an elicitor) of the organism.

Late research has given a more profound comprehension of this framework, and demonstrated that quality for-quality connections are normal over a scope of host–parasite blends, incorporating communications of plants with pathogenic organisms, microscopic organisms, infections, and nematodes, and even in some plant–insect associations. A direct protein–protein collaboration administered by R qualities and AVR qualities, and prompting the easily affected reaction, has been shown in two pathosystems: contamination of tomato by Pseudomonas syringae (pathovar syringae) and contamination of rice by Magnaporthe grisea (rice impact infection). In these cases the cooperation most likely includes the acknowledgment of surface-found proteins. Be that as it may, a direct protein–protein cooperation has not been found in some other host–pathogen frameworks to date. Rather, there is mounting proof that no less than a third (plant) protein is included in most quality for quality associations, and this third protein (or further proteins) intervenes the guard reaction. There are a few lines of proof for this (checked on by Luderer and Joosten 2001, Bogdanova 2002). For instance, the proteins required in these associations appear to be cytoplasmic proteins, not cell surface receptors included straightforwardly in pathogen acknowledgment, as was once thought. The cytoplasmic proteins constitute a family, or families, of related proteins with comparative properties. For instance, the "model" plant Arabidopsis thaliana (which is broadly utilized for atomic hereditary examinations) contains a quality that codes for imperviousness to two bacterial pathogens. The result of this quality is like the resistance quality results of a few different creatures – a quality from tobacco giving imperviousness to an infection, a quality from flax presenting imperviousness to flax rust, a quality from tomato giving imperviousness to the growth Fulvia fulvum, and a quality from sugar beet giving imperviousness to a nematode. All these resistance quality items have a locale of leucine-rich rehashes, and all have a nucleotide-restricting site that could start a flagging course prompting initiation of the plant's protection reaction.

SUMMARY

- Fungi is major disease causing agent in plants. About 70% of all main diseases of crops are due to fungus like organisms and fungi. One of the most disastrous example of fungi causing disease in plant is Phytophthora infestans that cause potato late blight disease. Potato was main food for people in Ireland. More than half million people died due to starvation.
- There are thousands of plants-pathogenic fungi, so to identify them they were divided in different groups, like seed rots pathogens, necrotrophic pathogens of compromised and immarture hosts, stalk rot pathogens, fruit pathogens, and host specialized pathogens.
- In order to prevent from these diseases plants have defensive system. These systems involve both fungi-toxic compounds and physical barriers in tissues of plants. Physical defensive system involves cuticle of stem and leaf surface having enzyme known as cutinase that has been thought important for fungal infection. Plants respond to the fungi by forming papillae at the sight of attempted invasion. Papillae is localized thickening of cell wall where attempts of invasions happened.

REFERENCES

1. Joseph, H. (2007). *Sex in Fungi*. ASM Press.
2. Agrios, G. (2014). *Plant Pathology*. Saint Louis: Elsevier Science.
3. Chen, R. S., & McDonald, B. A. (n.d.). *Sexual Reproduction Plays a Major Role in the Genetic Structure of Populations of the Fungus Mycosphaerella Graminicola*.
4. Pearson, MN, Beever, RE, Boine, B, & Arthur, K. (n.d.). *Mycoviruses of filamentous fungi and their relevance to plant pathology*. Blackwell Publishing Ltd.
5. Badouin, H., Giraud, T., Capy, P., Coelho, S., Glemin, S., Marais, G., De, M. S., Laboratoire Ecologie, systématique et évolution (Orsay, Essonne). (2016). *Génomique évolutive chez les champignons Microbotryum: Adaptation et chromosomes de types sexuels*.
6. S Pande, M Sharma, M Pathak, J Narayana Rao, P Anil Kumar, D

Madhusudan Reddy, VI Benagi, ... BS Eksinghe. (2006). *Outbreak of phytophthora blight of pigeonpea in the Deccan Plateau of India, 2005*. International Crops Research Institute for the Semi-Arid Tropics.

7. Rubin, R. (2014). *Clinical Approach to Infection in the Compromised Host*. Springer Verlag. .
8. Klastersky, J., & Staquet, M. J. (1982). *Combination antibiotic therapy in the compromised host*. New York: Raven Press.
9. Cunha, B. A., & Moellering, R. C. (2001). *Infections in the compromised host*. Philadelphia: W.B. Saunders Co.
10. Good, R. A. (1984). *Intravenous immune globulin and the compromised host: Proceedings of a symposium held September 21 to 23, 1983, in Dallas*. New York: The American Journal of Medicine.
11. In Hayden, R. T., In Carroll, K. C., In Tang, Y.-W., & In Wolk, D. M. (2009). *Diagnostic microbiology of the immunocompromised host*. Washington, D.C: ASM Press.
12. Shellito, J. (2005). *The compromised host: Virtual symposium*. New York, NY: American Thoracic Society.
13. Goyal, A., & Manoharachary, C. (2014). *Future Challenges in Crop Protection Against Fungal Pathogens*. New York, NY: Springer New York.
14. Hayden, K.J., Hardy, G.E.St.J., & Garbelotto, M. (2013). *Oomycete diseases*. CAB Publishing. .
15. International Plant Propagators' Society., & Plant Propagators' Society. (1999). *Combined proceedings, volume 48, 1998*. Mechanicsburg, Pa.: International Plant Propagators' Society. .
16. Gonthier, P., & Nicolotti, G. (2013). *Infectious forest diseases*. Wallingford, UK: CABI.
17. Giesler, Loren J. (2000). *NF00-411 Seed Treatment Fungicides for Soybeans (Revised January 2004)*. DigitalCommons@University of Nebraska - Lincoln.
18. Nahar, Sharfun. (2002). *PATHOGENIC EFFECTS AND PROFILES OF SECONDARY METABOLITES OF SEED-BORNE TOXIGENIC FUSARIUM SPECIES.*

19. Blancard, D., Laterrot, H., Marchoux, G., In Fletcher, J., Candresse, T., & McGee, D. (2012). *Tomato diseases: Identification, biology and control.*
20. Saha, B. C., Hayashi, K., American Chemical Society., & American Chemical Society. (2004). *Lignocellulose biodegradation.* Washington, DC: American Chemical Society.
21. Baminger, Ursula, Subramaniam, Sai S., Renganathan, V., & Haltrich, Dietmar. (n.d.). *Purification and Characterization of Cellobiose Dehydrogenase from the Plant Pathogen Sclerotium (Athelia) rolfsii.* American Society for Microbiology.
22. Harlton, Colleen E. (n.d.). *Genetic diversity and variation in Sclerotium (Athelia) rolfsii and related species.*
23. Kernkamp, M. F. (1952). *Investigations on physiologic specialization and parasitism of Rhizoctonia solani.* Minnesota: Univ. of Minnesota.
24. Durbin, R. D. (1958). *Factors influencing the ecology of clones of Rhizoctonia solani.*
25. Zarka, A. M. (1965). *Studies on Rhizoctonia solani Kühn, the cause of the black scurf disease of potato.* Wageningen: s.n.
26. Mishra, P. K., & Kumari, M. (2016). *Molecular characterization of Rhizoctonia solani: Rhizoctonia solani.*
27. Parmeter, J. R., & Symposium on Rhizoctonia Solani (1, 1965, Miami, Fla.). (1970). *Rhizoctonia solani, biology and pathology: based on an American Phytopathological Society (Lafayette, Ind.) Symposium on Rhizoctonia solani held at the Miami meeting of the Society, Oct. 1965.* Berkeley Calif: Univ. of California P.
28. Boogert, P. H. J. F. (1993). *The ecology of Verticillium biguttatum, its significance for the population dynamics of Rhizoctonia solani in potato =: De oecologie van Verticillium biguttatum, zijn betekenis voor de populatiedynamiek van Rhizoctonia solani in aardappel.*
29. Laysa, F. D. (1980). *Etiology and epidemiology of bacterial stalk rot of corn.* Los Banos: University of the Philippines.
30. Adeyanju, A. (2014). *Genetic study of resistance to charcoal rot and Fusarium stalk rot diseases of sorghum.* Manhattan, Kan: Kansas State University.

31. Christensen, S. A. (2011). *The function of the lipoxygenase ZmLOX10 in maize interactions with insects and pathogens*. College Station, Tex.: Texas A & M University.
32. Bushula, V. S. (2017). *Genetic diversity and pathogenicity of sorghum-associated Fusarium species*. Manhattan, Kan: Kansas State University.
33. United States., & United States. (2015). *Identification and characterization of glycosyltransferases involved in the synthesis of the side chains of the cell wall pectic polysaccharide rhamnogalacturonan II*. Washington, D.C: United States. Dept. of Energy. Office of Science.
34. Nagodawithana, T. W., & Reed, G. (1993). *Enzymes in food processing*.
35. University of Washington., United States., United States., United States., & Cleland, R. E. (1991). *Studies on the control of plant cell enlargement by cellular parameters*. Washington, D.C: United States. Dept. of Energy.
36. Society for General Microbiology., Howie, J. W., & O'Hea, A. J. (1955). *Mechanisms of microbial pathogenicity: Fifth symposium of the Society for General Microbiology held at the Royal Instiution, London, April 1955*. Cambridge: Published for the Society for General Microbiology at the University Press.
37. Ulvskov, P. (2010). *Annual Plant Reviews, Plant Polysaccharides: Biosynthesis and Bioengineering*. Somerset: Wiley.
38. Meister, A. (2009). *Advances in Enzymology and Related Areas of Molecular Biology, 11*. Hoboken: John Wiley & Sons.
39. Keen, N. T., & Staff, N. A. S. (1900). *(NAS Colloquium) Virulence and Defense in Host--Pathogen Interactions: Common Features Between Plants and Animals*. Washington: National Academies Press.
40. Society for General Microbiology., Howie, J. W., & O'Hea, A. J. (1955). *Mechanisms of microbial pathogenicity: Fifth symposium of the Society for General Microbiology held at the Royal Instiution, London, April 1955*. Cambridge: Published for the Society for General Microbiology at the University Press.

41. Leone, G.O.M. (n.d.). *Significance and role of polygalacturonase production by Botrytis cinerea in pathogenesis*. Leone.
42. Vidhyasekaran, P. (2002). *Bacterial disease resistance in plants: Molecular biology and biotechnological applications*. New York: Food Products Press.
43. Zadoks, J. C. (2008). *On the political economy of plant disease epidemics: Capita .a in historical epidemiology*. Wageningen, The Netherlands: Wageningen Academic Publishers.
44. Güssow, H. T., & Dominion Experimental Farms and Stations (Canada). (1917). *The black or stem rust of wheat: A popular account of the nature, cause and prevention of grain rust*. Ottawa: Dept. of Agriculture.
45. Hines, R. L., & Michigan State University. (2004). *Field crop pest management: A guide for commercial applicators*. East Lansing, Mich: Michigan State University Extension.
46. United States. (1934). *Technical bulletins, nos. 376-400, with contents*. Washington, D.C: U.S. G.P.O.
47. Voytyuk, S. O., & Volz, P. A. (2009). *Biodiversity of the powdery mildew fungi (Erysiphales, Ascomycota) of Israel*. Ruggell [Lichtenstein: A.R.A. Gantner Verlag.
48. *Taxonomy of powdery mildew fungi in Himachal Pradesh*. (1977).
49. Ballantyne, B. (1971). *Powdery mildew fungi on cucurbitaceae in relation to breeding disease-resistant cultivars*. Sydney: publisher not identified.
50. Sutherland, A. M. (2009). *The feasibility of using Psyllobora vigintimaculata (Coleoptera: Coccinellidae), a mycophagous ladybird beetle, for management of powdery mildew fungi (Erysiphales)*.
51. Reiner, T. (2015). *Receptor-like cytoplasmic kinases as downstream effectors of RAC/ROP GTPases in the interaction of Arabidopsis and barley with powdery mildew fungi*.
52. Angeli, D. (2013). *New insights into host specialization and mycoparasitic interaction between powdery mildew fungi and Ampelomyces quisqualis*. Zürich: ETH.
53. In Martin, F. (2014). *Fungi*.

54. In Dean, R. A., In Lichens-Park, A., & In Kole, C. (2014). *Genomics of plant-associated fungi: Monocot pathogens.*
55. Reed, G. M. (1920). *Varietal resistance and susceptibility of oats to powdery mildew, crown rust and smuts*. Columbia, Mo: University of Missouri, College of Agriculture, Agricultural Experiment Station.
56. Mains, E. B., Martini, M. L., & United States. (1932). *Susceptibility of barley to leaf rust (Puccinia anomala) and to powdery mildew (Erysiphe graminis Hordei)*. Washington: U.S. Dept. of Agriculture.
57. Shaw. & Lucas,. (1998). *Downy mildews.* Cambridge: Cambridge University Press.
58. Safeeulla, K. M., & Downy Mildew Research Laboratory. (1976). *Biology and control of the downy mildews of pearl millet, sorghum and finger millet.* Mysore: Downy Mildew Research Laboratory, Manasagangothri, Mysore University.
59. Downy Mildews Workshop, Spencer-Phillips, P. T. N., Gisi, U., Lebeda, A., & International Congress of Plant Pathology. (2011). *Advances in downy mildew research*. Dordrecht: Springer. .
60. Downy Mildews Workshop, Spencer-Phillips, P. T. N., & Jeger, M. J. (2011). *Advances in downy mildew research: Volume 2*. Dordrecht: Springer.
61. Minchinton, E. J., Minchinton, E. J., Horticultural Research & Development Corporation., Agriculture Victoria., & Institute for Horticultural Development (Vic.). (1998). *Downy mildews on nursery plants*. Knoxfield, Vic: Agriculture Victoria, Institute for Horticultural Development. .
62. Marx, G. J. (2010). *Downy mildews: Genetics, molecular biology and control.* Place of publication not identified: Springer.
63. Lebeda, A., & Spencer-Phillips, P. T. N. (2007). *Proceedings of the 2nd International Downy Mildews Symposium: July 2-6, 2007, Olomouc, Czech Republic*. Olomouc: Palacky University. .
64. Minchinton, E. J. (1998). *Control of downy mildews in nursery seedlings*. Gordon, N.S.W: Horticultural Research and Development Corp.
65. Preece, T. F., & British Mycological Society. (2002). *A checklist*

of the downy mildews (Peronosporaceae) of the British Isles. Richmond: British Mycological Society.

66. In McCarthy, J. E. G., & In Tuite, M. F. (1990). *Post-Transcriptional Control of Gene Expression.* .
67. Singh, D. P. (1986). *Breeding for Resistance to Diseases and Insect Pests*. Berlin, Heidelberg: Springer Berlin Heidelberg.
68. Prell, H. H., & Day, P. (2001). *Plant-Fungal Pathogen Interaction: A Classical and Molecular View*. Berlin, Heidelberg: Springer Berlin Heidelberg.
69. Cold Spring Harbor Laboratory. (1986). *Molecular biology of Homo sapiens*. Cold Spring Harbor, N.Y: Cold Spring Harbor Laboratory.
70. Park, R. (2014). *Biotic stress tolerance and resistance*. London: Henry Stewart Talks.
71. Mount, M. (1982). *Phytopathogenic Prokaryotes V2*. Oxford: Elsevier Science.
72. Vanderplank, J. E. (1978). *Genetic and molecular basis of plant pathogenesis*. Berlin: Springer-Verlag.
73. Century, K. S., & University of California, Berkeley. (1996). *Genetic and molecular analysis of disease resistance in Arabidopsis thaliana*.
74. Talbot, N. (2004). *Plant-Pathogen Interactions*. Oxford: Blackwell Pub.
75. Yang, Bing, Sugio, Akiko, & White, Frank F. (n.d.). *Os8N3 is a host disease-susceptibility gene for bacterial blight of rice*. National Academy of Sciences.

CHAPTER

9

MOLDS OF MAN

Just nearly 200–300 fungi are described to grounds diseases of different warm blooded animals and humans, in compare to the several thousands of fungus infecting plants illnesses that are jointly called as mycoses. The fungi that cause diseases in different warm blooded animals and humans are, for the utmost quantity, resourceful or show just slight signs in common, well individuals. However the state has altered significantly in current ages, by the growing usage of immune-suppressant drugs in the advance of AIDS/HIV, cancer therapy, transplant surgery. Fungal toxicities can be life intimidating in these circumstances, however there are few certainly suitable drugs to regulate them deprived of producing opposing adverse effects.

Notwithstanding the intrusive mycoses, organisms represent a risk to wellbeing by delivering mycotoxins in foodstuffs and creature encourages, and airborne contagious spores can be noteworthy reasons for asthma, feed fever, and more genuine word related maladies. Taking every one of these variables together, growths can significantly affect human and creature wellbeing. In this part we consider the significant organisms that taint people and some other warm-blooded creatures.

MAJOR CONTAGIOUS PATHOGENS OF PEOPLE AND DIFFERENT WARM BLOODED ANIMALS

The human-pathogenic parasitic fungi can be gathered into five classes in view of elements, for example, their essential course of passage into the host, the kind of illness that they cause, and their normal wellsprings of inoculum. [1] A layout of these gatherings is given beneath and will give the premise to more point by point treatment in later areas of this section:

- The dermatophytes, otherwise called ringworm parasitic fungi, develop in the dead, keratinized tissues of the skin, nails, and hair. They are exceptionally normal, and taint vast segments of the human and creature populaces. The maladies that they cause are shallow, being kept to the dead tissues, yet they can cause extreme bothering of the basic living tissues, prompting auxiliary attack by microbes. [2] Large portions of these organisms demonstrate a level of specialization for specific hosts (people, felines, steers, and so forth.) however can cross-taint different hosts. The wellspring of inoculum is generally the shed keratinized tissues (chips of skin, hairs, and so forth.) in which the growth can hold on in a torpid stage. As a gathering, the dermatophytes are fruitful parasitic fungi with a plainly characterized specialty.
- Several Candida species develop as typical inhabitant commensals on the mucosal films of solid people, yet can wind up plainly obtrusive in fitting conditions. The diploid yeast Candida albicans is the prime case of this. It is discovered regularly on the mucosa of the gut, the mouth, and the vulvovaginal tract. However, in light of ecological triggers the yeast cells can create hyphal outgrowths that attack the mucosa, prompting clinical or subclinical conditions. For instance, "thrush" is a white, intrusive dotting of the throat, which is normal in neonates and insusceptible traded off people, including AIDS and malignancy patients, and individuals in the propelled phases of diabetes. [3] Candida can likewise cause extraordinary aggravation of the vulvo-uterine tract of

ladies amid phases of feminine cycle or amid pregnancy, and it ordinarily causes stomatitis in individuals who wear dentures.

- A couple of growths with melanized hyphal dividers, for example, types of Phialophora, Cladosporium, and Sporothrix, can attack damaged tissues, for example, profound injuries and cause tissue-debasing sores. In spite of the fact that they can be vital reasons for contamination, particularly among farming specialists in poorer nations, they are basically pathogens of wounds or damaged tissues. They won't be viewed as further in this section. Essentially, some normal individuals from the Zygomycota cause awful contaminations of individuals who experience the ill effects of diabetes and ketoacidosis.
- A particular gathering of primitive growth like living beings, now delegated diverse types of Pneumocystis, are found in the lungs of people and an extensive variety of mammalian hosts. They appear to be have particular since DNA grouping examinations of Pneumocystis strains from various hosts indicate wide hereditary assorted qualities, yet close closeness between strains confined from any one sort of host. The organism that contaminates people, Pneumocystis jirovici (some time ago known as *P. carinii*), is by all accounts especially normal in youthful kids, in light of the fact that regularly over 80% of kids demonstrate a neutralizer reaction to the pneumocystis antigen in skin tests. Be that as it may, the organism appears to vanish as kid's age, and is discovered again in AIDS patients, where it causes a destructive pneumonia that can be transmitted from patient to tolerant via airborne cells. Pneumonia caused by P. jirovici in HIV patients is regularly viewed as one of the main "Guides characterizing" diseases. [4]
- A differing gathering of parasitic fungi that regularly develop as saprotrophs in soil or on plant or creature remains can build up contaminations in the lungs from breathed in spores. These contaminations constantly come from an ecological wellspring of spores as opposed to from guide persistent to-quiet transmission. Any parasite that produces spores sufficiently little (c. 3–4 μm) to achieve the alveoli, that can develop at 37°C, and that can withstand the host's cell resistances is a potential

risk to wellbeing. Be that as it may, these diseases are regularly limited to immunocompromised or immunosuppressed people, incorporating patients with cutting edge diabetes or tumors. [5] A remarkable element of a considerable lot of these diseases is that they are topographically restricted to particular locales – as such, they are endemic. For instance, Coccidioides immitis (a parasite known to have executed a lab specialist) is to a great extent restricted to the bone-dry forsake districts of southwestern USA and northern Mexico.

Table- 9.1: Main types of fungus causing mycoses in humans

Primary way of entry	Fungus	Sexual stage	Disease	Natural distribution
Skin	**Trichophyton (22 species), microsporium (19 species) however just 2 are involved in infections Epi-dermophyton**	***Arthroderma (Ascomycota)***	**Dermatomycosis: athlete's foot, tinea, ringworm, etc.**	**Keratinized tissues, wild or domesticated animals and human.**
Mucosa	***Candida Albi-cans***	**Recently reported**	**Candidosid: thrush, stomatitis, vulvo-vaginitis.**	**Commensal on mu-cosa.**
Lungs	***Aspergillus fumigatus***	**None**	**Aspergillosis: invasive (systemic) or aspergillomas of lungs**	**Saprotrophic in soil or organic matter (composts)**
	Blastomyces dermatitidis	***Ajellomyces***	**Blastomycosis: brain, skin lesions, bones.**	**Saprotrophic**
	Coccidioides immitis	**None**	**Coccidioidomycosis: Systemic, lungs.**	**Saprotrophic in soil**
	Histoplasma capsulatum	***Ajellomyces***	**Histoplasmosis: lungs, rarely sys-temic.**	**Bat and birds drop-pings**
Wounds	***Phialophora Cladosporium Sporothrix, etc.***	**Often none**	**Subcutaneous myco-ses: sporotrichosis, chromomycosis, etc.**	**Saprotrophic in dead plant material and soil.**
Lungs	**Pneumocytis species**	**None**	**Virulent pneumonia**	**Humans, and other mammals**

Dermatophytic Fungi

The dermatophytes are a plainly characterized gathering of around 40 species, generally doled out to three genera of mitosporic organisms such as Epidermophyton, Microsporum, and Trichophyton, and in light of elements of their conidial stages. [6] A few types of Trichophyton (and correspondingly of Microsporum) have been found to deliver a sexual stage when strains of various mating sorts are matched in lab culture. The sexual stage, named Arthroderma, is a cleistothecium containing asci and ascospores. However its part, assuming any, in the study of disease transmission is not known, thus most medicinal mycologists utilize the recognizable "agamic" names in light of the fact that these are related with particular sicknesses and land appropriations.

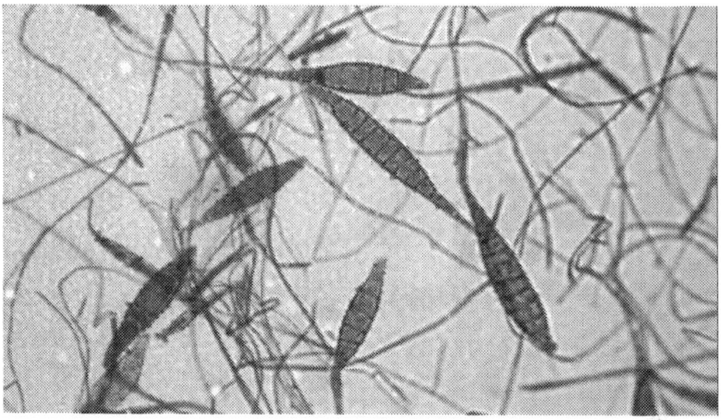

Figure- 9.1: Microsporum canis, a dermatophyte.

The trademark highlight of dermatophytes is their capacity to develop in the dead, keratinized tissues of the hair, nails and skin, where their metabolic items can actuate an inflammatory response. The living tissues are not attacked on the grounds that dermatophytes can't develop at 37°C. [7] By and by, the disturbance caused by dermatophytes frequently prompts scratching and further harm to the tissues, empowering microscopic organisms to attack. Scratching additionally prompts the shedding of dermatophyte-plagued skin and hair, so the mutual utilization of "scratching posts" by creatures can prompt transmission of the disease. Thus, in human populace's public offices, for example, showering territories, evolving rooms, and residential floor coverings can give a wellspring of inoculum. [9] This backhanded transmission is believed to be more imperative than guide have to-have transmission. Indeed, even the part of the conidial stages is easily proven wrong, in light of the fact that these are not generally found on tainted tissues while they are joined to the host, however pigmented hyphae and vegetative arthrospores (thickwalled spores framed by hyphal fracture) can be found in these tissues. [8]

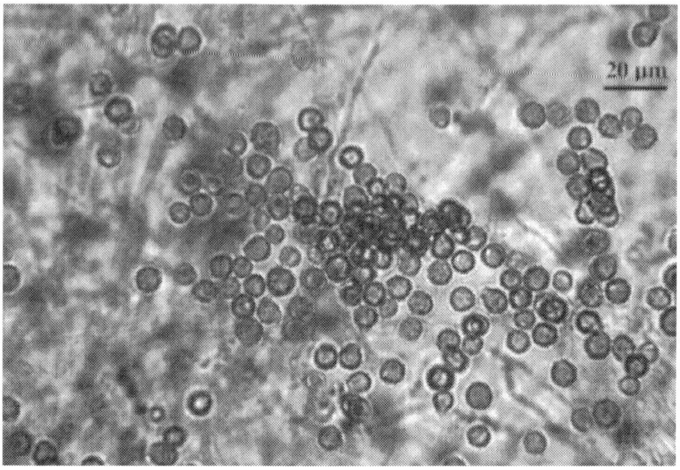

Figure- 9.2: Spores of dermatophytes.

The Virulence and Pathogenicity factors of Dermatophytes

Pathogenicity is the capacity to cause infection, while destructiveness variables decide the seriousness of an illness. Despite the fact that this refinement is clear on a fundamental level, it is not generally simple practically speaking without information of particular qualities and quality items that can be controlled to build up an immediate connection amongst infection and the parts of pathogenicity/harmfulness determinants. [10] Nonetheless, two key components are thought to add to ailments caused by dermatophytes:

- Their critical, despite the fact that not total, level of host specialization.
- Their capacity to develop on substrates rich in keratin.

Keratin is a noteworthy protein found in skin, nails, and hair. It is a troublesome substrate to corrupt, particularly when found as "hard keratin" (e.g. nails, feet) which contains various disulfide bonds between the sulfur-containing amino acids. In research center culture the dermatophytes can develop on keratin as their sole carbon and vitality source, by creating keratinolytic proteases named keratinases. This is maybe an extraordinary element of these organisms, on the grounds that no different parasite is accounted for to develop only on proteins. Keratinases could along these lines be harmfulness determinants of the dermatophytes, and a comprehension of these catalysts could prompt the improvement of antibodies. In a current report a metalloprotease (MEP) genome succession of the irrelevant organism, Aspergillus fumigatus, was utilized as a test to recognize three MEP qualities in a genomic library of Microsporum canis. One of these qualities, MEP3, was found to encode a 43.5-kDa keratinolytic metalloprotease, and both this protease and another (known as MEP2) were observed to be created in vivo amid test disease contemplates on guinea pigs. This work speaks to the principal report of a quality family encoding potential destructiveness determinants in dermatophytes. [11] Dermatophytes demonstrate a very noteworthy level of host-or living space specialization, so they are regularly assembled into three classes:

Anthropophilic (human-adoring) species are principally parasitic fungi of people and just seldom contaminate different species. They commonly cause diligent however moderately gentle contaminations. Cases of these parasitic fungi include:

- Epidermophyton floccosum, with an overall dissemination, causing disease of the crotch, the body, and feet (competitor's foot).
- Geophilic species live in soil and for the most part debase keratin-containing materials, for example, plumes and hair, however they can incidentally contaminate human hosts. A case is Microsporum gypseum which is discovered worldwide however normally in South America, where it is typically gained from soil yet periodically from creatures.
- Zoophilic (creature adoring) species normally taint wild or tamed creatures, for example, steers, felines, puppies, and steeds. They can likewise be transmitted to people who have close contact with creatures, yet these contaminations for the most part actuate a solid provocative reaction in the human host and afterward resolve suddenly. A great illustration is Microsporum canis, usually gained by people who have close contact with residential pets, for example, felines and canines.

- Trichophyton mentagrophytes and *T. rubrum* that presumably the commonest species, with overall dispersions, tainting many body parts. [12]

Quickly, in light of the fact that these organisms cause shallow contaminations they can be dealt with by topical utilization of oxidizing specialists or anti-microbials (e.g. amphotericin, B nystatin). In the event that the diseases are relentless they can be dealt with by oral organization of the anti-toxin griseofulvin or by fresher, less dangerous medications, e.g., terebinafine.

Table- 9.2: Few main dermatophytic fungi that infect human

Anthropophilic	Zoophilic	Geophilic
Epidermophyton floccosum	*Microsporum canis* (cats, dogs)	*Microsporum gypseum* (commonly infects humans)
Microsporum audouinii	*Microsporum equinum* (horses)	*Trichophyton terrestre*
Microsporum ferrugineum	*Microsporum nanum* (soil/pigs)	
Trichophyton mentagrophytes	*Microsporum persicolor* (rodents)	
Trichophyton rubrum	*Trichophyton equinum* (horses)	
Trichophyton tonsurans	*Trichophyton mentagrophytes* var. *mentagrophytes* (mice, rodents) *Trichophyton verrucosum* (cattle)	

Candida Albicans and different Candida Species

Candida albicans happens normally as a diploid growing yeast on the mucosal films of people and other warm-blooded creatures. [13] It is a typical commensal of people, being found on the mucosa of the mouth, gut, or vagina of over half of sound people. Typically it causes no mischief, yet an extensive variety of inclining elements can make it end up noticeably intrusive, prompting conditions on the whole known as candidosis. A couple of other Candida species, e.g., C. tropicalis and *C. glabrata*, cause comparable conditions, however less regularly.

The rundown of clinical appearances of *C. albicans* is broad. This organism can cause "thrush" in infants, when the parasite attacks the mucosa of the mouth and throat, creating speckly white pustules from which the illness picks up its name. This is regularly connected with conveyance through a tainted birth channel, and the parasite multiplies on the mucosa before a typical, adjusted microbial populace has created. Candida additionally causes cystitis, and it can cause irritation of the mouth (stomatitis) of many individuals who wear dentures. This is related with a few components, including scraped area caused by the dentures, grip of Candida cells to the dental plastic, and most likely the shut condition with absence of "flushing" underneath the denture plate. [14]

Individuals whose hands are as often as possible presented to water can create Candida diseases of the skin and fingernails. Also, Candida can be brought into the blood through catheters and other surgical methodology, yet the yeast populace soon decays when the catheter is evacuated. Candida disease of the gut is regularly connected with delayed antibacterial treatment, particularly with antibiotic medication anti-microbials which smother the bacterial populace. [15] Stress can be an extra figure Candida diseases, for instance, space explorers have been found to grow high populaces of *C. albicans* amid space flights. What's more, in outrageous instances of inclination, for example, propelled diabetes, neutrophil or macrophage issue, resistant scatters, and malignancies, Candida can develop systemically and progress toward becoming life-decline. This inventory of illustrations demonstrates that Candida is conceivably an ever-exhibit danger to human prosperity. A comprehension of its harmfulness determinants could make ready for new ways to deal with controlling these contaminations.

The Virulence causes of Candida Albicans

An extensive collection of confirmation proposes that there are two clear harmfulness determinants of *C. albicans,* its capacity to cling unequivocally to epithelia and a few different surfaces, and its capacity to experience a dimorphic switch between a yeast stage and a hyphal or pseudohyphal stage in light of natural components.

Adhesins

Regardless of numerous times of concentrate on the attachment of Candida cells, both in vitro and in vivo, it is as yet impractical to characterize unequivocally the part that grip plays in Candida contaminations, on the grounds that there appear to be many cooperating components. [16] Grip examines in vitro demonstrate that the yeast cells of *C. albicans* cling emphatically to shed epithelial cells of the vagina, and that attachment is more grounded to cells got from pregnant ladies. The two commonest cell sorts in the vaginal epithelium are the middle of the road cells and the shallow cells. Candida ties most emphatically to the middle of the road cells in vitro, and these phones prevail when there are elevated amounts of progesterone (for instance, amid in ladies who utilize oral contraceptives or in pregnancy). Progesterone appears to have an immediate impact, since Candida ties all the more emphatically to the epithelial cells of nonpregnant ladies if progesterone is included vitro. [17]

Furthermore, strains of *C. albicans* from dynamic vaginal contaminations are found to follow more emphatically in vitro examines than do disengages from sound individuals. This raises the likelihood that venereal spread of vaginal candidosis may include the transmission of emphatically disciple strains. These general relationships between's in vitro bond and the course of moderately gentle clinical diseases are bolstered by different examinations. For instance, *C. albicans* holds fast to cells of the buccal depression, to the methyl acrylate sap of dentures, and to the surfaces of catheters.

By utilizing buccal epithelia or denture tar as model frameworks in vitro, the grip of Candida was observed to be emphatically upgraded if the parasite had been developed on large amounts of galactose, maltose, or sucrose, as opposed to on glucose. This is related with the nearness of mannoprotein adhesins on the surface of the yeast cells. [18] By and by, the differential impacts of sugars appear to be strain-related, on the grounds that Candida strains from dynamic diseases indicate it every now and again, though strains from asymptomatic transporters can demonstrate it a great deal less regularly. Likewise, bond might be a particular component of *C. albicans* as opposed to of Candida species all in all, in light of the fact that the grip of nonpathogenic Candida spp. or, on the other hand of the nonpathogenic *Saccharomyces cerevisiae* is not uniquely adjusted when they are developed on various sugars. [19]

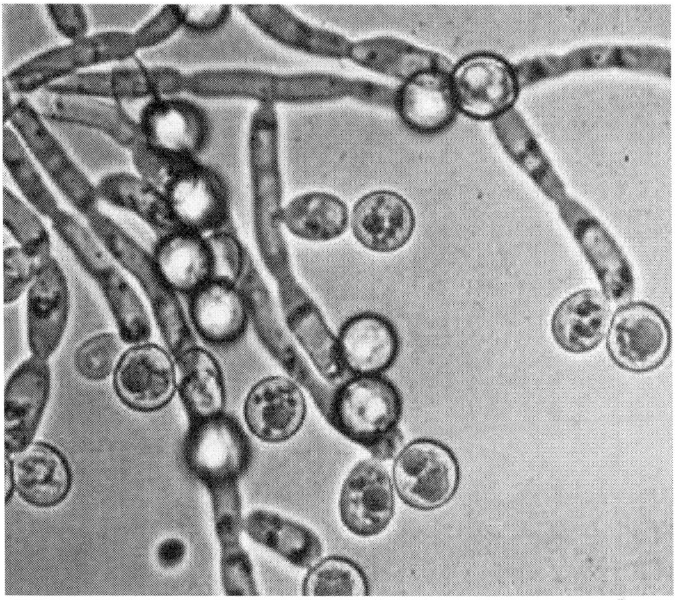

Figure- 9.3: Condition of Candida albicans in limited nutrient situations.

The Dimorphic Switch

Candida albicans is a dimorphic fungus parasite. It ordinarily develops as a sprouting yeast, yet because of supplement impediment the yeast cells can create a hyphal outgrowth. [20] This dimorphic change is fundamental to the pathogenicity of *C. albicans*, in light of the fact that attack of the mucosa is constantly accomplished by hyphal development. Actually, this recognizes *C. albicans* from all other (non-pathogenic) Candida spp. in clinical specimens, and it is utilized as a part of a basic demonstrative test: the yeast is refined on a specific medium, at that point a loopful of inoculum is exchanged to a vial containing horse serum and hatched at 37°C for 4–5 hours. Of all the Candida strains that may happen in clinical specimens just *C. albicans* will grow germ-tubes in these conditions, albeit a few other Candida spp. from nonhuman sources can do as such. The hyphal arrange is just brief, and brooding for longer circumstances brings about the creation of a pseudomycelium in which the hyphae comprise of strings of hotdog molded cells that deliver sprouting yeast cells at the septa. Advance in understanding the conduct of *C. albicans* has been hampered by the way that *C. albicans* is constitutively diploid, so mutants are hard to acquire. [21]

Some homozygous passive strains have been created by rehashed treatment with mutagens, particularly UV-illumination. Additionally, parasexual hereditary qualities has produced mutants. For this, strains with auxotrophic transformations are consolidated by protoplast combination, at that point permitted to recover dividers. A portion of the subsequent strains are tetraploid cross breeds or aneuploids which, when developed on negligible medium and subjected to warm stun or antimicrotubule operators, can return to a changed diploid state by loss of chromosomes. [22] The improvement of sub-atomic hereditary strategies for *C. albicans* has likewise been hampered. Numerous qualities from *C. albicans* can be changed into, and communicated in, *Saccharomyces cerevisiae*, yet not the other way around in light of the fact that *C. albicans* has a nonstandard codon utilization: the codon CUG encodes serine, though it encodes leucine in most different life forms. This is a huge confinement in light of the fact that a hefty portion of the putative destructiveness determinants of *C. albicans* can't be considered effectively out of sight of the pathogen itself.

The Innovation of Mating-Kind Genes

The mating arrangement of the ascomycetous yeast Saccharomyces cerevisiae includes a tape of three mating-sort qualities (Mata1, Matα1 and Mat α2) and the mating kind of a strain can change, contingent upon whether the MATa or MATα quality is moved into the expression locus by a particular change occasion. [23] By differentiate, no one had ever distinguished the comparable mating-sort qualities of Candida albicans, so this parasite had dependably been thought to be a clonal living being, with no limit with respect to sexual recombination. In any case, in 1998 a solitary mating sort like (MTL) locus was found in a research center strain of C. albicans. Since there is just a single MTL locus, the cell must end up noticeably homozygous for a α by mitotic recombination or quality transformation. [24] A strain can't change over to (α/α) on the grounds that each strain has lost the other MTL allele. This implies ordinary strains of *C. albicans* can't mate. Be that as it may, by erasing one of the two MTL alleles it is conceivable to deliver mating strains that are either a/ – or α/ – and that, in suitable conditions, can meld with each other. This was observed to be uncommon occasion, frequently prompting evident tetraploids that did not grow further. Yet, a couple a/ – and α/ – strains were found to change their phenotype from the ordinary white province appearance to a dark state shape, and these obscure strains have an expanded mating proficiency of around a million overlay. Hence, it is recommended that C. albicans can experience mating yet does as such just in the murky settlement frame. The murky stage cells are steady ay 25°C however not at 37°C, and they are significantly more proficient in colonizing skin surfaces. Thusly it is proposed that mating in this parasite must happen outside the body as on the skin surface or in natural stores. [25]

Universal Contagion through *C. albicans*

The uncommon instances of systemic spread of *C. albicans* in the blood and lymph tissues are constantly connected with extremely inclining elements, for example, propelled diabetes, delayed corticosteroid treatment, and leukemia, and so on. In any case, oddly, *C. albicans* at times develops systemically in AIDS patients, who have a tendency to create other systemic contagious diseases. Candida develops systemically as a yeast in the body liquids, yet dimorphism assumes a vital part in one regard. On the off chance that cells of *C. albicans* are blended with white platelets (macrophages and polymorphonuclear leucocytes) in vitro the yeast cells can be overwhelmed and pulverized, however some of them break out of the phagocyes by changing over to hyphae and afterward create a further populace of yeasts. In these conditions the force of the host guard framework will be vital in containing a contamination. [26] We will see that the real medications used to treat systemic candidosis (current subordinates of ketoconazole and related mixes) act synergistically with the host safeguards, in light of the fact that at even low fixations they smother the move from yeast to hyphal development. [41]

Resourceful and Secondary Pathogens

Hypothetically, any organism that can develop at 37°C could be a potential pathogen of people, yet by and by the range is much smaller than this. A couple of normal saprotrophic types of Sporothrix, Phialophora, Cladosporium, and Acremonium can contaminate wounds and cause harming subcutaneous mycoses. [27]

An alternate range of FUNGUS typically taint through the lungs in light of the fact that their airborne spores are sufficiently little to achieve the alveoli. Of every one of theseparasitic fungi, the hugest risk is postured by Aspergillus species, particularly *A. fumigatus* be that as it may, to a lesser degree, *A. flavus*. We will concentrate on these and related organisms in this area, since Aspergillus contaminations can be not kidding and regularly life undermining.

Aspergillosis

Aspergillus fumigatus, *A. niger* and *A. flavus* and are exceptionally normal saprotrophs on a scope of natural materials and deliver bottomless airborne conidia that are sufficiently little to enter the lungs. In individuals with impeded respiratory capacity these spores can sprout to deliver thick, confined saprotrophic provinces named aspergillomas. Normally these stay noninvasive, encompassed by stringy tissue of the host. Aspergillomas are very basic in poultry that are encouraged on formed grain, and can happen in cultivate specialists who frequently handle such materials.

Disease of the lungs is totally coincidental so far as the parasitic fungi are concerned in light of the fact that they develop as saprotrophs on plant natural issue and they have no normal methods for dispersing from the lungs to spread the contamination to different hosts. [28] Aspergillus fumigatus is by a long shot the most harming airborne contagious pathogen of people, and is in charge of around 90% of aspergillosis cases. Its spores are much of the time breathed in and enter the lungs, however typically are wrecked by the body's inborn safeguards. In any case, *A. fumigatus* can wind up plainly obtrusive and develop systemically in the body, either from contaminations in the respiratory tract of immunocompromised patients or after section through surgical injuries. It can be a potential issue in transplant surgery, when the patient's resistant framework is falsely smothered.

There are believed to be two primary courses of attack by *A. fumigatus*, through the ciliated epithelium that lines the upper districts of the respiratory tract or through the alveoli. [29] The cells of the ciliated epithelium can overwhelm conidia of *A. fumigatus*, and some of these inundated spores can make due inside the host cells. Corticosteroid medications are likewise known to be a hazard calculate intrusive aspergillosis, by decreasing the discharge or viability of antifungal peptides delivered by the covering epithelium. The spores that are not immersed by the covering epithelium enter the alveoli, where they can grow, however are decimated quickly by polymorphonuclear neutrophils. In any case, low neutrophil numbers coming about because of chemotherapy can diminish the effectiveness of this safeguard framework. [30] The spores can likewise be overwhelmed rapidly by alveolar macrophages where they are executed by receptive oxygen species. Be that as it may, the rate of killing is moderate, and it can take 2–3 days to clear a respiratory test. These elements can be essential in deciding the result of a disease.

To date there is no proof of a particular harmfulness figure related with *A. fumigatus*, and any strain of this growth is by all accounts ready to contaminate traded off people. The model framework most normally utilized for this is vaccination of mice (the alleged murine, i.e. mouse, show structure). Be that as it may, this may not be the most proper model for aspergillosis of people. [31] The entire genome arrangement of A. fumigatus is not accessible. This should empower correlations with be made of either the transcriptome (ambassador RNA) or the proteome (protein profile) of strains developed in vitro contrasted and in vivo, to recognize any potential destructiveness determinants that are communicated particularly in the host condition. [32]

WIDESPREAD DIMORPHIC FUNGI

Notwithstanding *C. albicans*, a little gathering of organisms can cause systemic diseases of solid or invulnerable traded off people. These parasitic fungi are especially intriguing in light of the fact that they are topographically confined (endemic) and they are dimorphic, changing starting with one development shape then onto the next because of temperature shifts. Every one of them start contamination from airborne spores that enter the lungs. [33] Individuals with weakened respiratory capacity can be especially inclined to lung contamination, and individuals in the propelled phases of leukemia, diabetes, and immune-suppressive issue can frequently build up the systemic malady. However, skin tests with antigens of some of these parasitic fungi demonstrate that a considerable extent of the populace in the endemic zones has been presented to contamination at some stage and maybe endured just gentle, influenza like side effects before the disease suddenly settled. In this manner, these organisms represent a noteworthy and never-ending risk to a segment of the populace. The four principle pathogens that cause these infections are talked about beneath.

Coccidioides immitis

Coccidioides immitis and a related animal groups, *C. posadasii*, are parasitic fungi that develop in the soluble soils of parched forsake areas of Arizona, California, and Texas, reaching out into parts of South and Central America. The hyphae of these growths piece to create little, thick-walled spores (arthroconidia or arthrospores) which can be scattered in dry, wind-blown tidy and in this manner enter the lungs. The majority of these cells are immersed and devastated by macrophages in the lungs, and cause an influenza like or pneumonia-like contamination which settle suddenly. [34] The frequency of this is very normal among individuals who live in these abandon territories, on the grounds that up to half of solid individuals demonstrate a positive reaction to a skin test with Coccidioides antigen, acquired from research center societies of the organism. In any case, in a few people like as particularly the individuals who are immune-compromised which is the growth is not contained by macrophages and it can spread inside the lungs and enter the circulatory system, causing a genuine summed up disease of the subcutaneous tissues, meninges, bones, and real organs. This malady is called coccidioidomycosis. [35]

Period of coccidiodomycosis, the arthroconidia swell and experience different atomic divisions, prompting the creation of multinucleate, swollen structures named spherules in the lungs. [36] The extensive spherules at that point experience cytoplasmic cleavage to create littler bodies (subspherules), which can rehash the contamination cycle inside the lung tissues. *Coccidioides immitis* has a few intriguing elements.

To begin with, at regular ecological temperatures, for example, 25°C it can develop as ordinary hyphae which create arthrospores for scattering, while it produces spherules at higher temperatures (37–40°C) average of the host condition. All things considered, a change of temperature alone is not adequate to deliver spherules, on the grounds that in vitro examines demonstrate that these are just created within the sight of high CO_2 fixations and in particular development media. *C. immitis* and the related species *C. posadasii* are the main dimorphic growths that deliver spherules, all other endemic contagious pathogens of people develop in a yeast-like maturing stage in the host tissues. In cutting edge phases of contamination, *C. immitis* moves toward becoming dispersed to a large portion of the body tissues. [37]

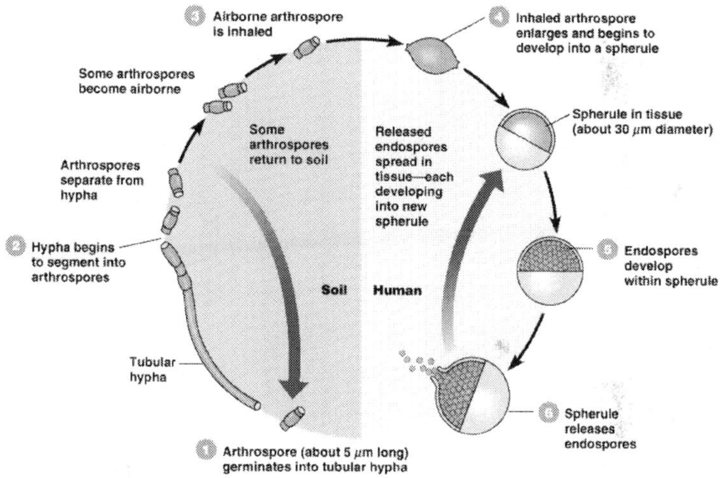

Figure- 9.4: Infection cycle of Coccidioldes immitis.

The geological appropriation of coccidioidomycosis is not completely known, and even inside the endemic districts it demonstrates a sketchy event. There are reports that it is most normal close to the tunnels of soil-abiding rodents. In controlled tests, mutts, rodents, felines, and other little creatures have been appeared to be vulnerable to disease (however so are bottlenose dolphins and stallions). [38] It has been recommended that wild leave creatures may give a characteristic repository to contamination of people. Most as of late, an atomic correlation of microsatellite DNA groupings from strains segregated from California contrasted and Texas, Central and South America uncovered that these speak to two particular types of Coccidioides (immitis and posadasii), both of which cause a similar infection. Maybe the most aggravating component is that Coccidioides can possibly be an inalienably destructive pathogen. [39] It is known to have executed a solid lab professional, and the Occupational

Safety and Health Administration of the US Government orders it as "an organic specialist with the possibility to represent a serious danger to general wellbeing and security that could conceivably be utilized by fear mongers." Put in basic terms, *C. immitis* is the main known parasite that is considered to speak to a noteworthy bioterrorism risk. [40]

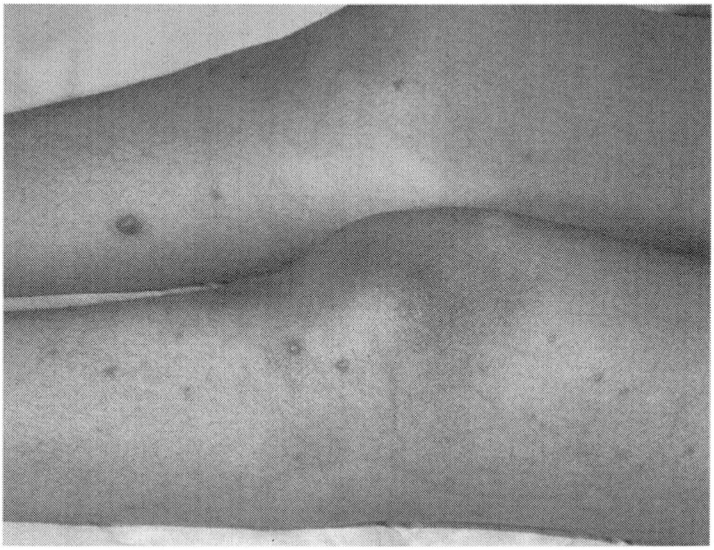

Figure- 9.5: Effects of disease caused by coccidioidomycosis in humans.

Histoplasma Capsulatum

Histoplasma is an organism with a "normal" form yeast dimorphism: it develops as hyphae on research center media at 25–30°C, yet as a sprouting yeast at 37°C on cysteinerich media. The hyphae deliver single conidia on the finishes of short branches. Some of these conidia are substantial, warty macroconidia (8–15 μm) though others are microconidia (2–4 μm) that are probably going to be the fundamental infective specialists in the lungs. The spores develop to frame a germ-tube which quickly offers ascend to a growing yeast stage in proper conditions. This yeast stage is constantly found over the span of disease, though the hyphae are included in the saprotrophic stage in dead body tissues or in common substrates. [42]

Histoplasma capsulatum has a wide geological appropriation, including a significant part of the eastern USA, the vast majority of

Latin America, and parts of Southeast Asia, Africa and Europe (e.g. Italy). It happens actually as a saprotroph in fecal-advanced soil around poultry houses, in bat droppings in caverns, and furthermore in starlings' droppings in towns and urban areas. The feathered creatures themselves are not contaminated, and the growth is not found in their new droppings. Nonetheless, bats have been observed to be actually contaminated by H. capsulatum, with intestinal injuries. Spores created from fecal materials are the probably wellspring of contamination of people, in light of the fact that there are recorded occasions of speliologists having built up the infection (histoplasmosis) in the wake of going by holes. Skin tests utilizing an antigen, histoplasmin, gotten from culture filtrates propose that a high extent of the human populace has been presented to contamination in the endemic ranges, one such examination evaluated that 20% of the considerable number of individuals in the USA had been tainted sooner or later from airborne spores. [43]

Indeed, even in gentle contaminations this parasite has been recognized in the pee, recommending that it may ordinarily be summed up, in spite of the fact that in by far most of cases the disease settle suddenly. Like Coccidioides, Histoplasma can be obtained in the lab on the grounds that there is a recorded example of understudies having tried positive subsequent to looking at societies in the classroom. The airborne spores can promptly enter the lungs and cause intense aspiratory histoplasmosis. Be that as it may, constant aspiratory contamination is kept to individuals with lung brokenness from different causes, and it can prompt spread to the next body tissues. This dispersal happens via carriage of the yeast cells in the lymph framework, and is most regular in resistant inadequate individuals or in the exceptionally youthful (under 1 year) or more established (50 years or more) segments of the populace. Magrini and Goldman (2001) depict the expanding scope of sub-atomic hereditary devices that have been utilized to research the pathobiology of *H. capsulatum*.

Blastomyces Dermatitidis

Blastomyces dermatitidis is an endemic dimorphic parasite found in the south focal locales and southeastern of the USA, along the Mississippi and Ohio streams, where it causes an illness named North American blastomycosis, or Chicago infection. It is likewise answered to be detached from people in parts of Africa and the Indian subcontinent, however these strains do not have the trademark antigens of *B. dermatitidis* and might speak to an alternate animal types. In North America contamination is related with damp soils improved with natural issue or decaying wood, yet little is thought about the fundamental environment of this parasite. [44] At the point when developed in research facility culture, it has limit hyphae (around 2 µm distance across) that deliver single globose conidia, 2–5 µm breadth, at the finishes of short hyphal branches, like (yet littler than) those of Thermomyces lanuginosus. It taints people from airborne conidia that enter the lungs and change into a thick-walled, yeast-like sprouting stage. Around half of contaminations are asymptomatic, yet different diseases can advance to an intense aspiratory stage after a hatching time of 30–45 days, creating indications in the lungs that emphatically look like a bacterial pneumonia. The contamination can likewise advance to an incessant stage that influences the lungs, skin, bones, genitourinary tract, and different organs, for instance, ulcerative injuries of the skin, granulomatous irritation of the lungs, and spread to numerous different tissues and organs.

Brandhorst *et al.* announced that thermally controlled dimorphism is the absolute most characterizing quality of Blastomyces, and this is connected to a stage particular quality of B. dermatitidis named BAD1 (Blastomyces adhesin), discovered just on the surface of the Y-stage and not on the hyphae that develop at bring down temperatures. The cell divider glucan of the yeast-stage cells comprises of 90% α-glucan, while mycelial stage cells have generally square with substance of α-glucan and β-glucan. This is likewise valid for Paracoccidioides brasiliensis and strains of Histoplasma capsulatum, in every one of the three cases the decreased substance of α-glucan connects with loss of destructiveness in creature models. In *B. dermatitidis* this connection between's α-glucan and destructiveness was affirmed by erasing the quality encoding BAD1. The erased strain had a lessened capacity to tie to macrophages or to mouse lung tissue, and it was exceedingly constricted in exploratory mice. Reclamation of the BAD1 quality by change reestablished the destructiveness of *B. dermatitidis* and reestablished the cement properties of the parasite. The BAD1 protein has been found to self-relate within the sight of calcium particles, and it is proposed that, in the states of the lung, chitin-bound BAD1 atoms could start to catch facilitate BAD1 atoms, adding further α-glucan layers to the yeast cell divider. [45]

Paracoccidioides Brasiliensis

Paracoccidioides brasiliensis is another endemic fungus, found mainly in subtropical forest regions of Central and South America, especially Brazil, Venezuela, and Columbia. It has been isolated from the digestive tract of some animals (e.g. armadillos) and from moist soil enriched with proteins, but the natural ecology of the fungus is poorly understood. It is a mitosporic fungus with no known sexual stage, and it grows very slowly in agar culture. *P. brasiliensis* is a dimorphic fungus that grows as mycelia at lower temperatures but as a budding yeast at 37°C or in body tissues. The yeast cells are oval or spherical and range in size from 2 to 10 μm or up to 30 μm or more. Often, they grow as a single large central cell with multiple buds attached to the central cell by narrow necks, resembling a mariner's wheel. The buds remain attached to the mother cell until they, themselves, bud. [46]

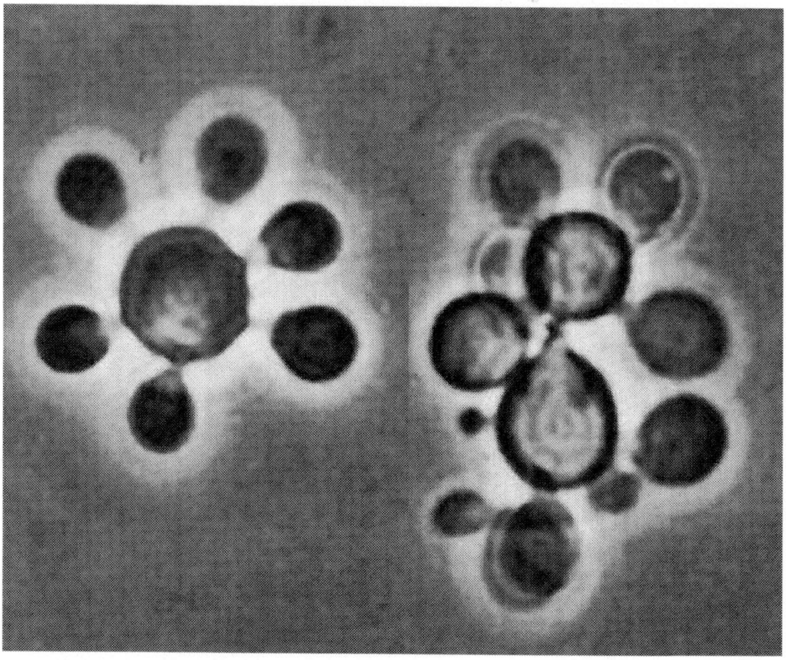

Figure- 9.6: Budding phase of Paracoccidioides brasiliensis.

P. brasiliensis causes the disease termed paracoccidioidomycosis, or South American blastomycosis. Infection is assumed to occur from airborne spores that enter the lungs, but most infections are asymptomatic, as evidenced by a skin test to the fungal antigen. In other cases the infection can develop several years after initial exposure, indicating a long latent period in which the fungus has remained dormant in the lymph nodes. Such infections are probably associated with immunodeficiency. [47] The characteristic form of the disease involves severe ulcerative lesions of the mouth, nose, larynx, and subcutaneous tissues, causing serious facial disfigurement. In extreme cases the fungus can also affect other organs such as the spleen, liver, bones, and central nervous system. A notable feature of paracoccidioidomycosis is that it occurs primarily in men – the ratio of infection is about 15 : 1 in men compared with women, but can be up to 78 : 1, whereas equal numbers of men and women show a positive skin test reaction to the paracoccidioides antigen. The reason is that estrogens inhibit the transition of cells from the conidial or mycelial form to the yeast phase which is necessary for colonization of the tissues.

Cryptococcus Neoformans

Cryptococcus neoformans contrasts from the growths examined above on the grounds that it is not endemic but rather has a wide worldwide conveyance. It is normal as a saprotroph and has been detached often from old, weathered pigeon droppings in urban areas and from soils advanced with feathered creature excreta. In any case, it doesn't contend well in wet droppings, where microscopic organisms can raise the pH to development inhibitory levels, and it doesn't taint the feathered creatures themselves. Concentrates in New York City demonstrate that a high extent of youthful youngsters have been presented to the organism, since they demonstrate a response when skintested with the Cryptococcus antigen, yet clearly don't build up any side effects. There are more than 30 types of the yeast Cryptococcus, however just a single animal groups, *C. neoformans*, is pathogenic to people, causing the ailment called cryptococcosis. [48]

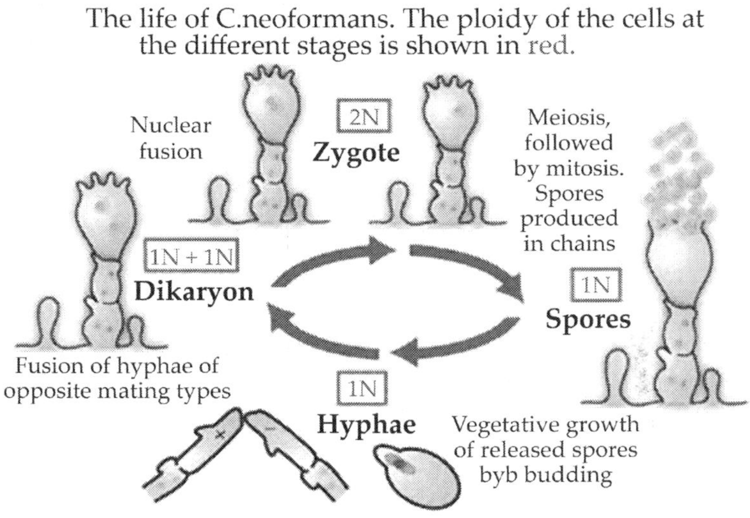

Figure- 9.7: Life cycle of C. neoformans.

Cryptococcosis used to be an uncommon sickness, discovered predominantly in individuals who were actually immunocompromised or who had experienced transplant surgery including immunosuppressive medications. Be that as it may, the expanding rate of AIDS/HIV in the 1990s prompted a noteworthy surge in cryptococcosis, influencing 7–10% of AIDS patients around the world. In spite of the fact that the illness can be treated with routine organization of antifungal specialists, this is from time to time conceivable in creating nations. The common course of disease is through the lungs, when spores or yeast cells are breathed in. After an underlying subclinical pneumonic disease, the parasite can cause ceaseless lung contamination and afterward scatter to the focal sensory system, where it demonstrates an inclination for development in the cerebral cortex, cerebrum stem, cerebellum, and meninges. This is constantly deadly if not treated, and it is regular in AIDS patients, maybe on the grounds that an essential injury in the lung has stayed calm for a long time and prompts optional spread when the resistant framework is debilitated. [49]

There are a few components of *C. neoformans* that still can't seem to be settled and that could help with averting or treating contaminations. Typically the organism is disconnected from ecological examples as a haploid yeast, and it was ordinarily expected that got dried out yeast cells are the essential wellspring of disease. With a measurement going from 2.5 to 10 μm, some of these cells would be sufficiently little to achieve the alveoli, rehydrate, and start contamination. In any case, more as of late it was found that *C. neoformans* has a sexual stage in the family Filobasidiella (Basidiomycota), prompting the generation of airborne basidiospores around 1.8–3.0 μm width. These future a perfect size to be kept in the alveoli. The basidiospores can be created in lab culture by matching of the two mating sorts, named "an" and "α". Yet, α strains alone can be actuated to create basidiospores (with a solitary haploid core in every cell) when refined on media that need nitrogen and in states of water-stretch. The α mating-sort locus (MATα) has been cloned and appeared to encode a pheromone-like peptide. The α strains are discovered considerably more as often as possible than are the a mating sorts in clinical examples, and in immunizations of mice they cause substantially more fast demise than do the a-matingtype strains. In this manner it appears that the quality result of α-mating-sort strains may add to destructiveness. Notwithstanding these focuses, there are two unmistakable types of *C. neoformans*, named assortments, *C. neoformans var. neoformans* and *C. neoformans var. gattii*. The first of these is ordinarily connected with AIDS disease in Europe and North America, and apparently starts from winged animal excreta. By differentiate, the assortment gattii is from time to time found in AIDS patients, rather it is related with endemic malady of nonimmunocompromised individuals in Papua New Guinea, Australia, and parts of India, Africa, south-east Asia and South and Central America. The ecological wellspring of *var. gattii* is by all accounts eucalypt trees and hollows in rotting wood.

Virulence and Pathogenicity determining factor of *C. neoformans*

As noted before in this section, pathogenicity determinants incorporate each one of those components that empower an obtrusive life form to live in a host domain. Harmfulness determinants incorporate the variables that decide the seriousness of ailment. For *C. neoformans* the significant pathogenicity determinants are the capacity to develop at 37°C of every a climate of around 5% CO_2 and at a pH of around 7.3. The way to this is by all accounts a quality that codes for calcineurin An, a particular sort of protein phosphatase that is initiated by Ca_2^+-calmodulin and that is included in stretch reactions in yeasts. Mutants that are blemished in calcineurin A. can develop at 24°C however not at 37°C, and they can't get by in 5% CO_2 or antacid pH.

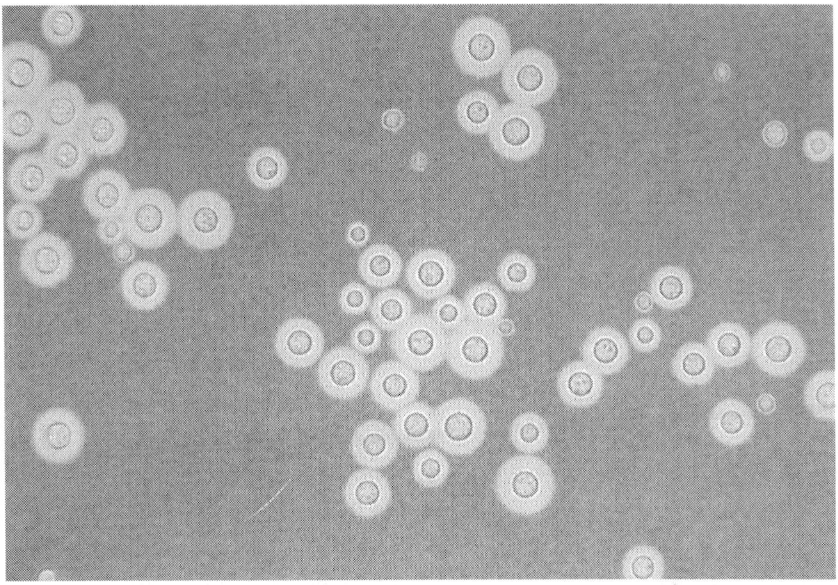

Figure- 9.8: Cells of a Cryptococcus albidus.

There are a few harmfulness determinants of *C. neoformans*. The most obvious of these are both in research facility culture and in clinical examples as the nearness of a thick, inflexible polysaccharide container around the yeast cells. The case comprises of a high sub-atomic weight polysaccharide, with a spine of α-1, 3-connected d-mannose units and single units of d-xylose and d-glucuronic corrosive. It is by all accounts a key harmfulness determinant of *C. neoformans*, in light of the fact that actually noncapsulate strains, or mutants that do not have the container, are nonvirulent. The typified cells are not as effectively phagocytized or slaughtered by neutrophils, monocytes or macrophages, contrasted and noncapsular cells, and their high net negative charge can lessen cell–cell connections required for leeway of the cryptococci. [50]

Pathogenic strains of *C. neoformans* additionally deliver dark colored or dark shades containing diphenolic mixes. Through the activity of phenoloxidase, these mixes are oxidized and polymerized to deliver melanin, which is stored in the yeast dividers, maybe ensuring against responsive oxidants in the host tissues. A comparative defensive part has been proposed for d-mannitol, which is created by Cryptococcus and is particularly powerful as a forager of hydroxyl radicals. In this way, general there appear to be a few potential harmfulness determinants in *C. neoformans*, regardless of the possibility that the exact parts of some of these are not completely comprehended at exhibit.

Pneumocystis Species

So far in this section, we have seen some very amazing cases of organisms that cause genuine illnesses of people, regularly as entrepreneurial or coincidental pathogens. They incorporate growths whose characteristic living space is eucalypt trees, or pigeon excreta, or the excreta of give in abiding bats, and soil organisms that are endemic to particular locales of the globe. We have likewise observed parasitic fungi that spread "quietly" in the tissues of people and that show inclinations for specific organs of the body. But then large portions of these forceful growths have no undeniable common methods for transmission starting with one host then onto the next. For a large portion of them, the disease of a human host is by all accounts simply coincidental.

Presently we swing to maybe the most bewildering fungus of all the numerous Pneumocystis species that cause destructive pneumonia-like manifestations related with AIDS and immunosuppressive treatments. These life forms were first portrayed as protozoa (protists), under the name *Pneumocystis carinii*, more than 100 years prior, and have an existence cycle that more intently takes after that of a protist than an ordinary organism. Phylogenetic investigation (in view of 18S ribosomal DNA arrangement) places them nearest to the early ascomycota, alongside the splitting yeast Schizosaccharomyces pombe. However, they have one of a kind element: they are the main growths that have cholesterol in the cell film, while most other parasitic fungi (except for some zygomycota) have ergosterol as their trademark layer sterol. One result of this is Pneumocystis is not powerless to the basic antifungal specialists that objective ergosterol, rather Pneumocystis is controlled by a portion of the antiprotozoal Agents. [54]

Usual Existence of *Pneumocystis* Species

Pneumocystis species are discovered worldwide in human hosts and in the lungs of an extensive variety of wild well evolved creatures. Be that as it may, before the across the board developmentof AIDS in the 1980s Pneumocystis was just discovered occasionally as a reason for pneumonitis, called Pneumocystis carinii pneumonia (PCP). It was discovered first in malnourished youngsters and after that in patients experiencing immunosuppressive treatments. Presently PCP has turned out to be one of the commonest reasons for death of AIDS patients, and there is an immediate connection between's the danger of creating PCP and a low number of circling CD_4^+ T lymphocytes. A cell number of 200 or less per microliter of blood is the basic lower confine. It is prominent that a high extent of kids in Europe and the USA demonstrate a positive serum test for Pneumocystis antigens by the age of 4 years, showing introduction to the growth. This is likely additionally valid for most different nations. Nonetheless, Pneumocystis at that point appears to vanish from the lungs and just returns as individual's age, or in light of immune-suppression.

Hence Pneumocystis disease is emphatically age-connected, and re-contaminations further down the road presumably result from across the board and consistent re-introduction to the growth. Not long after *Pneumocystis carinii* had been formally exchanged to the contagious kingdom, DNA succession examination demonstrated that strains acquired from various creature has were extremely differing and each was particular for the individual host species. [51] Utilizing the polymerase chain response, preliminaries were produced to intensify DNA from every single known specie of Pneumocystis, and each of these was observed to be unmistakable and have particular, with no confirmation of cross-disease starting with one host animal categories then onto the next. Affirmation of this in a few investigations has prompted the suggestion that all the host-particular types of Pneumocystis ought to be viewed as discrete species.

Along these lines, the humans have particular organism has been named *P. jiroveci*, and the first name *P. carinii* has been held for one of the two known Pneumocystis species found in rats. Pneumocystis life forms acquired from rats, mice, ferrets, pigs, and monkeys all show hostspecies specificity and new names for these are probably going to be proposed. Reliable with the co-advancement of parasitic fungi with their hosts, the Pneumocystis species from people are most like those of different primates.

Methods of Life Cycle, Pathology, and Transmission of *Pneumocystis*

Pneumocystis can't be kept up for long in research center culture, so it is by all accounts basically a commit parasite. Little measures of DNA of *P. jiroveci* have been recognized in natural specimens, including tests of airborne spores and lake water, yet the in all probability wellspring of contamination is either the enactment of a prior idle disease or re-disease from breathed in spores. The parasite gets away from the protections of the upper respiratory tract, and sets up contaminations in the alveoli, evidently by official to the sort 1 epithelial cells. The life cycle of Pneumocystis is muddled, and still not completely caught on. Abiogenetic trophic (sustaining) shapes multiply by mitosis in the lungs. There is additionally a sexual stage in which haploid cells conjugate to create a diploid pre-pimple. This experiences meiosis, trailed by mitosis, to deliver eight haploid cores in the "late stage sore." Mature growths are variable fit as a fiddle however frequently a saucer-like appearance. They are thought to break, discharging haploid vegetative cells.

In immune-compromised individuals the Pneumocystis cells multiply and can bring about levels of cells up to four layers thick in the alveolar lumen, making oxygen insufficiency. A frothy alveolar exudate is delivered, and in cutting edge phases of sickness the parasite can be recognized in every single significant organ of the body, particularly the lymph hubs, bone marrow, liver, and spleen. DNA grouping polymorphisms have frequently been found in segregates of the human pathogen *P. jiroveci*, showing that many strains exist inside this species. For instance, polymorphisms have been found in the mitochondrial 18S ribosomal RNA quality, the mitochondrial little subunit rRNA quality, the inward interpreted spacer (ITS) areas of the atomic rRNA quality, and the dihydropteroate synthase (DHPS) quality.

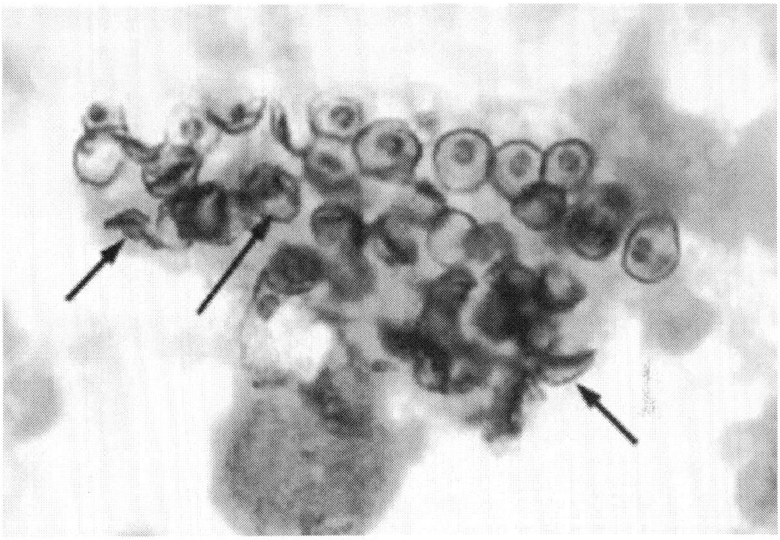

Figure- 9.9: Cysts of Pneumocystis jirovec.

These polymorphisms give the premise to exploring the study of disease transmission of Pneumocystis contaminations in people. Examinations of strains gathered earlier and after the start of the AIDS pandemic show that there has been no generous change in the Pneumocystis populace that could represent the monstrous increment in AIDS-related pneumonia. DNA succession polymorphisms have additionally been utilized to explore whether PCP (pneumonia) exists in a long haul idle shape in people or is re-procured amid nonstop or occasional introduction to the pathogen. The reactivation of long haul inert contamination can't be precluded, in light of the fact that hypothetically even a solitary dormant Pneumocystis cell could offer ascent to disease when a host is immune-compromised. Be that as it may, investigations of populace hereditary qualities have tended to this issue, by looking at the strains found in grown-up patients who have stayed near their origin against the strains found in patients who moved far from their origination. [52] The information show that grown-up patients procure strains that are more like those in their present place of habitation than in their origin, emphatically recommending that contamination is reacquired instead of being conveyed since early adolescence. The accompanying focuses can be made in synopsis of the present learning of Pneumocystis:

- In human populaces, youthful youngsters are every now and again tainted yet commonly display just a transient contamination and in this way create invulnerability. Kids might be the essential repository of disease.
- Pneumocystis is a family of overall appropriation. It exists the same number of various host-particular species, with no proof of cross-contamination between various hosts.
- Adults appear to secure contamination from natural sources near their zones of habitation, yet they just create infection on the off chance that they are immunosuppressed or immunocompromised. In these conditions the contamination can advance quickly, and is firmly corresponded with a low tally of flowing CD_4^+ T lymphocytes. [53]
- *Pneumocystis* is by all accounts a primitive growth like living being that speaks to one of the most punctual basal genealogies of organisms and has broadened as a commit parasite that contaminates numerous types of warm blooded animals, including people.

SUMMARY

- As compare to plants, which are affected by thousands of fungus, animals and humans are affected by only 200 to 300 fungi. These diseases are together known as mycoses. The fungi that effects humans and animals cause only mild symptoms in healthy individual. However due to the use of immunosuppressant drugs in cancer therapy and transplant surgery, fungal infections were proved life threatening.
- Fungi mainly effects the animals by producing mycotoxins in their feeds and spores of fungus are responsible for causing hay fever, asthama and other diseases related to respiration. Major fungal pathogens of animals and humans are followings: Dermatophytic fungi, Candida albicans, Aspergillosis, Coccidioides, Histoplasma capsulatum, Blastomyces dermatitidis, Paracoccidioides brasiliensis, Cryptococcus neoformans, and Pneumocystis species.

REFERENCES

1. Starr, C., In Taggart, R., In Evers, C. A., & In Starr, L. (2016). *Biology: The unity and diversity of life.*
2. Madigan, M. T., Martinko, J. M., Bender, K. S., Buckley, D. H., & Stahl, D. A. (2015). *Brock biology of microorganisms.*
3. Meyers, R. A. (2012). *Encyclopedia of sustainability science and technology*. New York: Springer. .
4. Anderson, R. P., & Young, L. (2012). *Case studies in microbiology: A personal approach*. Hoboken, NJ: Wiley. .
5. Marks, R., & Plewig, G. (1986). *Skin Models: Models to Study Function and Disease of Skin*. Berlin, Heidelberg: Springer Berlin Heidelberg.
6. Rojas, R. (2000). *Secondary metabolites of Vernonia gigantea roots: Isolation, antimicrobial, chemical and biotransformation studies.*
7. In Tosti, A., In Vlahovic, T. C., & In Arenas, R. (2017). *Onychomycosis: An illustrated guide to diagnosis and treatment.*
8. Dillavou, C. L. (1971). *Metabolic studies of dermatophytic fungi*. Columbia.
9. McCrea, A. (1934). *Fungicidal value of some common dyes against dermatophytic fungi.*
10. Davenport, R. A. F. (1978). *Detection and preliminary characterization of siderophores produced by dermatophytic fungi.*
11. Becskei, A. (2011). *Yeast genetic networks: Methods and protocols*. New York, N.Y: Humana Press.
12. International Conference on Environmental, Industrial and Applied Microbiology, & Méndez-Vilas, A. (2006). *Modern multidisciplinary applied microbiology: Exploiting microbes and their interactions*. Weinheim: Wiley-VCH.
13. Thompson, N. D. (2005). *Survival time from diagnosis of candidemia: An application of survival methods for epidemiology to the Mycoses Study Group multi-center observational study of hospitalized patients with candidemia.*
14. Witzany, G. (2012). *Biocommunication of fungi*. Dordrecht: Springer.

15. International Symposium on Bioluminescence and Chemiluminescence, & Tsuji, A. (2005). *Bioluminescence & chemiluminescence: Progress and perspectives*. Singapore: World Scientific.

16. Limjindaporn, T., & Georgetown University. (2001). *The transcription factor, GAT99P, regulates nitrogen utilization, hyphal development, and virulence in Candida albicans*. (Dissertation Abstracts International, 62-5.)

17. Emoődy, L., Pál, T., Hacker, J., & Blum-Oehler, G. (2002). *Genes and Proteins Underlying Microbial Urinary Tract Virulence: Basic Aspects and Applications*. Boston, MA: Springer US.

18. Sullivan, D. J., & Moran, G. P. (2014). *Human Pathogenic Fungi: Molecular Biology and Pathogenic Mechanisms*. Portland: Caister Academic Press.

19. Singaravelu, Kumala, Gácser, Attila, & Nosanchuk, Joshua D. (2014). *Genetic determinants of virulence: Candida parapsilosis*. Asociación Española de Micología.

20. Scherer, S, & Magee, P T. (n.d.). *Genetics of Candida albicans*.

21. Baloda, S. B. (1987). *Potential salmonella virulence factors: Studies on toxins and cell-surface adhesins*. Uppsala: Swedish University of Agricultural Sciences, Department of Veterinary Microbiology.

22. Schloß Ringberg Workshop on Type 4 Pili: Biogenesis, Adhesins, Protein Export and DNA Import. <1995, Kreuth>. & Manning, P. A. (1997). *Schloss-Ringberg Workshop on Type-4 pili: Biogenesis, Adhesins, Protein Export and DNA Import: Rottach-Egern, Germany 1995*. Amsterdam [u.a.: Elsevier.

23. Virkola, R. (1993). *Tissue-binding characteristics of Escherichia coli adhesins*. Helsinki: Yliopistopaino.

24. Kjærgaard, K. (2001). *Bacterial designer adhesins: FimH & Antigen 43*. Lyngby: Section of Molecular Microbiology, BioCentrum-DTU, Technical University of Denmark.

25. Paul, J. A. (2011). *Regulation of morphogenesis by ammonium transporters in Ustilgo maydis*.

26. Hughes, C. F. (2003). *Differential gene expression by Microbotryum violaceum during discrete stages of fungal development*.

27. Lokvam, J. (2000). *Chemo-ecological and evolutionary aspects of the pollinator reward system in Clusia grandiflora (Clusiaceae).*
28. Keigwin, B., Shepherd, G. R., Wuenschel, M. J., & Northeast Fisheries Science Center (U.S.),. (2016). *Geomorphometric analysis indicates overlap in body shape between sexes of black sea bass (Centropristis striata).*
29. Pöhlmann, Jennifer, & Fleig, Ursula. (n.d.). *Asp1, a Conserved 1/3 Inositol Polyphosphate Kinase, Regulates the Dimorphic Switch in Schizosaccharomyces pombe* ▿ . American Society for Microbiology (ASM.
30. Nigg, M., & Bernier, L. (2016). *From yeast to hypha: defining transcriptomic signatures of the morphological switch in the dimorphic fungal pathogen Ophiostoma novo-ulmi.* (BioMed Central Ltd.) BioMed Central Ltd.
31. Hunt, J, & Simmons, L W. (n.d.). *Status-dependent .ion in the dimorphic beetle Onthophagus taurus.*
32. Pérez-Campo, Flor Maria, & Domínguez Olavarri, Ángel. (2013). *Factors affecting the morphogenetic switch in Yarrowia lipolytica.* Springer.
33. Zill, O. (2010). *Evolutionary Genetic Studies of Mating Type and Silencing in Saccharomyces.* Berkeley, CA.
34. Scott, D. (2012). *Solutions manual for Introduction to genetic analysis, tenth edition.* New York: W H. Freeman & Co.
35. Sanders, M. F., & Bowman, J. L. (2015). *Genetic analysis: An integrated approach.*
36. Fedoroff, N. V., & Botstein, D. (1992). *The Dynamic genome: Barbara McClintock's ideas in the century of genetics.* Cold Spring Harbor, N.Y.: Cold Spring Harbor Laboratory Press.
37. In Martin, F., & In Uroz, S. (2016). *Microbial environmental genomics (MEG).*
38. Martin, S. H. (2013). *Mating type and pheromone genes in the Gibberella fujikuroi species complex: And evolutionary perspective.*
39. Wilken, P. M. (2013). *Identification and application of mating type gene sequences in Ophiostoma.*

40. Panwar, Sneh L., Legrand, Melanie, Dignard, Daniel, Whiteway, Malcolm, & Magee, Paul. T. (n.d.). *MF 1, the Gene Encoding the Mating Pheromone of Candida albicans†*. American Society for Microbiology.
41. Prasad, R. (2017). *Candida albicans*.
42. Kreier, J. P. (1993). *Parasitic Protozoa: Toxoplasma, Cryptosporidia, Pneumocystis, and Microsporidia*. Oxford: Elsevier Science.
43. Toledo, J. A. C., & Castro, M. R. (2001). *Pneumocystis carinii pneumonia, pulmonary tuberculosis and visceral Leishmaniasis in an adult HIV negative patient*. (Brazilian journal of infectious diseases, 5, 3, 154-157.)
44. Margolis, S. (1993). *The Johns Hopkins handbook of drugs: For the 100 major medical disorders of people over the age of 50 : with a special expanded table of contents organized by disorder*. New York: Rebus.
45. In Brogden, K. A., & In Guthmiller, J. M. (2002). *Polymicrobial diseases*. Washington, D.C: ASM Press.
46. Sullivan, D. J., & Moran, G. P. (2014). *Human Pathogenic Fungi: Molecular Biology and Pathogenic Mechanisms*. Portland: Caister Academic Press.
47. Tibayrenc, M. (2017). *Genetics and Evolution of Infectious Diseases*.
48. Parker, J. N., Parker, P. M., & Icon Health Publications. (2002). *The Official Patient's Sourcebook on Pneumocystis Carinii Infection*. ICON Group.
49. Coffey, S., Agins, B. D., & United States. (2011). *Guide for HIV/AIDS clinical care*. Rockville, MD: U.S. Dept. of Health and Human Services, Health Resources and Services Administration, HIV/AIDS Bureau.
50. United States., Coffey, S., Agins, B. D., & United States. (2011). *Guide for HIV/AIDS clinical care*. Rockville, MD: U.S. Dept. of Health and Human Services, Health Resources and Services Administration, HIV/AIDS Bureau.
51. DeVita, V. T., Hellman, S., & Rosenberg, S. A. (1997). *AIDS: Etiology, diagnosis, treatment, and prevention*. Philadelphia: Lippincott-Raven.

52. Tomlinson, D., & Kline, N. E. (2010). *Pediatric oncology nursing: Advanced clinical handbook.* Heidelberg: Springer.
53. Strelkauskas, A. J., Edwards, A., Fahnert, B., Pryor, G., & Strelkauskas, J. (2016). *Microbiology: A clinical approach.*
54. Tortora, G. J., Funke, B. R., & Case, C. L. (2010). *Microbiology: An introduction.*

CHAPTER

10

METHODS OF CONTROLLING FUNGAL GROWTH

We pass through many examples about controlling fungus by the controlling of environmental and physical factors. For instance, it is important to manage the water potential for storage of grain, and the grouping of water and temperature potential could be utilized to expect the not dangerous restrictions for avoiding development of mycotoxin-making fungus in store products related to food. The shipment and storage of fresh yield, comprising numerous fruits, is usually attained through organized atmosphere storing such as arrangement of low temperature and raised heights of carbon dioxide. This caue interruptions the onset of senescence and in ripening, and can be inexpensive than moreover of the single dealings only. Likewise, bananas are regularly transported to the European market as immature from Central America, unriped green fruits and are then artificially mature through contact to ethylene-producing chemicals. Green fruits live extensive-space transport nearly about 2–3 weeks, and although they are green the fruits will not change abrasions of banana anthracnose which is existing as hidden contaminations on the covering. Fruits rotting brings to the discharge of ethylene which is a wound hormone, and might bring to early invasion and ripening through fruit-decaying fungus. To stabilize this, some kinds of fruit are preserved with coatings of wax and fungicides, and detained in cold storing. [1] Numerous yeasts and bacterias are currently advertised commercially as protectants of fruit, by way of a substitute to the practice of fungicides. It might not be essential to treatment living organisms at all, since the fractions of cell wall of numerous yeasts, comprising *S. cerevisiae*, might action as influential elicitors defensing mechanism of plants.

So also, there is expanding enthusiasm for the utilization of chitosan to shield plants from pathogenic assault. At display, chitosan is acquired monetarily by substance deacetylation of crab shells. It upsets the development of a few organisms in vitro, causing over the top expanding, divider modification, and cytoplasmic disruption. Chitin is a trademark divider segment of organisms, and a few parasitic fungi – particularly Zygomycota, it has normally deacetylate chitin by the compound chitin deacetylase. They could be utilized as an option, inexhaustible wellspring of chitosan for activating plant resistance responses. Sanitation, quarantine, and similar biologically based strategies:

Product revolution is a conventional, exceptionally viable technique for controlling many plant maladies if the pathogen does not get by for long without a host trim. This is valid for the take-all growth Gaeumannomyces graminis, which for the most part decays to nondamaging levels inside 1 year if an alternate (noncereal) edit is sown. [2] The option is to develop oat edits ceaselessly, after a seemingly endless amount of time, until the point that the dirt turns out to be normally suppressive to take-all.

Yield administration practices can be modified to maintain a strategic distance from ailment. In numerous nations stinking filth of wheat is seed transmitted and should be controlled by seed-connected fungicides. In any case, in the Pacific northwest of the USA it is chiefly soil-borne, and if winter wheat is sown early it achieves the most helpless stage while the dirt dampness and temperature are unsatisfactory for germination of the filth spores. Other basic and compelling sickness shirking hones incorporate the liming of soil to anticipate genuine clubroot infection of cruciferous harvests caused by the dissent Plasmodiophora brassicae. Phosphate lack can be a noteworthy reason for yield misfortunes. The carmelizing root decay malady of wheat, caused by Pythium graminicola (Oomycota), was a standout amongst the most harming illnesses in the North American prairies until the 1940s, however was for all intents and purposes disposed of when phosphate manures turned out to be generally accessible.

Meteorological estimating has been utilized for a long time as a sickness administration instrument. It was produced at first for control of potato curse by Beaumont in 1947. By concentrate the connection amongst climate and curse plagues, he set up that, after a specific date which fluctuates from locale to district, a scourge pestilence will create inside 2–3 weeks following a 2-day time frame in which the temperature is at least 10°c and the relative mugginess is over 75%. [3] This 2-day term ended up noticeably known as a Beaumont period, and it empowered producers to time the use of fungicides so that significant scourges could be maintained a strategic distance from. This has now been refined further, and comparative guaging strategies are generally utilized as a part of other trimming frameworks. One case is the facial dermatitis cautioning framework for Pithomyces.

Sanitation can be exceedingly viable for controlling or keeping away from ailment. Two mycoparasites, *Trichoderma harzianum* and *Pythium oligandrum*, are best known for their capability to control plant pathogens however can likewise be an issue in business mushroom generation since they assault and obliterate the mycelium of *Agaricus bisporus*. [4] An unmistakable strain (or strains) of *T. harzianum* has turned into an issue in some British mushroom sheds, when spores defile the plate of fertilizer. P. oligandrum additionally has caused genuine trimming misfortunes in some mushroom-creation units. A current case was portrayed from New Zealand, where inward interpreted spacer (ITS) ribosomal DNA was utilized to recognize a particular strain of P. oligandrum in charge of a genuine flare-up of mushroom illness. [5]

The development of growths, for example, Amorphotheca resinae and Paecilomyces varioti in flying fuel stockpiling tanks has been especially hard to control. These organisms develop on the long-chain n-alkanes of aeronautics lamp fuel and cause issues by blocking channels and eroding the dividers of fuel-stockpiling tanks when they create acids as metabolic results. No less than one aircrash has been ascribed to A. resinae in the 1960s, preceding the earnestness of this issue was ecognized. Hypothetically, these growths ought not to develop in flight fuel since they require water. In any case, it is practically difficult to keep water from saturating fuel stockpiling tanks or gathering amid changes of air temperature. At that point the growths develop at the fuel–water interface. This issue is presently controlled by a mix of measures – fixing the tank dividers with rubber treated materials to avert erosion, standard checking and cleaning of channels to avoid blockages caused by contagious development in the fuel lines, and the expansion of biocides to the fuel, regularly blends of organoborates, or ethylene glycol monoethyl ether. [6]

Isolate is one of the foremost techniques for keeping the spread of pathogens, yet is ending up plainly progressively hard to authorize with the volume of global exchange. Most significant scourges are caused by growths that were presented coincidentally from different nations. Average cases incorporate potato scourge and the few noteworthy pandemics of Dutch elm ailment in North America and Western Europe, which have been connected to presentations of elm timber that was not de-yapped to evacuate the creepy crawly vectors. The demolition caused by chestnut scourge (*Cryphonectria parasitica*) in North America is by all accounts traceable to the presentation of a decorative Asian chestnut tree to the New York Zoological Garden in 1904, albeit as of late it has been recommended that the organism more likely than not been available sooner than this to have caused such broad decimation.

An exceptional strain of the human-pathogenic parasite *Cryptococcus neoformans* is progressively being separated from AIDS patients and has been followed to tropical sources. A similar strain was found to develop on specific sorts of eucalypt tree in Australia, and the locales where the clinical strain happened related to districts where these trees had been transported in – an affiliation that couldn't have been anticipated for isolate purposes. The spread of Panama malady of bananas to most banana-developing areas of the world was more likely than not caused by transport of planting material like bananas are built up from bits of corm at the base of the stem, since business bananas are sterile triploids, unfit to build up from seeds. This parasite is thought to have its focal point of inception in the Southeast Asian landmass and to have been scattered by brokers in the nineteenth century. The improvement of innovations for raising sterile "tissue-culture" plantlets (really inferred by meristem culture) should beat such issues later on.

Later cases can be refered to. A genuine root spoil illness of raspberry sticks caused by Phytophthora fragariae var. rubi is presently found in most raspberry generation zones of the world. It appears to have been spread in vegetative proliferating material (sticks) of new, enhanced cultivars. Reliable with the spread of this pathogen from a solitary source as of late is the finding that confinement section length polymorphisms (RFLPs) are for all intents and purposes indistinguishable in pathogenic strains over the world. The same was valid in the past for the spread of *P. fragariae var. fragariae*, the reason for red center ailment ofstrawberry plants, and for a specific frame of the crown annoy bacterium, Agrobacterium tumefaciens, on grape vines. This pathogen has been found to develop as a systemic, symptomless endophyte, so in the past it would have gotten away location in "obviously perfect" vegetative engendering material. [7]

This list of cases outlines a critical point: numerous ailments can be controlled effectively and economically, without the requirement for harmful chemicals, yet essentially by "great housekeeping."

BIOLOGICAL AND INTEGRATED CONTROL

A few cases of natural control have been talked about in this book, so here we survey the subject in more extensive terms. Natural control (biocontrol) can be characterized as the training in which, or process whereby, an undesirable living being is controlled by the exercises of different living beings. [8] This definition covers both actually happening biocontrol and the commonsense strategies used to accomplish control, for example,

- The intentional acquaintance of one life form with control another.
- The intentional misuse and administration of normally happening bio-control frameworks.
- Any mix of these, in light of the fact that frequently it is important to change some ecological figure request to support a presented biocontrol operator.
- The intentional change of an ecological component to advance the exercises of regular bio-control operators.

Cases of the deliberate presentation of a control operator incorporate the utilization of Phlebiopsis gigantea to control root decay of pine, the utilization of hypovirulence to control chestnut scourge, the utilization of myco parasitic fungi, for example, Trichoderma spp. as seedling protectants, and the utilization of different organisms to control creepy crawly bothers. The intentional abuse of actually happening biocontrol was found in the decrease of grain blister nematode, the misuse of soils normally suppressive to Pythium illnesses of cotton, and the control of turf pH to control the takeall fix malady of turf grasses.

The third way to deal with biocontrol includes changing an ecological element to advance the exercises of actually happening biocontrol specialists. Plant-pathogenic parasitic fungi tend to develop over the span of an editing season in glasshouses, so the dirt should be dealt with to demolish pathogens before another product is sown.

Customarily this has been finished by raising the dirt temperature to 65–70°C for 30 minutes, utilizing steam–air blends (circulated air through steam) instead of by treatment with steam alone (100°C). The utilization of steam–air blends is less expensive as well as more compelling, as appeared by the exploratory outcomes in Table. [9] In this investigation, plate of soil actually defiled with the seedling pathogen Pythium ultimum were dealt with at various temperatures at that point sown with a thick product of ringer pepper seedlings. A little inoculum of another pathogen, Rhizoctonia solani, was put in one corner of every plate, to mimic a surviving pocket of inoculum that may be discovered just past a treated range in a business glasshouse. The region of ailing plants in each case was then surveyed.

Table- 10.1: Disease caused by Rhizoctonia solani

Disease caused by Rhizoctonia solani			
Temperature	*Area (cm^2)*	*Linear spread (cm)*	*Area of disease (cm^2) caused by resident Pythium species*
No treatment	None	<1	103
100°C	253	18	0
71°C	65	9	0
60°C	3	2	0

Treatment of the dirt for 30 minutes at any temperature (61, 70 or 100°C) disposed of the inhabitant Pythium populace. In any case, after treatment at 100°C the dirt was sterile and profoundly helpful for the spread of Rhizoctonia, which can become quickly through cleaned soil. By differentiate, soil treated at 61°C was free from any malady. This was clarified by the way that Bacillus spp., which happen normally as torpid spores in soil, can survive sanitization as well as are activated to grow at this temperature. These normally happening biocontrol operators develop quickly to fill the organic "vacuum" and they deliver a few antifungal anti-microbials that secure the recently sown harvest. The utilization of steam–air blends in glasshouse crops has now been supplanted by the utilization of Trichoderma spp. Furthermore, comparable opponents. Yet, sun based warming of soil (solarization) has turned out to be basic practice for bringing profitable field manifests in a large number of the hotter parts of the world. [10]

It is accomplished by covering the dirt with thin polyethylene sheeting for a 20-to 30-day time frame. This makes a "nursery impact," raising the temperature at the dirt surface to at least 40°c, and regularly to at least 30°c at even 50 cm soil profundity. The spores of most soil-borne pathogens can survive these temperatures in vitro, yet in clammy, sun oriented warmed soil the pathogens are dynamically wiped out by the exercises of contending and opposing microorganisms that twist at the higher temperatures. [11]

Chemical Control of Fungi

Fungi-toxic or fungicidal chemicals are frequently used to control parasitic development, however a refinement must be made between general toxicants and specifically lethal specialists.

GENERAL TOXICANTS

Fungitoxic chemicals assume a noteworthy part in forestalling rot of a scope of items including makeup, starch-based materials, cotton textures, and auxiliary timbers. For each situation the decision of additive relies upon particular needs. For instance, the additives utilized as a part of beautifiers and treatments must be nontoxic, nonirritant, and good with the item definition. [12] Frequently, low-atomic weight alcohols and esters of p-hydroxybenzoic corrosive (parabens) are utilized as a part of these cases. At the other outrageous, marine timbers, fence posts, and utility poles should be treated with determined aggravates that will secure against rot and creepy crawly assault for quite a few years. The three most normally utilized wood additives are coal-tar creosote, pentachlorophenol, and inorganic arsenicals, for example, chromated copper arsenate. Timbers are regularly weight treated with these mixes to guarantee high infiltration and maintenance in the wood. Unmistakably, a number of these wood-treatment items posture potential wellbeing risks. Pentachlorophenol is a known cancer-causing agent that is consumed by the lungs and skin, and it can contain follow measures of dioxin. Arsenic-based wood medications are generally known to be toxic, and the Environmental Protection Agency of the USA as of late declared (2003) that arsenic will be pulled back from utilize. The outrageous lethality of huge numbers of these wood additives has impelled enthusiasm for the conceivable utilization of natural control specialists for wood assurance. Among the potential applicants are strains of Trichoderma that as often as possible colonize the cut surfaces of logs and that irritate wood-rot growths in research facility tests. [13]

Control of Human and Crop Pathogens: the Principle of Selective Toxicity

Rather than general toxicants, chemicals used to control growths in the tissues of another living being must show specific harmfulness. Whatever remains of this part will be committed to this theme, managing first with the control of plant infections and after that with the control of human mycoses. These zones are more firmly related than one may might suspect, in light of the fact that similar sorts of substance are utilized to control both plant-and human-pathogenic organisms. For instance, the azole fungicides (imidazoles and triazoles) were first created to control plant illnesses, however in adjusted structures they are presently generally used to control human mycoses. [14]

Additionally, the actually happening antifungal anti-infection griseofulvin was first found as a "twisting variable" that caused the germ-containers of a plant-pathogenic parasite, Botrytis allii, to develop in a mutilated winding manner, yet it was created financially as an orally regulated anti-microbial to control contaminations caused by the dermatophytic organisms. It acts by disturbing parasitic microtubules, and this clarifies its morphogenetic impact since microtubules are included in the conveyance of cell parts to the developing hyphal tip. [15]

Principal Cellular targets of Antifungal Agents

The fundamental cell targets right now used to control plant or human ailments. At first sight it may appear that there are countless focuses on that could be misused for illness control. In any case, practically speaking the range is constrained. [16]

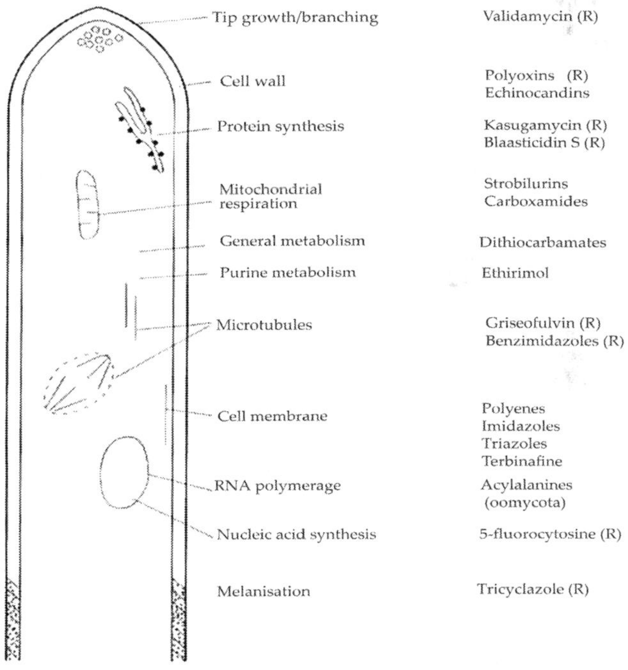

Figure- 10.1: Major cellular targets for chemical control of fungi that cause disease in plants and animals.

Huge numbers of the mixes have an exceptionally confined utilization (appeared as "R") and are utilized fundamentally in Japanese farming. In the event that we prohibit these mixes then we are left with only five fundamental sorts of antifungal target:

- The cell filın, since organisms are interesting in having ergosterol as their trademark layer sterol.

- Mitochondrial breath, which is focused by some plant fungicides.
- Various parts of general digestion.[17]
- Fungal cell divider segments, particularly β1-3 glucans, for which another gathering of medications, the echinocandins, has as of late come into utilization.
- The microtubules and microtubule-related proteins, which are disturbed by the anti-toxin griseofulvin, and by benzimidazole fungicides (which have now been pulled back).

There is an urgent need to find new chemicals (with novel modes of action) and new cellular targets, to provide a greater range of options for controlling fungal diseases. [18]

FUNGICIDES UTILIZED FOR PLANT INFECTION CONTROL

The expression "fungicide" is utilized as a part of a wide sense for any intensify that murders or inactivates growths. The vast majority of these mixes are artificially blended, yet some are changed subordinates of normally happening mixes. All fungicides must be authoritatively endorsed for use by the important national administrative body, which additionally indicates the products on which the fungicide can be utilized and any "withholding period" between the last use of a fungicide and the reaping of a harvest. As of now, there are around 80 diverse concoction gatherings of fungicides (where a gathering is characterized as a particular sort of compound, for example, the pyrimidines, the imidazoles, the triazoles, and so forth.). These gatherings can be consolidated into 12 classes as per their essential cell target or method of activity. By and by, fungicides can be gathered into three more extensive classes, mirroring their diverse parts in plant illness control – the inorganic fungicides, the natural contact (protectant) fungicides, and the systemic fungicides. [19]

Inorganic Fungicides

Inorganic fungicides were the first to be utilized to control plant illnesses, starting with natural sulfur in 1846, copper salts in 1882, and mercury around 1920 (in spite of the fact that its utilization has been prohibited since 1992 on account of its outrageous poisonous quality, and Britain holds the refinement of being the last industrialized nation to acknowledge this boycott!). Inorganic sulfur dependably had a restricted utilize in view of its insolubility, yet it can be tidied onto leaves to control fine buildup organisms, on the grounds that the hyphae and spores of these growths create on or close to the leaf surface. Sulfur is the main fungicide accessible to natural cultivators since it is a completely characteristic item.[20]

Fungicides in view of different copper salts have been utilized widely for their expansive antifungal and antibacterial range. They were fantastically fruitful against wool buildup of grapevines (Plasmopara viticola, Oomycota) in the late 1800s, when "Burgundy blend" and "Bordeaux blend" were created. Ordinarily, the copper fungicides are buildings of copper sulfate with lime (Bordeaux blend) or they depend on copper hydroxide or copper oxychloride. They control many leaf and natural product infections, yet are possibly phytotoxic. Their method of activity depends on the interruption of a few essential metabolic procedures, so organisms don't create imperviousness to them effectively.[21]

Organic Contact (Protectant) Fungicides

Natural fungicides were produced in the 1930s and immediately substituted the inorganic fungicides for generally purposes. They are named "protectant" or "contact" fungicides since they act just close to the site where they are connected in order to secure the plant surface or to control a built up contamination. Accordingly, they should be connected over the entire plant surface, and should be reapplied to secure any new development. [22]They are, be that as it may, amazingly sturdy fungicides and a large portion of them are still being used today. Parasitic fungi don't create imperviousness to them effortlessly on the grounds that they meddle with fundamental metabolic procedures and regularly have numerous destinations of activity in contagious cells. Cases of these protectant fungicides incorporate the dithiocarbamates, for example, maneb and thiram. These contain at least one thiol (sulfur-containing) gatherings and they inactivate sulfydryl (- SH) bunches in amino acids, proteins, and chemicals. They are utilized to control many leaf infections.

Other protectant fungicides incorporate the dinitrophenols, for example, dinocap, which uncouples breath from ATP blend. [23]

Systemic Fungicides

Systemic fungicides are consumed by plants and are then circulated inside, more often than not by upward development in the xylem, where they can ensure new development. The advancement of these mixes in the 1960s altered harvest assurance on the grounds that these fungicides can annihilate existing diseases and ensure against resulting contaminations. They are currently utilized generally, albeit the vast majority of them don't move downwards in the phloem, so they give little assurance against the significant soil-borne pathogens. Their utilization has additionally uncovered another issue: the dynamic fixings are not general toxicants but rather have profoundly particular locales or methods of activity, for instance, on a solitary stride in an enzymic pathway or by official to a particular segment of a parasitic cell. Along these lines, organisms can create imperviousness to them effortlessly, frequently by a solitary quality change. [24] At the point when this happens a growth regularly demonstrates cross-imperviousness to every other individual from that fungicide gathering. Thus, the systemic fungicides should be utilized as a part of conjunction with a resistance administration methodology. This can include the exchanging utilization of fungicides with various methods of activity, or the mix of systemic and wide range protectant fungicides. We think of some as agent cases underneath.[25]

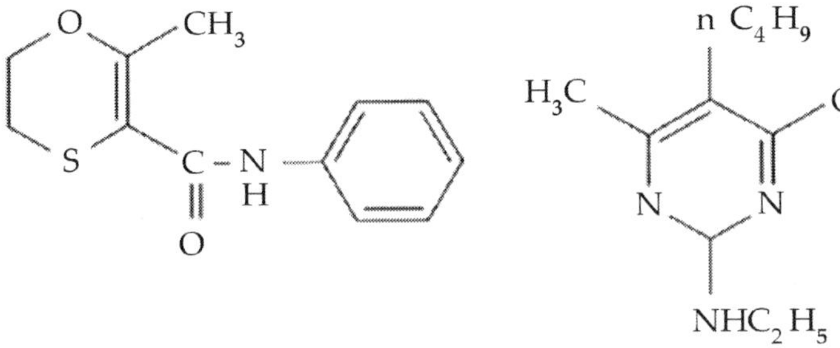

Carboxin **(Carboxamide)** Ethirimol **(2-aminophyrimidine)**

Figure- 10.2: Examples of two systemic fungicides.

Benzimidazole Fungicides

The benzimidazole mixes were the principal systemic fungicides to be presented as thiabendazole in 1964, benomyl in 1967, and afterward others, for example, carbendazim (1970) and thiophanate-methyl. Amusingly, thiabendazole was created and discharged in 1962 as an antihelminthic, and just later was its fungicidal part found. This exact way to deal with fungicide improvement is as yet normal. It includes the screening of thousands of possibly dynamic mixes against a couple of significant pathogens, normally including a grain rust parasite, a fine mold organism, a wool buildup pathogen (e.g. Phytophthora infestans), and a rice pathogen (Magnaporthe grisea). Aggravates that show guarantee in this essential screen are synthetically changed to create a scope of isomers for additionally testing, to upgrade the fungicidal action, and to limit undesirable symptoms. This sort of approach prompted the different protecting of benomyl and carbendazim from the underlying compound, thiabendazole. [26] It is currently trusted that they have comparable or indistinguishable methods of activity: they are changed over to methyl benzimidazole-2-yl-carbamate (MBC) which ties to axle microtubules and squares mitosis. They likewise influence parasitic tip development, by authoritative to the cytoplasmic microtubules which convey parts to the developing hyphal tip. [27]

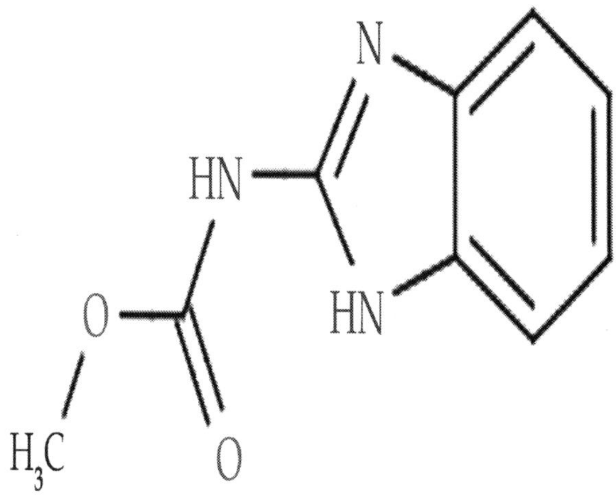

Figure- 10.3: Example of Benzimidazole fungicides.

Benzimidazole fungicides are specifically antifungal, with practically no impact on plant and creature cells, notwithstanding the way that plants and creatures additionally have microtubules. The purpose behind this was appeared by in vitro thinks about. Microtubules are framed when β-tubulin shapes a dimer with α-tubulin, and the microtubule develops by progressive increments of these dimers. MBC ties unequivocally to the β-tubulin ofparasitic fungi, keeping its relationship with α-tubulin, and consequently keeping the self-get together of microtubules. MBC does not tie firmly to the tubulins of higher creature or plant cells, nor to the tubulins of Oomycota. So it is a particular antifungal operator.[28]

The benzimidazoles have been utilized effectively to control a few plant pathogens, yet organisms can create resilience (resistance) to them in the event that they are utilized over and again. This is by all accounts caused by point changes at different destinations in the β-tubulin quality, in some cases including decreased official of the fungicide to β-tubulin and some of the time influencing the collaboration of β-tubulin with α-tubulin or with the tubulin-related proteins, all of which are vital for practical microtubules.[29]

Notwithstanding the utilization of benomyl (Benlate) as a fungicide for over 30 years, it has been connected to a birth deformity in which moms presented to benomyl in the early phases of pregnancy can bring forth kids with discharge eye attachments. As per The Observer daily paper, "over 40% of pregnant rats bolstered abnormal amounts of benomyl delivered embryos with extreme eye surrenders." The producer as of late pulled back Benlate from the worldwide market. [30]

Sterol Formation Inhibitors

Sterols are found in the cell layers of all eukaryotes and of some archaea (the methanotrophs), however are missing from bacterial layers. Sterols embed into the phospholipid bilayer and help to keep up layer steadiness and smoothness. Diverse gatherings of living beings have distinctive sorts of layer sterol. Organisms naturally have ergosterol as their film sterol, while cholesterol is the trademark sterol of creatures, and sitosterol and comparable phytosterols are found in plants and Oomycota. [31]

All sterols are combined by a complex, multistep pathway. They are gotten from the buildup of three particles of acetyl coenzyme A, to frame isoprene units (5-carbon mixes), at that point three isoprene units gather to shape the 15-carbon compound, farnesyl pyrophosphate. Two atoms of this consolidate to frame the 30-carbon compound, squalene, which at that point experiences a progression of cyclization responses and ring terminations, bringing about the sterol, lanosterol. This is the antecedent sterol from which every other sterol are delivered. There are a few middle of the road ventures from lanosterol to the last sterol, however the key stride prompting ergosterol is the expulsion of a methyl amass from the C-14 position of the particle. This progression is catalyzed by the catalyst lanosterol 14 α-demethylase, which has an iron-containing cytochrome P-450 as its coenzyme. This demethylation step happens just amid combination of the contagious sterol, not amid the union of plant or creature sterols, so it gives a perfect focus to antifungal specialists.[32]

A few systemic fungicides act by restraining sterol demethylation, with the outcome that the organism can't orchestrate its typical sterols, and rather different sterols, for example, lanosterol are consolidated in the parasitic film. This prompts film spillage and at last to cell passing. The imidazoles (e.g. imazalil) and triazoles (e.g. propiconazole) are critical cases of fungicides that demonstration along these lines. Their regular element is the ownership of a five-membered heterocyclic ring containing either two (imidazoles) or three (triazoles) nitrogen iotas. These azole fungicides are thought to act in an indistinguishable path from the azole drugs used to treat contagious contaminations of people (see beneath). The lipophilic piece of the fungicide is thought to tie to the demethylase catalyst, while nitrogen in the heterocyclic ring partners with an iron-containing coenzyme, hindering demethylation. [33]

Group-Specific Systemic Fungicides

A few systemic fungicides act pretty much particularly on specific contagious gatherings. For instance, the carboxamide fungicides, for example, carboxin act against Basidiomycota (rusts, mucks, Rhizoctonia solani) by meddling with breath, they repress the progression in the TCA cycle where succinate is dehydrogenated to fumarate. The 2-aminopyrimidine fungicides, for example, ethirimol act particularly against fine mold growths by restraining a protein (adenosine deaminase) included in purine digestion.[34] For this situation, fungicide-tolerant mutants regularly demonstrate no change of the catalyst yet are thought to evade the hindrance by acquiring purines from the host plant. The acylalanine fungicides (e.g. metalaxyl) act particularly against Oomycota by hindering the atomic RNA polymerase, in this manner blocking RNA blend. They have no impact on the genuine organisms yet are exceedingly compelling in controlling Pythium and Phytophthora spp., including Phytophthora infestans.[35]

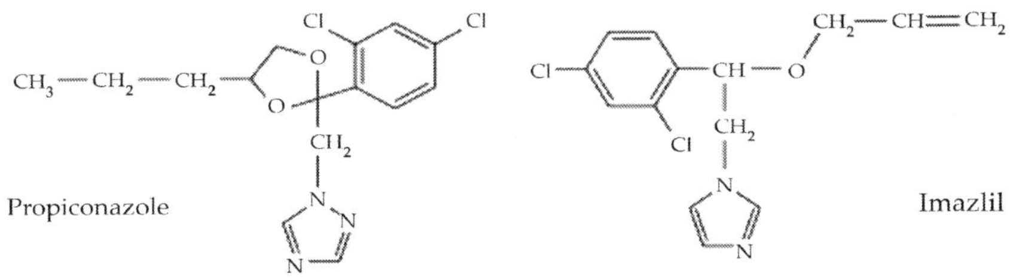

Figure- 10.4: Examples of azole fungicides.

Imperviousness to the acylalanines can grow quickly in field conditions, so these fungicides are utilized as a part of mix with protectant fungicides, for example, the dithiocarbamates. Fosetyl-aluminum is another aggravate that demonstrations particularly against Oomycota, and it has the striking property of being completely versatile in phloem. So it can be connected to the shoots and it moves into the roots, making it to a great degree profitable for control of root-tainting Phytophthora spp, for example, and Phytophthora cinnamomi.[36]

The method of activity of fosetyl-Al is as yet misty. It indicates practically zero movement against Oomycota in research facility culture, and this drove at first to the view that it may act by instigating host resistance. Ensuing work demonstrated that fosetyl-Al breaks down promptly in plants to yield phosphorous corrosive (H_3PO_3) and this straightforward mineral corrosive can smother the development of Oomycota in culture media of low phosphate content. [37]

A couple of fungicides go about as melanin blend inhibitors. They have no impact on contagious development in culture, yet meddle with the contamination procedure, particularly when parasitic fungi taint from melanized appressoria. These fungicides have discovered a constrained application in the control of Colletotrichum spp. Furthermore, particularly of Magnaporthe grisea (the reason for rice impact). Appressoria join firmly to a host surface, at that point progress toward becoming melanized, empowering them to develop an amazing osmotic weight equal to around 8 airs (8 MegaPascals). [38]This is fundamental for "centering" of the thin entrance peg which creates from a little nonmelanized locale of the divider in contact with the host surface. On the off chance that the organism can't blend melanin at that point there is an expansive zone of contact with the host surface, and the endeavored entrance comes up short. Melanin-lacking mutants of these pathogens demonstrate a comparable failure to enter the host. The fundamental target pathogens, including rice impact, can quickly create imperviousness to these fungicides, whose method of activity depends on a particular protein – either a reductase or a dehydratase in the melanin biosynthetic pathway. [39]

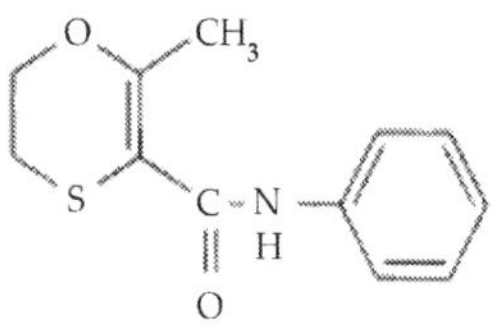

Carboxin **(Carboxamide)**

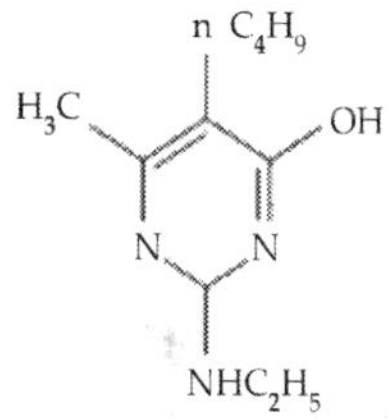

Ethirimol **(2-aminopyrimidine)**

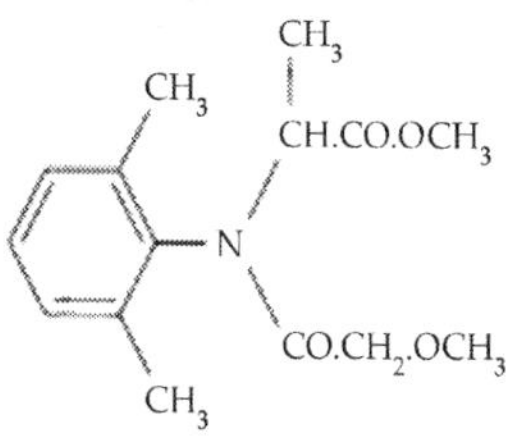

Metalaxyl **(acylalamine)**

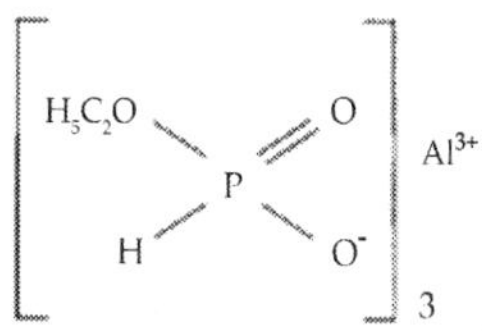

Fosetyl aluminum

Figure- 10.5: Systemic fungicides (Ethirimol, and Carboxin) and Metalaxyl and Fosetyl-aluminum.

Strobilurin Fungicides

Strobilurin A and Oudemansin A are normal items found in the toadstools of two wood-decaying Basidiomycota, Strobilurus tenacellus and Oudemansiella mucida. They were found freely however later were observed to be synthetically indistinguishable, with a typical method of activity as they repress mitochondrial breath inparasitic fungi, by hindering the oxidation of ubiquinol at a particular site of the cytochrome bc1 complex in the electron-transport chain. [40] Along these lines, they repress the era of ATP. In spite of the fact that these mixes had been known for quite a while, they just begun to be created financially in the mid-1980s when agrochemical organizations delivered engineered analogs, known as the strobilurins, which were photochemically steady and had other attractive properties, for example, low mammalian danger, suitable portability inside plants and worthy product security. The principal business strobilurins like azoxystrobin and kresoxim methyl, were propelled in 1996, and there are currently more than 700 licenses recorded on these mixes by the major agrochemical organizations.[41]

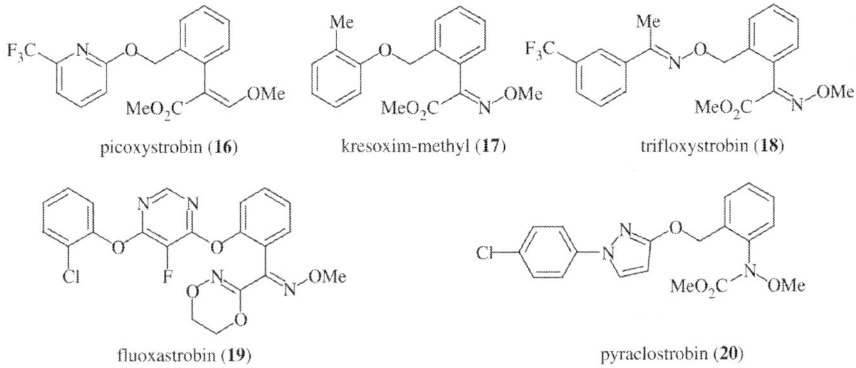

Figure- 10.6: Examples of strobilurin fungicides.

The strobilurin fungicides demonstrate an incredibly extensive variety of movement against all the major scientific classifications of plant pathogens. For instance, azoxystrobin is enrolled for use against more than 400 plant pathogens. It firmly restrains spore germination, indicates amazing precaution movement, and has eradicant and antisporulation properties. Particular details are promoted for a wide range of sorts of product, including all the real maladies of grains. The inquiry emerges with reference to why these mixes are harmful to organisms however have little impact on plant has, which have the same mitochondrial target locales. The proposed clarification is that there could be differential entrance and corruption of these fungicides in plants contrasted and in organisms. Like all fungicides that follow up on particular metabolic focuses, there is solid proof that organisms can create imperviousness to the strobilurins. So they should be utilized as a part of blends or as exchanging medications with different fungicides, as a major aspect of a resistance-administration procedure.[42]

ANTIFUNGAL ANTI-INFECTION AGENTS UTILIZED FOR PLANT ILLNESS CONTROL

There are a few antifungal anti-infection agents however just a couple are sufficiently specific to be utilized for illness control. For instance, cycloheximide is a broadspectrum antifungal specialist and is generally utilized as an exploratory instrument in research facility ponders, yet it acts by blocking protein amalgamation on 80S ribosomes and is in this manner lethal to all eukaryotes. It is not utilized for sickness control. The couple of anti-infection agents that have been utilized financially for plant sickness control were created in Japan, for the most part for the control of rice illnesses (rice impact caused by Magnaporthe grisea, and rice sheath curse, caused by Rhizoctonia solani). [43]

Table- 10.2: Few antifungal antibiotics for the control of human and plant mycoses

Antibiotic	**Produced by**	**Fungi affected**	**Site/mode action**
Griseofulvin	*Penicillium gerisefulvum*	Many (not oomycota)	Fungal tubulins
Poly macrolide	*Streptomyces spp.*	Many (not oomycota)	Cell membrane
Polyoxins	*Strep. Cacaoi*	Many (not oomycota)	Chitin synthesis
Viladamycin A	*Strep. Hygroscopicus*	Some	Morphogen
Blasticidin-S	*Strep. Griseochromogenes*	Some	Protein synthesis
Kasugamysin	*Strep. Kasugaensis*	Some	Protein synthesis
Streptomycin	*Strep. Griseus*	Oomycota	Calcium?

Pyrrolnitrin	*Pseudomonas spp.*	Various plant pathogens	Implicated in biocontrol by nutrient competition, antibiosis, parasitism of other fungi, etc
Pyoluteorin	*Pseudomonas spp.*		
Gliotoxin	*Trichoderma virens*		
Gliovirin	*T. virens*		
Viridian	*T. virens*		
Viridiol	*T. virens*		
Heptelidic acid	*T. virens*		
Terichodermin	*Trichoderma spp.*		
6-pentyl-α-pyrone	*Trichoderma spp.*		
Suzukacillin	*Trichoderma spp.*		
Alamethicine	*Trichoderma spp.*		

Notwithstanding these fermenter-created anti-infection agents, a few natural control specialists are known to deliver antifungal anti-toxins. Yet, these mixes are misused in a roundabout way by advertising the biocontrol specialists as microbial inoculants which is a methodology that can keep away from the need to embrace point by point and costly toxicological testing.[44]

The Japanese Agricultural Antibiotics

Five novel sorts of antifungal anti-microbial have been found and popularized in Japan. They are examined quickly underneath, and their structures.

Polyoxins and Nikkomycin

The polyoxins, including polyoxin D, are particular inhibitors of chitin combination and subsequently speak to an exceptional method of activity. Their revelation in 1965 raised expectations that they could have huge parts in controlling parasitic fungi and creepy crawlies – the two noteworthy gatherings of chitin-containing life forms – while having no impact on higher creatures or plants.[45] Nikkomycin is a comparable exacerbate that demonstrations similarly. Both mixes tie firmly to the dynamic site of chitin synthase, and rival the ordinary substrate, UDP-N-acetylglucosamine from which chitin is integrated. They have been utilized as a part of training to control some plant maladies, particularly sheath scourge of rice (Rhizoctonia) and dark spot of pear (Alternaria) in Japan. Be that as it may, these pathogens quickly create resilience/imperviousness to the anti-infection agents in field conditions, since mutant strains of the organisms indicate decreased anti-microbial take-up. Endeavors to utilize these anti-microbials to control maladies of human and creature has have likewise been unsuccessful. In vitro considers in lab media demonstrate that the polyoxins are intense inhibitors of chitin union. In any case, these mixes are taken into contagious cells through layer proteins that are typically utilized for the take-up of dipeptides, and the wealth of little peptides in human or creature tissues aggressively obstructs the take-up of these anti-infection agents.[46]

Blasticidin-S, Kasugamycin, and Validamycin

These three anti-infection agents have discovered constrained parts in Japanese farming. Blasticidin-S hinders protein blend in microorganisms and a couple of growths, by official to the vast subunit of 80S ribosomes. It was created for control of rice impact (*M. grisea*) however later it was supplanted by another protein amalgamation inhibitor, kasugamycin, which has bring down mammalian danger. Resistance has created to both of these mixes in field conditions. It depends on lessened take-up of blasticidin-S, and decreased official of kasugamycin to the ribosome. Validamycin An is utilized to control a few maladies caused by Rhizoctonia, including seedling illnesses, dark scurf of potato, and sheath scourge of rice, yet it is idle against most parasitic fungi and microbes. It has no impact on Rhizoctonia in rich culture media, yet it causes anomalous stretching and discontinuance of development in feeble media. Validamycin has been observed to be changed over to validoxylamine in growths, and this compound is a solid inhibitor of trehalase, the protein that severs the contagious disaccharide trehalose to glucose.[47]

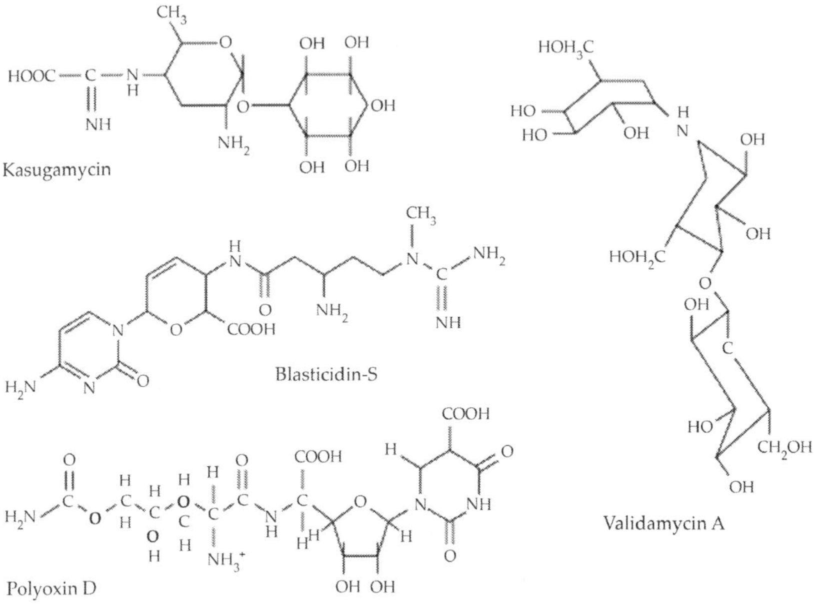

Figure- 10.7: Japanese agriculture antiotics.

Antibiotics implicated in Biocontrol

We consider here just the commonsense utilizations of these mixes in sickness control. A few new biocontrol items are experiencing field trials or have been showcased over the most recent couple of years. One of these, advertised as GlioGard™ however now as SoilGard, comprises of little, air-dried alginate globules containing fermenter-developed biomass of Trichoderma virens and wheat grain as a sustenance base. [48]

This item is thought to act principally by creating the anti-infection gliotoxin. The dried granules are blended with soil-less establishing media in glasshouses and left for a few days after the medium is dampened, before seedlings or established cuttings are transplanted into the beds. The planning of these operations is essential, on the grounds that the organism must develop on the wheat grain substrate and create gliotoxin which diffuses into the establishing medium, shielding the plants from pathogenic assault. In the event that there is a deferral in planting then the anti-microbial levels begin to decay and the level of assurance is lessened. [49]

Rhizosphere microscopic organisms additionally create numerous antifungal mixes. Table demonstrates the aftereffects of "nonselective" screening (on Trypticase Soy agar) of the prevailing microorganisms from a scope of plants by a business organization. The most widely recognized microorganisms with antifungal properties included Pseudomonas spp. (some of which deliver the anti-infection agents pyrrolnitrin, pyoluteorin, phenazines, and diacetylphloroglucinol), Xanthomonas maltophilia (obscure anti-microbials), Bacillus subtilis and different Bacillus spp. (some of which create iturins, mycosubtilins, bacillomycin, fengymycin, mycobacillin, and mycocerein), and Erwinia herbicola (which produces herbicolins). Some of these microbes are utilized as a part of business biocontrol definitions. [50]

For instance, Bacillus subtilis is showcased for seed treatment of cotton and soybeans, to ensure against damping-off growths. Others have been embroiled in characteristic biocontrol in field conditions, a great case being the part of fluorescent pseudomonads (*P. putida* and *P. fluorescens*) in stifling the take-all illness of grains. [51]

Control of Fungal Infections of Humans

The control of parasitic diseases of people presents significant troubles, on the grounds that regularly the patient is extremely traded off or immunosuppressed. It is significant that even the regular antibacterial anti-infection agents, for example, penicillins and cephalosporins don't work alone, they act in conjunction with the host's typical safeguards. There are, be that as it may, a few antimycotic operators that can be utilized effectively to treat human mycoses. We consider these beneath. [52]

Griseofulvin

Griseofulvin is a normally happening antifungal anti-toxin, created by the organism Penicillium griseofulvum. It is insoluble in water yet can be taken orally in a microcrystalline frame and it at that point gathers in the keratin forerunner cells of the skin, nails, and hair. It is utilized only to control the dermatophyte contaminations of the keratinized tissues, however it is fungistatic instead of fungicidal, so for genuine diseases the treatment must be drawn out until the point when all the tainted tissues have been shed.

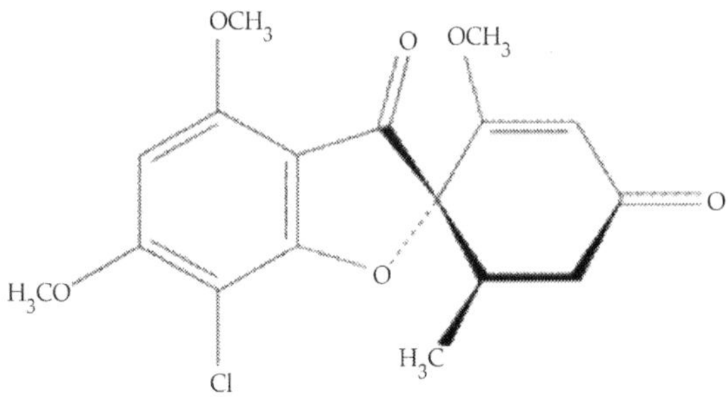

Figure- 10.8: Griseofulvin.

As noted before, griseofulvin causes development bends of contagious hyphae in spore germination tests, including thickening of the hyphal dividers, related with overwhelming testimony of glucans. In any case, this is currently known to be an optional impact, and the essential part of griseofulvin is to meddle with the get together of microtubules or of the tubulin-related proteins. It causes unusual atomic division by upsetting the axle microtubules, and strange development, apparently by disturbing the microtubule-related transport of materials to the hyphal tip. In these regards, the part of griseofulvin is like that of the benzimidazole fungicides utilized for plant infection control, on the grounds that these additionally square microtubule get together. Yet, the benzimidazoles obviously act at an alternate site on microtubules since mutants that create imperviousness to the benzimidazoles are not impervious to griseofulvin, and the other way around.

Griseofulvin has been the primary medication used to treat dermatophyte contaminations for a long time, and in spite of the fact that imperviousness to the anti-toxin can happen every now and again in vitro, it has not been a noteworthy issue in clinical practice. In any case, griseofulvin is presently being continuously supplanted by more current manufactured medications, including terbinafine and itraconazole, which are less dangerous and require shorter treatment times.

Terbinafine

Like griseofulvin, terbinafine can be taken orally (or topically) and it has a tendency to gather in the skin, nails, hair, and greasy tissues. It represses ergosterol union by meddling with an early stride in sterol biosynthesis, where the protein squalene epoxidase begins the procedure of cyclization prompting lanosterol.

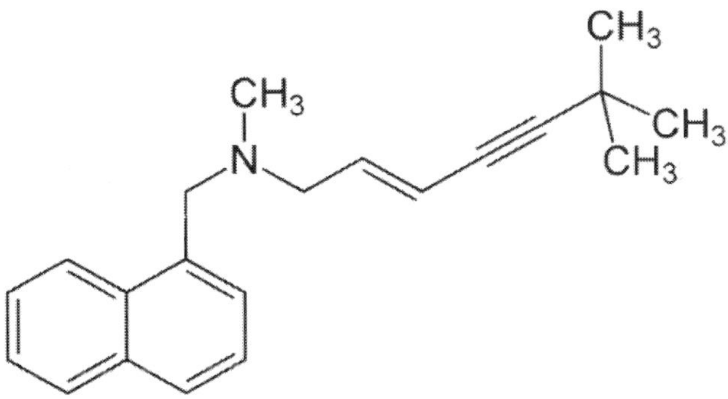

Figure- 10.9: Terbinafine.

Polyene Macrolide Antibiotics

The polyene macrolide anti-toxins are created by a few Streptomyces spp. Their ordinary structure is exemplified by amphotericin B. An aminosugar gather is appended to a vast lactone ring with a progression of twofold bonds on one face (subsequently the term polyene) and a progression of hydroxyl bunches on the other face. So the particle has both a hydrophobic and a hydrophilic face. By X-beam crystallography the particle apparently has a bar like structure, around 2.1 nm long on account of amphotericin B, like a layer phospholipid. The polyenes embed in the phone film by partner with sterols (the hydrophobic face) and are thought to cause adjustment of the sterols so that a gathering of eight polyene atoms frames a ring with the hydrophilic faces in the inside. Along these lines they shape a polar pore through which little particles (K^+, H^+) can pass uninhibitedly, disturbing the cell's ionic control.

All eukaryotic cells have sterols in the films, albeit some Oomycota don't create them or even need them for development. So the polyene anti-infection agents would not, at first sight, appear to be specific antifungal operators. Notwithstanding, there are a wide range of polyenes that vary in the ring size and nearness or nature of the aminosugar gathering, and they demonstrate particular official to various sterols. Three of these mixes are amphotericin B, nystatin, and pimaricin – have a substantially higher fondness for ergosterol than for cholesterol, the mammalian layer sterol, so they are specifically antifungal. Nystatin and pimaricin can be connected topically as creams to control Candida contamination of the vagina, or as powder definitions to control competitor's foot.

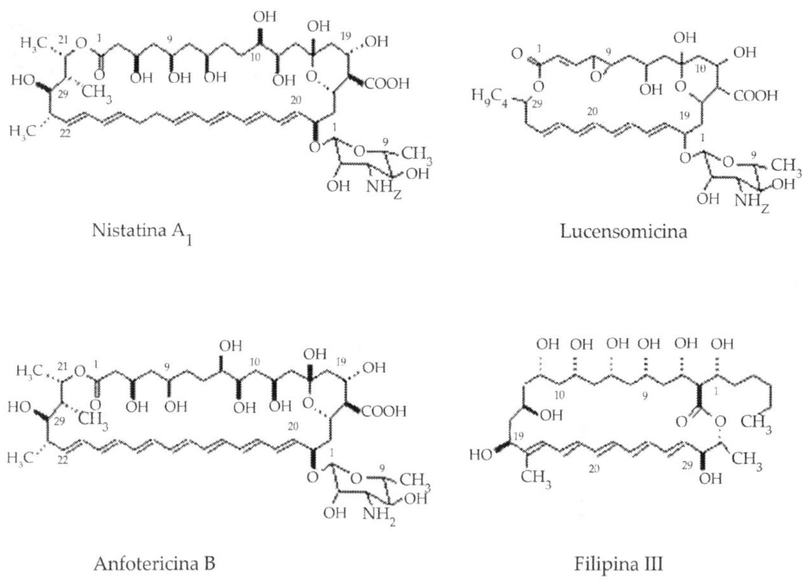

Figure- 10.10: Examples of Polyene macrolide antibiotics.

Amphotericin B has bring down mammalian danger than nystatin or pimaricin, and it additionally has a few (low) level of water solvency. Starting improvement work in the 1960s demonstrated that amphotericin B could be directed orally, with some achievement. In any case, all the polyene anti-infection agents have antagonistic symptoms, so the central purpose of research has been to create details that limit these impacts, for example, constant renal harmfulness. There are presently a few lipid details of amphotericin B, either endorsed for human utilize or in phases of advancement. For instance, the business definition named AmBisome comprises of little (<100 nm breadth) liposomes comprising of a solitary layer of phospholipids that fuse amphotericin B. Single dosage intravenous poisonous quality tests on mice demonstrated that AmBisome had a LD50 esteem (the measurements bringing about half mortality) of progressively that 175 mg/kg, contrasted and a LD50 of 2–3 mg/kg for a standard colloidal definition of amphotericin B blended with sodium deoxycholate. Amphotericin in the new details can possibly control a few systemic contagious pathogens, including Candida, Cryptococcus, Aspergillus, Blastomyces, Coccidioides, and Histoplasma.

Azole Drugs

The azole drugs are ergosterol biosynthesis inhibitors like those utilized for plant infection control, yet with structures intended for the treatment of systemic mycoses of people. They act by blocking sterol demethylation at the C-14 position, amid the pathway from lanosterol to ergosterol.

Contingent upon whether these medications have a few nitrogens in the five-membered heterocyclic ring, they are named imidazoles or triazoles. Ketoconazole was the first of these medications to be produced for treatment of people. It has solid movement against Candida and Cryptococcus neoformans, and furthermore can be utilized against other systemic pathogens, for example, Blastomyces dermatitidis and Coccidioides immitis, yet it has little action against Aspergillus. In any case, ketoconazole has a few unfriendly symptoms, and it has been to a great extent supplanted by the later and less dangerous triazole medications, for example, fluconazole and itraconazole.

In spite of the fact that ketoconazole is as yet accessible, it is utilized essentially as a moment line tranquilize for diseases that don't react to the triazoles. Fluconazole is for the most part utilized against Candida and Cryptococcus species, and is accessible as both oral and intravenous details. It likewise has constrained movement against Histoplasma and Blastomyces, yet no action against other human pathogens. It is a generally utilized bleeding edge medicate against numerous Candida albicans contaminations, however is not viable against a portion of the less basic Candida spp. for example, *C. glabrata* and *C. krusei*.

Itraconazole is accessible in oral or intravenous structures and is dynamic against numerous organisms including yeasts, dimorphic parasitic fungi, and mycelial growths. Its range incorporates Candida spp., Cryptococcus, Aspergillus, Histoplasma, Blastomyces, Coccidioides, and dermatophytes. In any case, imperviousness to itraconazole has been accounted for among numerousparasitic fungi, and this occasionally prompts cross-imperviousness to drugs with a comparative method of activity. Three new triazoles with various spectra of movement have as of late been produced. In one case, voriconazole, the compound showed up available in 2002, and the other two mixes, ravuconazole and posaconazole, are in the last phases of clinical trials. In this manner, contrasted and the circumstance just 10 years prior, when ketoconazole was the medication of decision, there is currently an extensive variety of azoles accessible for the treatment of particular mycoses. For all these azole drugs, the result of hindering the demethylation venture of the ergosterol biosynthetic pathway is to exhaust the convergence of ergosterol with the goal that parasitic fungi use different sterols in the film.

The impact of this can be examined in a model framework, by misusing the way that Saccharomyces cerevisiae can become both anaerobically and vigorously. At the point when developed vigorously, Saccharomyces blends ergosterol, yet when developed anaerobically Saccharomyces should be provided with sterols, in light of the fact that these are results of oxygen consuming biosynthetic pathway. In the event that anaerobic cells of Saccharomyces are provided with an alternate sterol, for example, lanosterol, they wind up noticeably cracked and can't gather the divider in the ordinary way. In this regard it will be reviewed that the significant divider manufactured catalysts, chitin synthase and glucan synthase, are indispensable film proteins, so probably any adjustment in the layer creation could influence these chemical capacities. In Candida, one of the impacts of the azole drugs is to keep the stage move from yeasts to hyphae. As we noted in, this stage move is imperative for intrusion of tissues and for enabling Candida to break out of phagocytes.

5-Flucytosine (5-FC)

5-FC is a fluorine-substituted nucleoside, grew initially as an antitumour operator yet it has an extra part as an antimycotic specialist for the treatment of systemic Candida and Cryptococcus. It is taken into a contagious cell by the cytosine permease of the cell layer, at that point deaminated to 5-fluorouracil which, through a progression of steps, is consolidated into RNA set up of uracil, causing disabled RNA work.

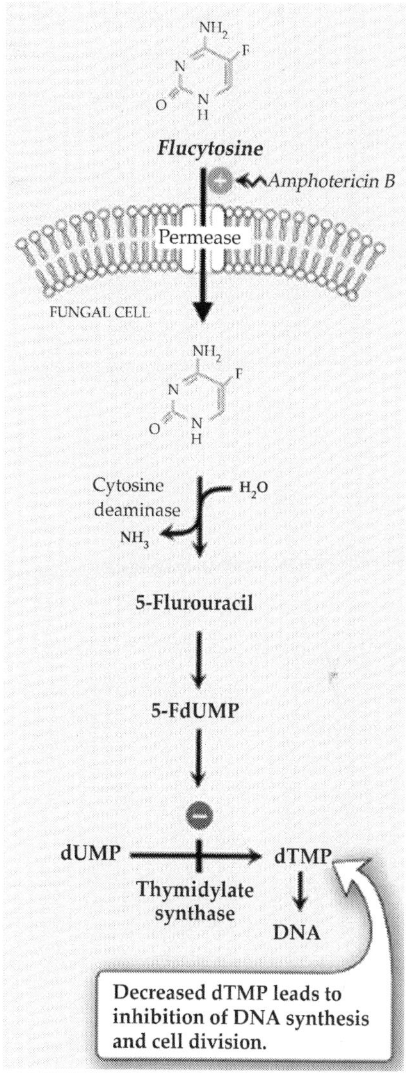

Figure- 10.11: Pathway of synthetic pyrimidine 5-flucytosine.

It additionally debilitates DNA union since one of the results of 5-fluorouracil restrains the protein thymidylate synthase in the pathway that creates thymidine nucleotides. 5-FC is specifically harmful to parasitic fungi since mammalian cells have a low rate of take-up of the atom and have practically zero capacity to change over it to 5-fluorouracil since they do not have the catalyst cytosine deaminase. In any case, 5-FC influences just few growths, strikingly Candida, Cryptococcus, and

Aspergillus fumigatus, and even these parasitic fungi effectively create imperviousness to it in clinical practice. The reason for this resistance can be examined in vitro by providing cells with the different intermediates along the pathway. Most safe strains from clinical examples are observed to be insufficient in uridylate pyrophosphorylase movement, however a couple of do not have a cytosine permease or deaminase. In view of the fast improvement of resistance, 5-FC is not utilized alone but rather in conjunction with different medications, particularly amphotericin B, with which it has a huge synergistic impact.

Echinocandins

Echinocandins are actually happening auxiliary metabolites of fungal organisms, which are presently in a propelled phase of improvement as novel antifungal specialists since they restrain the blend of β1-3 glucan, one of the real divider polymers ofparasitic fungi. Caspofungin, a polypeptide antifungal, was authorized for clinical use in 2002. It is dynamic against some Candida spp. What's more, Aspergillus fumigatus? Two different mixes of the same basic class, anidulafungin and mycafungin, are relied upon to be authorized soon.

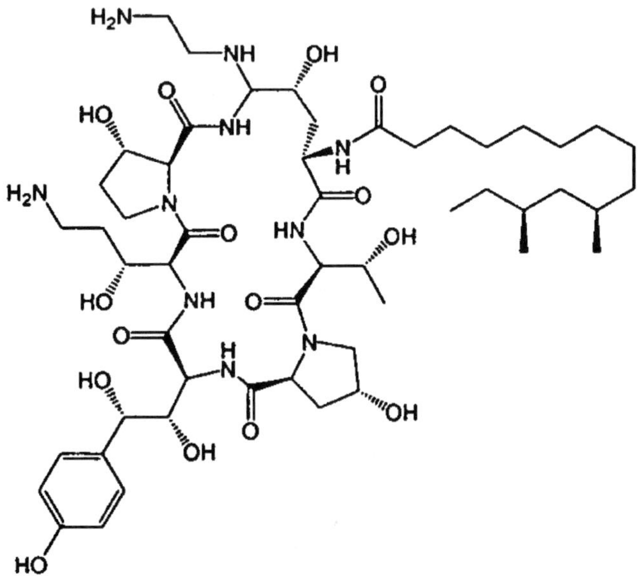

Figure- 10.12: Structure of Caspofungin.

Regulator of Pneumocystis jirovici

P. jirovici is an irregular parasite which for a long time was delegated a protozoan. The vital methods for controlling it is with the medication pentamidine which is thought to follow up on DNA. It is likewise touchy to the antibacterial operators trimethoprim and sulfomethoxazole, which upset the folate pathway required in exchange of gatherings from the amino corrosive serine to a scope of different mixes. *P. jirovici* is not influenced by the antifungal polyenes or azoles in light of the fact that it has cholesterol, not ergosterol, in its cell layer. Be that as it may, every one of the three of the echinocandins specified above are dynamic against *P. jirovici*.

The Imminent

Notwithstanding the great advances that have been made in controlling parasitic diseases of people, and the current presentation of the echinocandins, there is still worry that the pace of improvement of medications with novel methods of activity is moderate. In parallel with the improvement of new antifungal specialists, there is expanding enthusiasm for the advancement of antibodies to ensure against a portion of the endemic human-pathogenic parasitic fungi. These investigations are at a rel-atively early stage, yet potential antigens have been distinguished for both Coccidioides immitis (and the related organism *C. posadasii*) and Paracoccidioides brasiliensis.

SUMMARY

- Human infection through fungus has been increasing throughout the world. Most of the infections were caused through new fungal agents, and also involves traditional fungal agents with new process of infection or by new virulence aspects. Infections of fungi are not easily detectable and hence option of treatment are also poor. Fungi cause numerous diseases like rusts and smuts in plants, and it also cause damage to cosmetics.
- Prevention is better than to attain diseases. Fungi can be found in the presence of sunlight and UV radiations but becomes weak and fragile, that is why full sunlight is natural fungal agent. Clean indoor environments with ventilation. Never

keep garbage or decomposed materials in house. Cleanliness is basic necessity to prevent from fungal growth like, mycotoxin forming fungi in stored products.

- Different types of fungicides are used to control or kill fungi. Many of these compounds are synthesized chemically, and some are derivatives of naturally occurring compounds. Before use all fungicides should be approved officialy through national controlling organization, which also told timeperiod among harvesting crop and the last fungicide application. There are 80 different fungicides currently that are determined via specific chemical group, such as, triazoles, pyrimidines, and imidazoles.
- It is difficult to control fungal infections in humans, due to the fact that the victum is rigorously immunosuppressed. Even the usual antibacterial antibiotics like, cphalosporins and penicillin work by conjuction with their hosts common defense system. However, human mycoses is successfully being treated with the help of numerous antimycotic agents.

REFERENCES

1. Kroeck, S., & Langer, J. (2011). *Crop rotation and cover cropping: Soil resiliency and health on the organic farm*. White River Junction, VT: Chelsea Green Pub.
2. Osborn, W. M., Mathews, O. R., & United States. (1955). *Crop rotation, tillage, and fertility experiments at the Lawton (Okla.) Field Station, 1917-49*. Washington, D.C: U.S. Dept. of Agriculture.
3. Kirsch, O. C., & Wörz, J. F. G. (1985). *Cooperative promotion of production in Egypt: The failure of unified crop rotation within agrarian reform programmes for old lands*. Saarbrücken: Breitenbach.
4. Van, D. D. M., Triplett, G. B., & Henry, J. E. (1977). *Influence of long-term tillage and crop rotation combinations on crop yields and .ed soil parameters for an Aeric Ochraqualf soil*. Wooster: Ohio Agricultural Research and Development Center.
5. United States. & United States. (1900). *Pest management practices*. Washington, D.C.: National Agricultural Statistics Service,

Agricultural Statistics Board, U.S. Dept. of Agriculture.

6. Schroeder, R. A. (1999). *Shady practices: Agroforestry and gender politics in the Gambia*. Berkeley: University of California Press.
7. Pannell, D. J., & Vanclay, F. (2011). *Changing land management: Adoption of new practices by rural landholders*. Collingwood, Vic: CSIRO Pub.
8. Salazar, A., & Rios, I. (2010). *Sustainable agriculture: Technology, planning and management*. New York: Nova Science Publishers.
9. Ali, M. H. (2011). *Practices of irrigation & on-farm water management*. New York: Springer.
10. Bosart, L. F., & Sanders, F. (2008). *Synoptic-dynamic meteorology and weather analysis and forecasting: A tribute to Fred Sanders*. Boston, Mass: American Meterological Society.
11. Ramsay, A. C., & United States. (1970). *A model of ionospheric total electron content*. Scott Air Force Base, Illinois: United States Air Force, Air Weather Service.
12. Guerra-Ramon, J. C., & University of Durham. (2012). *Analysis of the astroclimate parameters associated with the tropospheric conditions and their time evolution at Roque de Los Muchachos Observatory*. Durham University.
13. Marriott, Norman G. (2013). *Principles of Food Sanitation*. Springer Verlag.
14. Guthrie, R. K. (1988). *Food sanitation*. New York, N.Y: Van Nostrand Reinhold.
15. United States. Hanford Site (Wash., United States. JOHNSON, A. R., & FH (US). (2003). *Integrated Biological Control*. Washington, D.C: United States. Dept. of Energy. Office of Environmental Management.
16. Eilenberg, J., & Hokkanen, H. M. T. (2006). *An ecological and societal approach to biological control*. Dordrecht: Springer.
17. Neuenschwander, P., Borgemeister, C., & Langewald, J. (2003). *Biological Control in IPM Systems in Africa*. Wallingford: CAB International.
18. Van, D. R. (2016). *Integrating biological control into conservation practice*.

19. Van, D. R., Hoddle, M., & Center, T. D. (2008). *Control of pests and weeds by natural enemies: An introduction to biological control*. Malden, MA: Blackwell Pub.
20. Gates, J. P., & National Agricultural Library (U.S.). (1993). *IPM and biological control of plant pests: Field crops: January 1991 - July 1993*. Beltsville, Md: National Agricultural Library.
21. Lacey, L. A., & Kaya, H. K. (2007). *Field manual of techniques in invertebrate pathology: Application and evaluation of pathogens for control of insects and other invertebrate pests*. Dordrecht: Springer.
22. Langerak, C. J. (1985). *The Pathogenesis of Fusarium oxysporum f.sp.narcissi and the role of antagonistic bulb-borne fungi in the chemical control of basal rot*. Wageningen.
23. West, J. (1994). *Chemical control of Armillaria root rot*. University of Reading.
24. Lippmann, M. (2009). *Environmental toxicants: Human exposures and their health effects*. Hoboken, N.J: John Wiley & Sons.
25. Stadler, R. H., & Lineback, D. R. (2009). *Process-induced food toxicants: Occurrence, formation, mitigation, and health risks*. Wiley: Hoboken, N.J.
26. Staff, N. R. C. (1900). *Assessment of Exposure-Response Functions for Rocket-Emission Toxicants*. Washington: National Academies Press.
27. Cheeke, P. R., & Cheeke, P. R. (1998). *Natural toxicants in feeds, forages, and poisonous plants*. Danville, Ill: Interstate Publishers.
28. Wang, R. G. M., Knaak, J. B., & Maibach, H. I. (1993). *Health risk assessment: Dermal and inhalation exposure and absorption of toxicants*. Boca Raton: CRC Press.
29. United States., & United States. (1991). *Reproductive and developmental toxicants: Regulatory actions provide uncertain protection: report to the Chairman, Committee on Governmental Affairs, U.S. Senate*. Washington, D.C: The Office.
30. Parker, Philip M. (2006). *The 2007-2012 World Outlook for Inorganic Compounds of Agricultural and Commercial Fungicides*. ICON Group.
31. Parker, P. M. (2005). *2006-2011 World Outlook for Inorganic*

Compounds of Agricultural and Commercial Fungicides. ICON Group International Inc.

32. Parker, P. M. (2005). *2006-2011 World Outlook for Inorganic Compounds of Lawn and Garden Fungicides*. ICON Group International Inc.
33. Parker, Philip M. (2006). *The 2007-2012 World Outlook for Inorganic or Organic Compounds of Agricultural and Commercial Fungicides*. ICON Group.
34. Parker, P. M., & Parker, P. M. (2006). *2007 report on inorganic compounds of agricultural and commercial fungicides, the: world market segmentation by city*. ICON Group.
35. Lebold, S. A. (1990). *A general synthetic approach to the C-terminal amino acid residue of the polyoxins: The total synthesis of Nikkomycin B*.
36. Li, Jine, Li, Lei, Feng, Chi, Chen, Yihua, & Tan, Huarong. (2012). *Novel polyoxins generated by heterologously expressing polyoxin biosynthetic gene cluster in the sanN inactivated mutant of Streptomyces ansochromogenes*. (BioMed Central Ltd.) BioMed Central Ltd.
37. McCarthy, P J, Newman, D J, Nisbet, L J, & Kingsbury, W D. (n.d.). *Relative rates of transport of peptidyl drugs by Candida albicans*.
38. Mukherjee, Prasun K., Latha, Jagannathan, Hadar, Ruthi, & Horwitz, Benjamin A. (n.d.). *TmkA, a Mitogen-Activated Protein Kinase of Trichoderma virens, Is Involved in Biocontrol Properties and Repression of Conidiation in the Dark*. American Society for Microbiology.
39. Pierre de Wit, Jos Raaijmakers, & Bergsma-Vlami, M. (n.d.). *Genotypic diversity and rhizosphere competence of antibiotic-producing Pseudomonas species*. S.n.
40. Grinyer, J., Macquarie University., & Macquarie University. (2007). *Proteomic analysis of the biological control fungus Trichoderma*. Australia: Macquarie University.
41. Morris, J. G., & Potter, M. E. (2013). *Foodborne infections and intoxications*. Amsterdam: Elsevier/Academic.

42. Topazian, R. G., & Goldberg, M. H. (1994). *Oral and maxillofacial infections*. Philadelphia: W.B. Saunders Co.
43. Surawicz, C., & Owen, R. L. (1995). *Gastrointestinal and hepatic infections*. Philadelphia: Saunders.
44. In Scheld, W. M., In Armstrong, D., Hughes, J. M., & American Society for Microbiology. (1998). *Emerging infections: 1*.
45. Olsen, L. A., & Institute of Medicine (U.S.). (2011). *Fungal diseases: An emerging threat to human, animal, and plant health : workshop summary*. Washington, D.C: National Academies Press.
46. In Donelli, G. (2015). *Biofilm-based healthcare-associated infections: Volume II*.
47. COID, C. R. E. (1977). *Infections and pregnancy*. New York: Academic Press.
48. Engelkirk, P. G., Burton, G. R. W., & Burton, G. R. W. (2007). *Burton's microbiology for the health sciences*. Baltimore, MD: Lippincott Williams & Wilkins.
49. Talaro, K. P., & Talaro, A. (2002). *Foundations in microbiology*. Boston: McGraw-Hill.
50. International Symposium on Coccidioidomycosis, Clemons, K. V., Laniado-Laborin, R., Stevens, D. A., & New York Academy of Sciences. (2007). *Coccidioidomycosis: Sixth International Symposium*. Malden, MA: Published by Blackwell Pub. on behalf of the New York Academy of Sciences.
51. Centers for Disease Control and Prevention (U.S.). (2000). *You can prevent PCP: A guide for people with HIV infection*. Atlanta, GA: U.S. Dept. of Health and Human Services, Public Health Service, Centers for Disease Control and Prevention.
52. Kreier, J. P. (1993). *Parasitic Protozoa: Toxoplasma, Cryptosporidia, Pneumocystis, and Microsporidia*. Oxford: Elsevier Science.

Index